Student Study Guide
to accompany

Calculus

Fifth Edition

by Robert Ellis & Denny Gulick

SANDRA Z. KEITH

St. Cloud State University

SAUNDERS COLLEGE PUBLISHING
Harcourt Brace College Publishers

Fort Worth Philadelphia San Diego New York
Orlando Austin San Antonio Toronto
Montreal London Sydney Tokyo

Sandra Z. Keith: Student Study Guide to accompany
CALCULUS, 5/e, by Ellis/Gulick

ISBN 0-03-098116-6

7 8 9 0 1 2 3 4 5 023 12 11 10 9 8 7 6 5 4

PREFACE

ॐ Teachers and students today face new pressures that create problems in and out of the calculus classroom . Teachers do not have time to teach the requisite skills and background at a pace that suits every student . Students, on the other hand, do not have the hours of free time to struggle with individual problems. If they cannot do the first few problems in a problem set, they frequently give up on the entire days' assignment, thereby creating a situation of "deficit learning". In fact, since students' main interactive contribution to the learning of calculus is usually the problem sets, in their minds, basic skills may assume more importance in their success in calculus than they probably should. As a result, vital elements of understanding may be lost, leading to frustration on everyone's part.

ॐ This Guide attempts to lead students through the material in the text, paralleling it, providing many worked examples similar to the exercises. Only the most basic of problems are stressed. This Guide in no way replaces the text; it is intended rather to help students "get going".

ॐ The Guide also provides self-assessment exercises for students to help them come to learn *when they know something*. These exercises provide overviews of the material, encourage the student to find similarities and differences in concepts, to reconstruct the development of the theory, and to learn to summarize vital points of the theory. Some cross-checks and self-tests are offered along the way. Students are also encouraged to create their own examples, grade themselves on a problem, and construct tests for themselves. Tips for studying are provided, as well as answers to some common complaints of students, such as why they may feel they understand the material, but perform poorly on tests. Suggestions are offered to help students find ways of learning and performing more effectively.

ॐ Students often digest the complicated language of mathematics by creating a language of their own. This Guide attempts to frame some of the theory in the "learning" language of students, with many illustrations and diagrams, and emphasis on the more important points. The Guide rehearses common student misunderstandings and errors. It then attempts to return to the student an understanding of why a precise mathematical language is important. In this way, it is hoped students will develop a more intrinsic understanding and appreciation of the methods of mathematics and its special symbolism.

Sandra Keith is a Professor at St. Cloud State University. Enormous gratitude goes to Eric Bibelenieks for his excellent advice and patient proof-reading.

TABLE - OF - CONTENTS

CALCULUS ADVISING

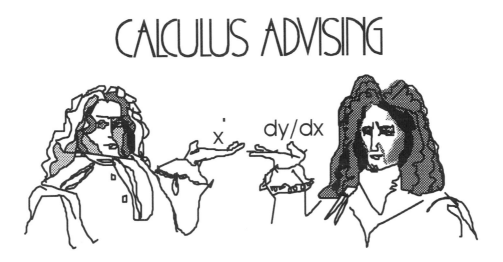

For many students, the problems encountered in the calculus sequence are not related as much to mastering the concepts of calculus as to handling the algebra and trigonometry. Before embarking on the calculus sequence, you need to know if your algebra and trigonometry background is adequate. Many schools offer a pre-test. Talk to your instructor, and follow his or her advice. If you had algebra or trigonometry a while ago, it is probably as good as lost -- it rarely comes back until too late. It is far better to enter a course with a solid background and do well, than to try to master the subject and the background at the same time. The subject of calculus is not difficult, but the pace will probably be faster than you are accustomed to. It is important now, even if it wasn't before, to keep up with the material on a daily basis, study for tests, and do all the problems assigned, even if they are not collected.

HOW TO USE THIS GUIDE

Nobody learns from watching a machine perform tricks -- to use this guide as a "crib" for solving problems is not useful. The purpose of this guide is to help you get going when you might be stuck. This guide concentrates only on the most basic of skills, but this is far from saying that this is all you will need. Thought problems help you get a grasp of concepts that will benefit your understanding of calculus in ways that skill problems alone cannot do. Nevertheless, it is often the case that with time pressures of instructors and students, skills go unaddressed. And because of the *lack* of help with basic skills, they can take up more time and importance than they probably should. We hope you will balance the skill and rote drill exercises with the conceptual development and assessment exercises scattered throughout this guide. Please do not disdain these exercises as unimportant. Of crucial importance in learning calculus is to know, *before* a test, whether your knowledge is solid-- to know, in effect, *when* you understand something. We also hope you will remember that the casual, informal mathematical language in this guide in no way replaces the precision and clarity and thoroughness of the language of the text--this guide should merely be thought of as a way-station to help you gain some confidence as you embark on the adventures of the calculus sequence.

HOW TO STUDY

HOMEWORK

Keep a special notebook for homework. Work with a study group if possible. If you do homework with a group right after class or before, you will be able to ask your questions in the next class period. Try not to leaf forwards and backwards in the text as you do homework. Don't rely on copying from worked examples in the text or working from the answers. Do a few problems, then check your results. Mark the problems that gave you trouble, and be sure you've mastered them before the exam. Many students feel that to do as much as they can is good enough, but if you skip any problems at all, they should be the ones you are fully confident about. Do the "long" problems--while these may or may not be on the tests, they improve your ability to work with the mathematics and retain it, and they develop your reasoning ability in the subject.

TEXT AND CLASS

Read the text before coming to class--your questions will be informed, and will probably be welcomed by your teacher. So much can be covered in class, that your in-class notes might not seem well organized. So rewrite your notes, and incorporate material from the text in your own words. Summarize what you can and leave messages for yourself. Try to become aware of what the teacher feels is important. Constantly ask yourself: why am I being given this problem; how does it relate to the theory, why is it important? Be prepared to memorize in calculus, even if you never had to before. We memorize so that we will have the resources to work with the material.

TESTS

In studying for tests (and you will have to study, even if you never did before!), write up a review sheet. This is a one or two page summary of important concepts, things to memorize, reminders of errors you are likely to make, sample problems, important concepts, etc. Work with a study group before an exam. The more you can verbalize the material, speaking or writing it, the more it will be "owned" by you and the more in-control you will feel. This guide has many self-testing sections to help you anticipate what teachers might feel is important.

CHAPTER 1

"The beginning is the most important part of the work"...Plato

This chapter serves as a review of topics that you have probably seen before. The section on logarithmic and exponential functions may be new, however. These skills will be of special importance throughout the course.

SECTION 1.1

"Newton was the greatest genius that ever existed and the most fortunate, for we cannot find more than once a system of the world to establish."...Lagrange

THE REAL NUMBERS

THE REAL NUMBERS are located on the REAL LINE. Any quantity not located here will not be recognized in calculus as a "number". For example, although ∞ has significance in calculus, it is not a "number", to us. Neither are imaginary numbers, such as i (where $i^2 = -1$). Thus, given $f(x) = x^2 + 1$, the roots of this function are $x = \pm i$, but these roots will not show up on the graph-- the graph will never cross the x-axis.

Common Errors

- The following description of an interval is wrong: $-3 > x > 3$. No x satisfies this relation, because the inequalities are meant to apply at the same time. The proper way to describe the set of points intended is in "two pieces"; namely, "x > 3 or x < -3". Another way is "x $\varepsilon(3, \infty) \cup(-\infty, 3)$" where "$\varepsilon$" means, "is an element of".

- The expression $\sqrt{2}$ means the positive square root of 2. If we want the negative square root, we write $-\sqrt{2}$.

- If $a^2 > b^2$, it is not necessarily true that a > b. However, |a| > |b|.

- The symbol $\varnothing$ means "the empty set". For example, to say $\sqrt{-4} = \varnothing$ is not good form, because we really mean that $\sqrt{-4}$ "does not exist". Many computers use a zero with a slash to represent "zero" as opposed to the letter --"oh", but this can be confusing. Just avoid using $\varnothing$ unless you are really describing the empty set.

- Cube roots and other odd roots of negative numbers **are** defined: thus $\sqrt[3]{-27} = -3$.

♦ A negative exponent does not necessarily mean the result is negative. $(27)^{-2/3} = (1/9)$.

♦ The definition of a **function** may not be what you have been told in the past. To define a function as "a set of ordered pairs..." is actually to define the "graph" of the function.

♦ $-x^2$ means $-(x^2)$. E.g., $-4^2 = -16$.

♦ Avoid putting two operations together without parentheses : E.g., x · -3 is bad notation. Instead, write -3 x.

♦ Too many parentheses may be confusing stylistically , but are better than too few.

♦ 3^{4^2} is not the same number as $(3^4)^2 = 81^2 = 6561$. Rather, it is $3^{16} = 43046721$.

♦ Expressions such as $\dfrac{0}{0}$, $\dfrac{2}{0}$, etc., **have no meaning**. Division by 0 is not allowed on the Real Line. In fact, we should not even write such expressions.

♦ $\sin x^2$ means $\sin(x^2)$, whereas $\sin^2 x$ means $(\sin x)^2$.

♦ $\sin xy$ will be taken to mean $\sin(xy)$. To avoid confusion, $(\sin x)y$ is usually written as $y \sin x$.

You Do The Explaining

Students often write $\sqrt{x^2 + 4} = x + 2$. But it is not true that $\sqrt{a^2 + b^2} = \sqrt{a^2} + \sqrt{b^2} = a + b$. The square root does not have an "additive" property. Rather, it has a "multiplicative" property: $\sqrt{ab} = \sqrt{a} \cdot \sqrt{b}$. Try to explain *why* this statement is true--don't simply give examples. Explain further why $\sqrt{(x+2)^2} = |x + 2|$.

SOLVING INEQUALITIES

EXAMPLE: Find all x such that $f(x) = \dfrac{(x+1)(3-x)}{(x+2)x^2} \geq 0$.

1. Find all "critical numbers", values of x for which the numerator or denominator is 0. In this problem, they are x = -1, 3, -2, or 0. Label these points on a line . (Scale is not important.) Critical numbers divide the line into intervals.

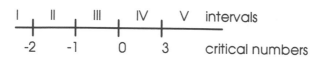

The intervals above are:

I	II	III	IV	V
$(-\infty, -2)$	$(-2, -1]$	$[-1, 0)$	$(0, 3]$	$[3, \infty)$
x may not equal - 2		x may not equal 0		

2. The sign of f will not change within an interval. That is, it will remain + or - within each interval. When f is a rational function (a quotient of polynomials) , we often employ one of the following two methods for determining the sign of f within the intervals determined by the critical points.

Method (1): Test Points

Select a "test-point" in each interval. This is an arbitrary point in each interval that we "plug into" f simply to test the sign of f. For example, in interval **I** we might use the point $x = -3$. The sign of f(-3) is "+" and so $\dfrac{(x+1)(3-x)}{(x+2)x^2} > 0$

on **the entire interior of interval I.**   On **interval II,** we might use the test point x = -3/2. And so, we obtain the sign chart below. Note that the signs do not automatically alternate from interval to interval. Here the solid vertical line is our way of showing that x may not equal that number.

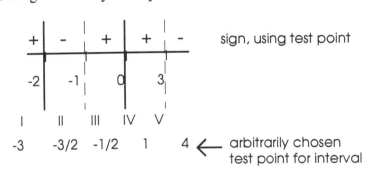

Method (2) : Charting Signs of Factors

Many students begin with the "test point" method, but switch to this method, because it requires less computation. We chart the signs of all **factors**, and multiply to obtain the sign of our quotient, f.

+	--	+	+	--	sign of quotient (multiply signs below)
-2	-1	0	3		
--	--	+	+	+	sign of factor (x+1)
+	+	+	+	--	sign of factor (3-x)
--	+	+	+	+	sign of factor (x+2)
					factor x^2 is always + or 0
I	II	III	IV	V	

(Note: the symbol x^2 means x^2.)

Either method gives the conclusion: $f(x) = \dfrac{(x+1)(3-x)}{(x+2)x^2} \geq 0$ for x in $(-\infty,-2)$, $[-1,0)$, or

$(0,3]$. (Notice we have included points where the numerator is 0 and excluded points where the denominator is 0.)

DIVIDING POLYNOMIALS

When a polynomial $p(x)$ is factored, say, as $p(x) = (x-a)(x-b)(x-c)$, the factors are found by asking: where is $p(x)$ zero? I.e., in order to factor a polynomial we need to find the roots of the polynomial. There are formulas for this, but chances are, the problems you will see are "rigged". Guess a number that makes the polynomial 0; we usually try $x = 0$, $x = \pm1$, or $x = \pm2$. Given the polynomial $x^3 - 13x + 12$, for example, we notice that $x = 1$ makes the polynomial 0. So $(x-1)$ is a factor. To find the other factor, we divide:

First arrange polynomials with highest degree terms first. Then ask:

$$\begin{array}{r} x^2 \\ x-1\overline{)x^3 - 13x + 12} \end{array}$$ How many times does leading term "x" in x - 1 go into x^3?

$$\underline{x^3 - x^2}$$ Multiply x^2 by $(x-1)$

$$x^2 - 13x + 12$$ Find remainder and repeat

$$\begin{array}{r} x^2 + x - 12 \\ x-1\overline{)x^3 - 13x + 12} \\ \underline{x^3 - x^2} \\ x^2 - 13x + 12 \\ \underline{x^2 - x} \\ -12x + 12 \\ \underline{-12x + 12} \end{array}$$

Now we can factor our polynomial into linear factors. ("Linear" factors have degree 1).

$$x^3 - 13x + 12 = (x-1)(x^2 + x - 12) = (x-1)(x+4)(x-3).$$

THE ABSOLUTE VALUE FUNCTION

The absolute value of x, denoted |x| is defined as: $|x| = \begin{cases} x & \text{if } x \geq 0 \\ -x & \text{if } x < 0 \end{cases}$. Some

students at first think this definition gives negative numbers. However, by itself, "$-x$" is not always a negative number any more than "x" is always a positive one. Try the definition out with $x = -3$. Then $|-3| = -(-3) = 3$.

Another definition for absolute value is: $|x| = \sqrt{x^2}$.

NOTE: When working with absolute value, $|x - a| < b$ means "the distance from x to a is less than b". Similarly, $|x - a| \geq b$ means, "the distance from x to a is greater than or equal to b."   The absolute value is a *distance measure*. Note that | x - a | is the same as | a - x | because "the distance from x to a" is the same as "the distance from a to x ".

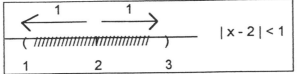

All x-values whose distance from 2 is less than 1: the interval (1,3).
THIS IS A ONE-PIECE ANSWER: we demand that x be "close to" 2.

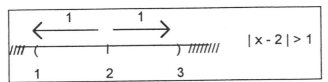

All x-values whose distance from 2 is greater than 1: the intervals (3, ∞) and (-∞,1).
THIS IS A TWO-PIECE ANSWER: we demand that x be "far" from 2.

EXAMPLE: Solving with absolute values. Find all x-values such that:

1.) | x - 3 | ≤ 2.
Answer: This is a **one-piece** answer because of the " < " -sign: x lies in the interval [1, 5]. The point x = 3 is the center of the interval, the "radius" of the interval is 2.

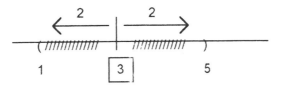

2.) | x + 4 | = 2.
Answer: The equals sign tells us our answer will be points. Interpret this equation to mean: the distance from x to -4 equals 2. We have x = -2 or x = -6.

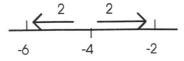

3.) |3x + 5 | ≥ 2.
Answer: This is a **two-piece** answer because of the " > " sign. Rewrite as | x + 5/3 | ≥ 2/3. The distance from x to -5/3 is greater than or equal to 2/3, so from the diagram below, we see that x is in (-∞, -7/3] or [-1, ∞).

SECTION 1.2

"There are two methods in which we acquire knowledge--argument and experiment." ...Roger Bacon

CHECK WITH YOUR TEACHER! Does your teacher have a feeling about

1. How sets of numbers on the real line are expressed, using set notation, intervals, etc.?
2. Rationalized denominators; expressing $\dfrac{1}{\sqrt{2}}$ as $\dfrac{\sqrt{2}}{2}$? Some teachers (and the text) do not prefer the latter notation.
3. Decimal approximations to answers?

4. Does your teacher have a policy about GRAPHING CALCULATORS?

EXAMPLE: Determine the distance between (6,-2) and (-1, 3).
Answer:

$$D = \sqrt{(\text{change in x })^2 + (\text{change in y })^2} =$$

$$D = \sqrt{(6-(-1))^2 + (-2-3)^2} = \sqrt{49 + 25} = \sqrt{74}$$

EXAMPLE: Determine the slope of the line through (2, -1) and (3,4). Slope is given as

$$\text{Slope} = \frac{\text{rise}}{\text{run}} = \frac{\text{change in y}}{\text{change in x}} = \frac{\Delta y}{\Delta x}.$$

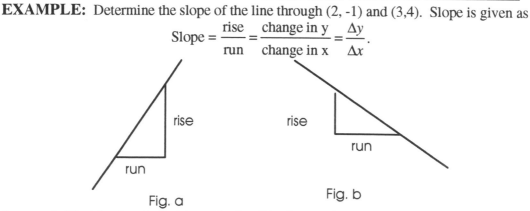

Fig. a   Fig. b

Answer: In Fig.(a), slope is positive. The "run" is always taken in the positive direction, and the "rise" is positive. In Fig. (b), the "run" is positive, but the "rise" is negative; this creates a negative slope.

EXAMPLE: Determine the equation of the line through (2, -1) with slope 4.

Answer: $\dfrac{y - y_0}{x - x_0} = m$, so $\dfrac{y - (-1)}{x - 2} = 4$, or $y + 1 = 4(x - 2)$. Rewrite in a form acceptable to your teacher.

EXAMPLE: Determine the equation of the line through (2, -1) and (-3, 5).

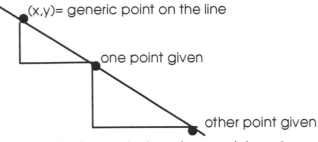

(x,y)= generic point on the line

one point given

other point given

The slope between (x,y) and one point must equal
the slope between the two given points

Answer: $\dfrac{y - y_0}{x - x_0} = \dfrac{y_1 - y_0}{x_1 - x_0}$, so $\dfrac{y \text{ - (-1)}}{x \text{ - } 2} = \dfrac{(-1) - (5)}{2 \text{ - (-3)}}$, or $y + 1 = -\dfrac{6}{5}(x \text{ - } 2)$. Rewrite in a form acceptable to your teacher.

EXAMPLE: Determine the equation of the line perpendicular to y = 3x -4 passing through (1,0).

Answer: The line will have slope m = -1/3. Then $\dfrac{y - y_0}{x - x_0} = m,$ so $\dfrac{y \text{ - } 0}{x \text{ - } 1} = -1/3,$

or $y = -\dfrac{1}{3}(x \text{ - } 1).$

SECTION 1.3

" 'We called him Tortoise because he taught us, ' said the Mock Turtle. 'Really, you are very dull!'"...Lewis Carroll

EXPONENTS

EXAMPLE: Find $(27)^{-2/3}$:

Answer:

$$27^{-2/3} = 1/9$$

(1) take cube root of 27

(2) square that

(3) put that in the denominator

EXAMPLE: Find $\dfrac{16^{-1/4}}{5^{-2}\,64^{2/3}}.$

Answer: Put every term with negative exponent on the opposite "level" of the

fraction: $\dfrac{5^2}{64^{2/3}16^{1/4}}$. Now take the roots indicated by denominators of the fractional

exponents; for instance, $64^{1/3} = \sqrt[3]{64} = 4$ and $16^{1/4} = 2$. Finally raise roots to the powers

indicated by the numerators of the exponents. Answer: $\dfrac{25}{4^2 2^1} = \dfrac{25}{32}$.

FUNCTIONS: WHAT IS AND WHAT IS NOT A FUNCTION

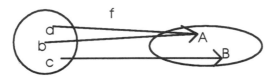

A function: defined on entire domain,
and each variable goes to only one variable.

Two questions can have
one answer.

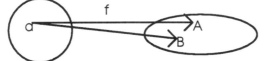

NOT a function: variable goes to more than
one variable

One question should not
have two answers.

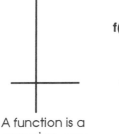

A function is a
mapping--you
cannot see it.

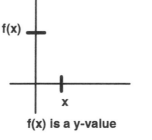

f(x) is a y-value

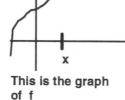

**This is the graph
of f**

Vertical Line Test:
f is not a function
if vertical line intersects
the graph more than
once.

 **DOMAIN OF A FUNCTION**

In finding the domain, ask where the function is NOT defined. Watch out for
1. Division by zero.
2. Square roots and positive roots of negative numbers.
3. Trigonometric functions on certain values, such as $\tan \pi/2$.

EXAMPLE: Find the domain of the function f given by : $f(x) = \dfrac{x+1}{x^2 + 3x + 2}$.

Answer: We can rewrite f(x) as $f(x) = \dfrac{x+1}{(x+1)(x+2)}$, but the factor (x + 1) does not

cancel when x = -1. Cancellation is division, and to cancel would be to divide by 0. This function is defined for all reals except x = -1 and x = -2 , where we would be dividing by 0. We can write the domain of f as $\{x \in R \mid x \neq -1, -2\}$. This is not the same function as

$g(x) = \dfrac{1}{x+2}$ whose domain is $\{x \in R \mid x \neq -2\}$.

EXAMPLE: Determine if f is a function where $f(x) = \begin{cases} x+1 & \text{if } x \geq 0 \\ 3 & \text{if } x \leq 0 \end{cases}$.

Answer: f is NOT a function because of the double definition at x = 0; f(0) = 1 by the top definition, and f(0) = 3 by the bottom definition.

EXAMPLE: Determine whether f, g, h, or k are the same, where:

$f(x) = \sqrt{1+2x}$, $g(x) = \dfrac{1+2x}{\sqrt{1+2x}}$, $h(x) = \sqrt{(1+2x)^2}$, $k(x) = \mid 1+2x \mid$.

Answer: The functions h and k are the same, using a definition of absolute value. However f is defined only for $1 + 2x \geq 0$, and g is defined only for $1 + 2x > 0$. We conclude that only h and k are the same.

SECTION 1.4

"My life is like a stroll upon the beach, As near the ocean's edge as I can go"...Henry David Thoreau

GRAPHING: IMPORTANT TYPES OF FUNCTIONS TO RECOGNIZE

1. Polynomials. A polynomial over the real numbers is an expression of the form $f(x) = a_n x^n + a_{n-1} x^{n-1} + \ldots a_1 x + a_0$ where the a_i are real numbers and n is a positive integer. In other words: **coefficients are real and exponents are positive integers.**

Which of the following are polynomials?

(a) $3x^{-4} + 5x^2$ (b) $4\sin x + x + 1$ (c) $\sqrt{x} + 21$ (d) $x^{1/3} + 34x^2$ (e) 5

(f) $\pi x^7 - \sqrt{2}\, x^4$

Answer: only (e) and (f) are polynomials.

In graphing a polynomial, watch for even or odd- leading degree (as well as sign of leading coefficient). The leading term governs a polynomial's behavior when x is a large positive or negative number.

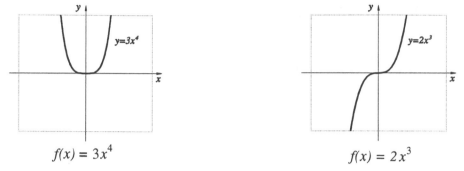

$$f(x) = 3x^4 \qquad\qquad f(x) = 2x^3$$

2. Square Roots: These are defined on "half" of the real line, and they "open" to the left or right. The sign in front of the "x" under the radical sign is the clue.

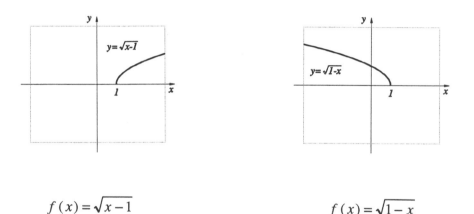

$$f(x) = \sqrt{x-1} \qquad\qquad f(x) = \sqrt{1-x}$$

3. Absolute Value: To graph a function like $y = |x - 5|$, first graph the function inside the absolute value, then reflect it across the x-axis to make it positive or zero.

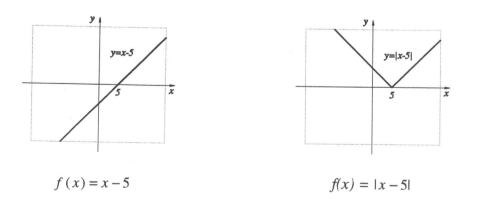

$$f(x) = x - 5 \qquad\qquad f(x) = |x - 5|$$

4. "Greatest Integer Less Than": This is the "round down" or "floor" function, denoted $[x]$ or sometimes, $\lfloor x \rfloor$. Round down to the nearest integer below (or equal) the number. For example, $[3/2] = 1$, $[2] = 2$, and $[-3/2] = -2$. To graph this function, first graph the function inside the brackets, then flatten it between integer values of y (not x). The graph of $y = [3x+1]$ is found in two stages below.

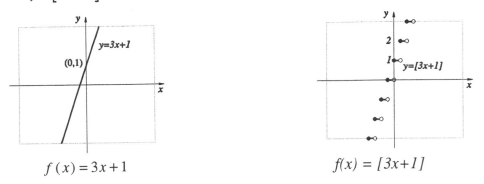

$$f(x) = 3x + 1 \qquad\qquad f(x) = [3x+1]$$

5. "Piecewise-defined" Functions These functions have "broken" definitions, such as

$$f(x) = \begin{cases} x^2 & \text{if } x > 0 \\ -x + 2 & \text{if } x \leq 0 \end{cases}.$$ To graph such a function, graph both pieces, and "jump

tracks" as indicated. Here we jump from f(x) = -x + 2 to f(x) = x^2 at x = 0.

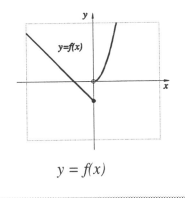

$$y = f(x)$$

You Construct The Problem

When you know a subject, you can almost anticipate the test. It helps if you imagine yourself writing a test. Many teachers like multi-step reasoning problems.

1. Write a problem that combines these notions: a parabola, a straight line, and the distance between two points.
2. Write a problem that combines solving an inequality with finding the domain of a function.
3. Write a problem that combines these notions: a circle and a line perpendicular to a given line.

Improve Your Graphing Skills!

☎ Without naming the function, describe the graph of a function to a friend, as if you were speaking over the phone. Your friend should try to construct the graph from your description. For example, you might need to say such things as, "as x goes to negative infinity, y approaches 0 from below...", or, "in the second quadrant, y is negative,"...etc.).

If your teacher will cooperate, try this game with your teacher as well: the class describes a graph to the teacher--will the teacher be able to guess the function from your description? (This also works well if one side of the room plays against the other side.)

SECTION 1.5

"Recently there has been a good deal of news about strange giant machines that can handle information with vast speed and skill."...Edmund Berkeley, 1949

AIDS TO GRAPHING

EXAMPLE: Finding intercepts. Find the intercepts , where $f(x) = 3x^2 + 6x + 2$.
Answer:

♦ **x-intercepts** occur where y = 0. Set $3x^2 + 6x + 2 = 0$. By the quadratic formula,

$x = \dfrac{-b \pm \sqrt{b^2 - 4ac}}{2a} = \dfrac{-6 \pm \sqrt{36 - 24}}{6} = \dfrac{-6 \pm 2\sqrt{3}}{6} - 1 \pm \dfrac{\sqrt{3}}{3}$. The x - intercepts are

$x = -1 \pm \dfrac{1}{\sqrt{3}}$. They are located on the graph at $(-1 \pm \dfrac{1}{\sqrt{3}}, 0)$.

♦ **y-intercepts** occur where x = 0. We obtain y = 3(0) + 6(0) + 2, so the y-intercept is 2, located on the graph at (0,2).

EXAMPLE: Completing the square. For the parabola given by $f(x) = 3x^2 + 6x + 2$, locate the vertex, and graph.
Answer:

1. Factor out the leading coefficient (the coefficient of the x^2 - term) rewriting the coefficient of the x-term.

$$f(x) = 3(x^2 + \boxed{2}\,x + \,) + 2$$

2. Compute one-half-the-coefficient-of-x-squared: $\left(\dfrac{1}{2}\boxed{2}\right)^2 = \boxed{1}$

3. Add this to the terms in parentheses , and subtract what you have added from the constant term outside the parentheses :

$3(x^2 + 2x + \boxed{1}) + 2 - 3(\boxed{1})$

      ↑   ↑

we added this   and subtract this

4. Now we have completed the square:

$$f(x) = 3(x + 1)^2 - 1.$$

5. The **vertex** of the parabola given by $f(x) = A(x + B)^2 + C$ is located at $(-B, C)$. In our example, vertex is (-1,-1). We can tell this parabola opens "upwards" because of the positive coefficient of x^2, i.e. $A > 0$. In general, if the parabola is given by $f(x) = ax^2 + bx + c$, the vertex will occur where $x = -\dfrac{b}{2a}$. To find the y-value of the vertex, plug $x = -\dfrac{b}{2a}$ into the equation for the parabola.

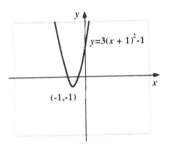

$$f(x) = 3(x + 1)^2 - 1$$

EXAMPLE: Determine the intercepts of $x^2 = 3y^4 + 1$.

Answer: In this problem, y is not a function of x, but we can nevertheless find the intercepts. When x = 0, y is not defined. There are no y-intercepts. When y = 0, x = ±1. so (±1,0) are the x-intercepts.

SYMMETRY

The graph of **y = f(x) is symmetric about the y-axis, or f is EVEN**, if   positive or negative x produces the same y; that is, if f(x) = f(-x).

The graph of **x = g(y) is symmetric about the x-axis** if positive and negative y-values produce the same x-values; that is, if g(y) = g(-y). This type of symmetry implies that y is not a function of x. The functions will not pass the "vertical line test".

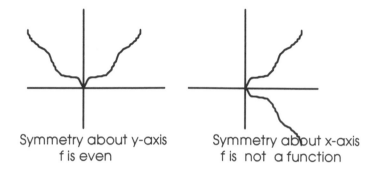

Symmetry about y-axis   Symmetry about x-axis
      f is even   f is not a function

The graph of a function given by y =f(x) is symmetric about the origin, or f is ODD, if a change in the sign of x produces a change in the sign of y.

The graph of an odd function is symmetric about a 180° degree rotation; the curve looks the same if you admire it upside down. This is the same as giving the graph a horizontal flip and a vertical flip.

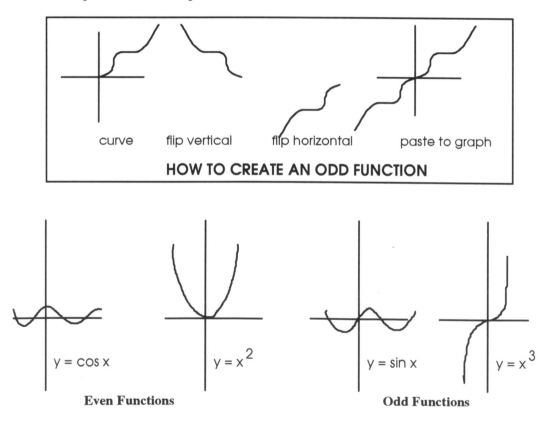

curve   flip vertical   flip horizontal   paste to graph

HOW TO CREATE AN ODD FUNCTION

$y = \cos x$ $y = x^2$ $y = \sin x$ $y = x^3$

Even Functions   **Odd Functions**

Graphing Tips

❶ Frequently, **even** functions have expressions with even powers of x, or cos x, or | x | terms. The function f given by $f(x) = x^4 - x^2 + |x| + \cos x$ is even. (Plug in x or -x and you get the same y!)

❷ Often **odd** functions have expressions with odd powers of x, or sin x terms. The function f given by $f(x) = x^3 - x^5 + x - \sin x$ is odd. (Plug in -x and you get -y). **BUT BE CAREFUL:** f given by $f(x) = x^3 - x^5 + 1$ is neither even nor odd!

❸ An odd function times an even function is odd. (The rule works with division, too.)

For example, f given by $f(x) = \dfrac{x}{1 - x^2}$ is odd. The numerator is odd; denominator is even.

❹ An odd function times an odd function is even. For example, f given by $f(x) = x \sin x$ is even. Both factors represent odd functions.

TRANSLATION OF FUNCTIONS

Suppose a function f is given by y = f(x). The graph of y-b = f(x-a) translates every point on the graph of y = f(x) "a" units **to the right** if a is positive, and "b" units **up** if b is positive. If a is negative, translate left, and if b is negative, translate down.

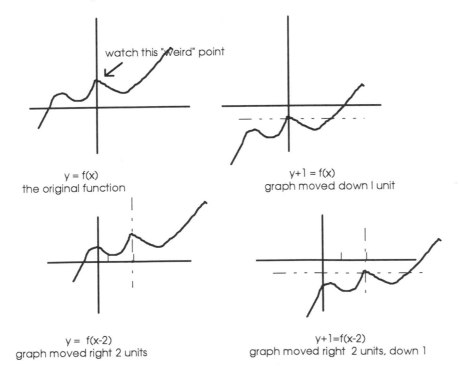

watch this "weird" point

y = f(x)
the original function

y+1 = f(x)
graph moved down 1 unit

y = f(x-2)
graph moved right 2 units

y+1=f(x-2)
graph moved right 2 units, down 1

NOTE: Think of the coordinate axes as having moved oppositely to the graph. The dashed coordinate system at bottom right is "2 units too far right" and "1 unit too far down". So we add (-2,1) to all x- and y-values to make the correction.

EXAMPLE: Find the vertex of the parabola given by $y = 2(x+4)^2 - 5$.

Answer: The shape is essentially that of $y = 2x^2$, which opens upward (positive coefficient of x^2). But where (x, y) =(0,0) is the vertex of the parabola $y = 2x^2$, the vertex of our parabola has moved 4 to the left and 5 down. The vertex of our parabola is (-4, -5).

OR: we can rewrite this parabola as $y + \boxed{5} = 2(x + \boxed{4})^2$ to see that the vertex is (-4,-5).

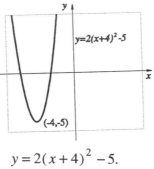

$$y = 2(x+4)^2 - 5.$$

EXAMPLE: Find the center and radius of the circle $2x^2 + 4x + 2y^2 - y = 1$.

Answer: We complete the square on x and y at the same time. Because this is an *equation*, the procedure for completing the square will look a little different from before.

$$2(x^2 + 2x \quad) + 2(y^2 - y/2 \quad) = 1$$

$$2(x^2 + 2x + 1) + 2(y^2 - y/2 + 1/16) = 1 + 2 + 2/16 = 25/8$$

(where we add to the right side what we have added to the left side)

$$2(x+1)^2 + 2(y - 1/4)^2 = 25/8$$

$$(x+1)^2 + (y - 1/4)^2 = 25/16.$$

This is the equation of a circle with radius $5/(2\sqrt{2})$ and center (-1, 1/4).

SECTION 1.6

"Sixty four millions and fifty thousand of pounds, which, estimating the whole at the medium price of thirty sols the pound, makes the sum of ninety-six millions and seventy five thousand livres tournois...An immense sum! that the city of Paris might save every year, by the economy of using sunshine instead of candles!"...Benjamin Franklin

COMPOSITION OF FUNCTIONS

Not only can we compose functions, we can add, subtract, multiply, divide functions. We describe the function $f + g$ by **"adding pointwise"**: $(f + g)(x) = f(x) + g(x)$.

EXAMPLE: Let f and g be given by $f(x) = \sqrt{x+1}$ and $g(x) = \dfrac{1}{x^2 - 4}$.

1. Find the domains of f and g.

Answer: The domain of f is x such that $x \geq -1$ and the domain of g is all x such that $x \neq \pm 2$.

2. Find f + g.

Answer: This is given by f + g (x) = $\sqrt{x+1} + \dfrac{1}{x^2 - 4}$. The domain is the intersection of the domains of f and g; namely $x \geq -1$ but $x \neq 2$. Or $[-1,2) \cup (2,\infty)$.

3. Find f · g.

Answer: This is given by the pointwise multiplication of the equations representing f and g:

$(f \cdot g)(x) = f(x)\, g(x) = \dfrac{\sqrt{x+1}}{x^2 - 4}$. The domain is the same as above.

4. Find f ∘ g.

Answer: This will be described by y = f(g(x)). If you are uncertain about how to find this, take the outside function, f (x) and put a box around the x. This is where g goes.
Recall: $f(x) = \sqrt{x+1}$.

$$f(\boxed{x}) = \sqrt{\boxed{x}+1}$$

$$f \circ g(x) = f(\boxed{g(x)}) = \sqrt{\boxed{g(x)}+1} = \sqrt{\boxed{\dfrac{1}{x^2-4}}+1} = \sqrt{\dfrac{x^2-3}{x^2-4}}$$

To find the domain of $f \circ g$ we must consider the following:

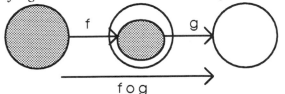

f ∘ g
The domain of the composition consists
of x for which f(x) is in the domain of g.

Thus if $f \circ g(x) = \sqrt{\dfrac{x^2-3}{x^2-4}}$, it is not enough to require that $\dfrac{x^2-3}{x^2-4} \geq 0$. It is advisable to go back in our series of equations to see where g(x) was used:

$$f \circ g(x) = \sqrt{\boxed{\dfrac{1}{x^2-4}}+1} = \sqrt{\dfrac{x^2-3}{x^2-4}}$$

$\uparrow$ $\qquad\qquad$ $\uparrow$

watch x here $\qquad$ and here!

We see from the term $\boxed{\dfrac{1}{x^2-4}}$, that $x \neq \pm 2$. And with the aid of the sign chart below,

$\dfrac{x^2-3}{x^2-4} \geq 0$ for $x \, \varepsilon \, [-\infty,-2) \cup [-\sqrt{3},\sqrt{3}\,] \cup (2,\infty]$. Since this domain already

excludes $x = \pm 2$, our domain consists of these x values.

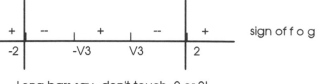

sign of f o g

Long bars say--don't touch -2 or 2!

5. Find g∘f.

Answer:

$$g \circ f(x) = g(f(x)) = \frac{1}{f(x)^2-4} \;=\; \boxed{\frac{1}{(\sqrt{x+1})^2-4}} \;=\; \frac{1}{x+1-4} \;=\; \frac{1}{x-3}.$$

watch this term

Therefore the domain of consists of x in $[-1,3) \cup (3,\infty)$.

f o g
(we require x > -1 and x not 3)

NOTE: We can see from this example: f∘g and g ∘f   are in general not the same functions. (Any more than to wash the dishes and dry them is the same as to pour out the dry the dishes and wash them.)

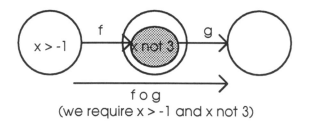

SECTION 1.7

"What can I wish to the youth of my country who devote themselves to science? Firstly, gradualness. About this most important condition of fruitful scientific work I never can speak without emotion. Gradualness, gradualness, and gradualness...."...Ivan Pavlov

Trigonometry is frequently a stumbling block for students in calculus. Be sure to review your trigonometry!

REVIEW OF TRIGONOMETRIC VALUES

θ	$\pi/6$	$\pi/4$	$\pi/3$	$\pi/2$	$2\pi/3$	$3\pi/4$	$5\pi/6$	π
cos θ								
sin θ								
tan θ								
sec θ								
csc θ								
cot θ								

QUIZ: Find the angles θ satisfying the following, where θ lies in the interval $[0, 2\pi)$.
(a) $\cos \theta = -1$　(b) $\sin \theta = 1/2$　(c) $\tan \theta = -\sqrt{3}$　(d) $\cos \theta = -\sqrt{3}/2$　(e) $\sec \theta = -2$
(f) $\csc \theta = 0$　(g) $\sin \theta = -1$　(h) $\cos \theta = 1$　(i) $\sin \theta = -1/\sqrt{2}$　(j) $\cot \theta = 1$

ANSWERS: (a) $\theta = \pi$　(b) $\theta = \pi/6, 5\pi/6$　(c) $\theta = 2\pi/3, 5\pi/3$　(d) $\theta = 5\pi/6, 7\pi/6$
(e) $\theta = 2\pi/3, 4\pi/3$　(f) not possible, (g) $\theta = 3\pi/2$, (h) $\theta = 0$, (i) $\theta = 5\pi/4, 7\pi/4$
(j) $\theta = \pi/4, 5\pi/4$.

TRIGONOMETRIC FUNCTIONS

Hints for graphing trigonometric functions. Describe the graph of the following functions (They are graphed on the following page.)

(a) $y = \cos(x - \pi/4)$. **Hint:** this is just the graph of the cos function shifted right by $\pi/4$.

(b) $y = \sin(\pi - x)$. **Hint:** rewrite as $y = -\sin(x - \pi)$. (Recall that $\sin(x) = -\sin(-x)$.) Shift the graph of the sine function to the right by π, then flip upside down (because of the minus sign).

(c) $y = 3 \sin 2x$. **Hint:** this is the graph of the sine function with *frequency* 2 and *amplitude*

3. It goes twice as fast as the graph of y = sin x, finishing a period in the interval $[0, \pi]$, and it reaches "heights" of y = ±3.

(d) y = |cos x| . **Hint:** graph y = cos x, and reflect all negative y-values across the x-axis.

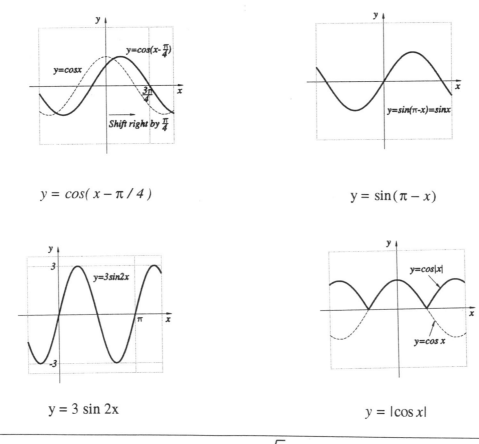

$$y = cos(x - \pi / 4)$$ $$y = sin(\pi - x)$$

$$y = 3 \sin 2x$$ $$y = |\cos x|$$

EXAMPLE: Solve the inequality for x: $\sin x > \dfrac{\sqrt{3}}{2}$.

Answer: We know $\sin \dfrac{\pi}{3} = \sin \dfrac{2\pi}{3} = \dfrac{\sqrt{3}}{2}$, and if we restrict x to lie in $[0, 2\pi)$, we require that $x \varepsilon (\pi / 3, 2\pi / 3)$. Since sine is periodic with period 2π, the complete set of x satisfying the inequality is given by

$$x \varepsilon (\pi / 3 + 2n\pi, \ 2\pi / 3 + 2n\pi) \ , \ n = 0, \ \pm 1, \ \pm 2, \text{ etc.}$$

SECTION 1.8

"Not a log in this buildin' but its memories has got/ And not a nail in this old floor but touches a tender spot."....Will Carleton

EXPONENTIAL AND LOGARITHMIC FUNCTIONS

You will see these functions again, in Chapters 5 and 6. An introduction is worthwhile because these functions are used in many other courses.

Properties of the Natural Logarithm Function
When b and c > 0 and r is a real number (*)

$$\ln bc \quad = \ln b + \ln c$$

$$\ln (b/c) = \ln b - \ln c$$

$$\ln b^r \quad = r \ln b$$

(*) We know the result only if r is rational, now.

Since these properties are so easily confused, think of them this way:

Difficult Operation Inside ln ( )	Easy Operation Outside ln ( )
$\times$	+
$\div$	-

NOTE: When we had the equation $y^2 = x$, we wanted to express y in terms of x, and for introduced a new function (positive square root), and new notation so that we might write: $y = "\sqrt{x}"$.

Now we have the function $a^y = x$ and need a new function and new notation to express y in terms of x. we want to say:

y = the number such that when the base a is raised to y we get the number x

y = exponent_{base is a} (we get x)

y = exponent$_{\text{base is a}}$ (we get x)

y = logarithm$_{\text{base is a}}$ (we get x)

y = $\log_a x$

The word "log*arithm*" means "logical/rightful *number*". The "logarithm" is the exponent.

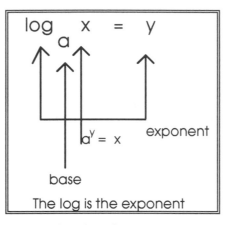

NOTE: For every exponential equation there is a corresponding logarithmic equation We can go back and forth between these equations.

EXPONENTIAL FORM	LOGARITHMIC FORM
$a^y = x$	$y = \log_a x$

Study: College Grads earn $1,039 more per month

The Star Tribune (Minneapolis) reports on Jan. 28, 1993 that on average, persons with a bachelor's degree earn $2,116 a month, and high school graduates earn an average of $1,077 a month; the gap in earning grows with further education. Forty-three percent of adults earned their degrees within 4 years of high school graduation. Females were more likely than males to finish college within 4 years. The article concludes that the payoff comes several years after graduation from college, as college graduates are promoted past their less-educated colleagues.

SELF-QUIZ: Which of the following are true,
where a, b > 0 and r is a real number?

$$\ln(a - b) = \frac{\ln a}{\ln b}$$

$$\ln (a - b) = \ln \frac{a}{b}$$

$$\ln a \ln b = \ln ab$$

$$\ln a \ln b = \ln (a + b)$$

$$\frac{\ln a}{\ln b} = \ln a - \ln b$$

$$\ln a \ln b = \ln a + \ln b$$

$$\ln ab^r = r \ln ab$$

$$(\ln a)^b = b \ln (a)$$

$$\frac{1}{\ln(a / b)} = \ln(b / a)$$

Answer: None of these

NOTE: $e^{(\ )}$ and $\ln (\ )$ are "inverses". Like matter and anti-matter, they cancel each other out when they are in direct proximity:

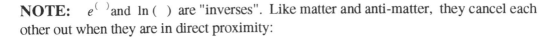

$$x \xrightarrow{\ e^{(\ )}\ } e^x \xrightarrow{\ \ln(\ )\ } \ln(e^x) = x \ , \text{ back to the start!}$$

$$x \xrightarrow{\ \ln(\ )\ } \ln x \xrightarrow{\ e^{(\ )}\ } e^{\ln x} = x \ , \text{ back to the start!}$$

OR:
$$e^{\ln x} = x \quad \text{and} \quad \ln e^x = x$$

CAUTION: Notice the above rule does not allow us to say :

$e^{3\ln x} = 3\ln x$. Instead, $e^{3\ln x} = e^{\ln x^3} = x^3$. Similarly $\ln (3e^x) \neq 3e^x$. Instead, $\ln(3e^x) = \ln 3 + \ln(e^x) = \ln 3 + x$. "Ln" and "e" must "touch" in order to use the rules above.

NOTE: We can use any $a > 0$ as a base. If we allowed negative numbers as bases, we would have expressions such as $f(x) = (-3)^x$, (where $a = -3$). Here f(x) could be positive and negative erratically and there would be zillions of values of x (such as x = 1/2), where f would not be defined. This would be a mess!

USEFUL PROPERTIES TO MEMORIZE

$$a^x = e^{\ln(a^x)} = e^{x \ln a}$$

$$\log_a x = \frac{\ln x}{\ln a}$$

$$a^{\log_a x} = x$$

$$\log_a a^x = x$$

NOTE: Somehow students often miss the fact that $\log_e x = \ln x$. The "ln" function is called the "natural logarithm". You will see why this is a good definition later. Can you find ln3, for example, on your calculator? Can you find e? Try the inverse function (or second function) of ln on the number one.

📖 Keep a Diary of Mistakes: Reserve a section of your notebook for a record of the mistakes you tend to make--not only algebraic errors, but also confusions that have prevented you from solving exercises from the homework, quizzes and exams. (Be sure to review this before taking another quiz or exam!)

EXERCISE: Fill in the chart.

EXPONENTIAL EQUATION	LOGARITHMIC EQUIVALENT	SOLVE FOR x
$3^x = 9$	$\log_3 9 = x$	from exponential form , x = 2
$x^3 = 9$	$\log_x 9 = 3$	from exponential form , x = $\sqrt[3]{9}$
$3^2 = x$	$\log_2 x = 3$	from exponential form , x = 9
$2^x = 9$	**(a)**	**(b)**
(c)	$\log_x 512 = 9$	**(d)**
(e)	$\log_2 x = 9$	**(f)**
$4^{x+1} = 16$	**(g)**	**(h)**

Answers: (a) $\log_2 9 = x$ (b) solved from the logarithmic form where we use $\log_2 9 = \dfrac{\ln 9}{\ln 2} =$ (from calculator) 3.16993. (c) $x^9 = 512$ (d) $x = \sqrt[9]{512} = 2$ (e) $2^9 = x$ (f) x = 512.   (g) $\log_4 16 = x+1$ (h) solved from the exponential equation. Since x +1=2,   x = 1.

EXAMPLE: Simplify the following:

1. $e^{\ln 5x} = ?$ **Answer:** $5x$

2. $\log_5 3 = ?$ **Answer:** $\log_5 3 = \ln 3 / \ln 5 = .6826$

3. $\log_3 \dfrac{1}{81} = ?$ **Answer :** -4, since $\log_3 \dfrac{1}{81} = \log_3 3^{-4}$, and log is the exponent.

4. $e^{3-\ln 2x} = ?$ **Answer :** $e^{3-\ln 2x} = \dfrac{e^3}{e^{\ln 2x}} = \dfrac{e^3}{2x}$

Here' s another method:

5. $\log_{16} 4 = ?$ **Answer:** Let $\log_{16} 4 = y$. Then $16^y = 4$ or $(4^2)^y = 4$, or $4^{2y} = 4^1$ so
   $2y = 1$ and $y = 1/2$.

6. $\log_5 5x = ?$ **Answer:** $\log_5 5x = \log_5 5 + \log_5 x = 1 + \log_5 x$

7. $\log_5 5^x = ?$ **Answer :** x ($\log_5(\ )$ and $5^{(\ )}$ touch and annihilate each other.)

8. $\log_{1/9} 3^x = ?$ **Answer:** Let' s say $\log_{1/9} 3^x = y$. Then $(1/9)^y = 3^x$ or $3^{-2y} = 3^x$, so
   $2y = -x$. Then $y = \log_{1/9} 3^x = -x/2$.

9. $6^{\log_6 36} = ?$ **Answer:** 36 ($6^{(\ )}$ and $\log_6(\ )$ touch and annihilate each other .)

SOLVING EQUATIONS

❖**VARIABLE IN THE EXPONENT:**   If an equation is difficult to solve because of a variable in the exponent, take the natural logarithm of both sides.

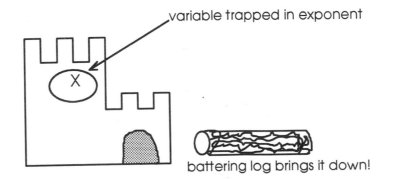

variable trapped in exponent

battering log brings it down!

EXAMPLE: Solve for x: $e^x = 3$.
Answer:
(1) We might recognize immediately that the logarithm form of the exponential equation $e^x = 3$ is $x = \ln 3$.
(2) Or we might take the natural logarithm of both sides. Then $\ln e^x = \ln 3$ or $x = \ln 3$.
(3) Or we could say $e^x = 3 = e^{\ln 3}$. So $x = \ln 3$.

EXAMPLE: Solve for x: $e^x = 2e^{2x-1}$.
Answer: Take the natural logarithm carefully, on both sides of the equation.
$\ln e^x = \ln 2e^{2x-1} = \ln 2 + \ln e^{2x-1}$. Thus $\ln e^x = x = \ln 2 + (2x-1)$ and $-x = \ln 2 - 1$, or $x = 1 - \ln 2$.

EXAMPLE: Solve for x: $3^x = 4^{x+1}$.
Answer: Take the natural logarithm of both sides of the equation: $x \ln 3 = (x+1) \ln 4$. So $x \ln 3 = x \ln 4 + \ln 4$, and gathering x-terms to the left, $x(\ln 3 - \ln 4) = \ln 4$. Thus $x = \ln 4 / \ln(3/4)$.

❖**VARIABLE INSIDE THE LOGARITHM:** If an equation has the variable inside the natural logarithm function use properties of the logarithm. Often exponentiation is the trick.

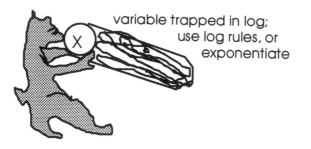

variable trapped in log; use log rules, or exponentiate

EXAMPLE: Solve $\ln x = 5$.
Answer: Exponentiate. Then $e^{\ln x} = e^5$ and $x = e^5$.

EXAMPLE: Solve $\ln x + \ln(x+4) = \ln 45$.
Answer: We could exponentiate, but here's another way.

$\ln(x(x+4)) = \ln 45$, so $= x(x+4) = x^2 + 4x = 45$ or $x^2 + 4x - 45 = 0$ and $(x+9)(x-5) = 45$. Thus, it would appear that $x = -9$ or $x = 5$. But there's more to the story....

Frequently, using log rules, we get extraneous roots. These roots must be checked in the original equation: $\ln x + \ln(x+4) = \ln 45$. Since $\ln x$ is not defined when $x = -9$, the only solution is $x = 5$.

EXAMPLE: Solve for x: $\log_2 x = \log_2 (x+1) + 3$.

Answer: Using log rules, $\log_2 x - \log_2(x+1) = 3$, so $\log_2 \dfrac{x}{x+1} = 3$ and $2^3 = \dfrac{x}{x+1}$. Thus

$8(x+1) = x$, or $x = -8/7$. Since log is not defined when x is negative , there are no

solutions.

EXAMPLE: Where do the graphs intersect?

1. $f(x) = e^{3x}, g(x) = e^{x+1}$.

Answer: Graphs intersect where $3x = x + 1$, or $x = 1/2$ and $y = e^{3/2}$.

2. $f(x) = e^x, g(x) = 4e^{-x}$.

Answer: Write as $e^x = 4e^{-x}$. To bring the exponent down, take ln ( ) of both sides:

$\ln e^x = \ln(4e^{-x}) = \ln 4 + \ln e^{-x}$.

Thus $\ln e^x = x = \ln 4 - x$, or $2x = \ln 4$, or $x = \dfrac{\ln 4}{2} = \ln 4^{1/2} = \ln 2$. So the graphs

intersect at $(\ln 2, 2)$.

3. $f(x) = \log_3 x, \quad g(x) = \log_4 x$.

Answer: Let $y = \log_3 x = \log_4 x$. Then $x = 3^y = 4^y$. Take ln( ) of both sides:

$y \ln 3 = y \ln 4$. But this impossible unless $y = 0$. And when $y = 0$, $x = 1$.

So the graphs intersect at $(1, 0)$.

REVIEW QUESTIONS TO ASK YOURSELF

Are you comfortable with ...
Solving an inequality? Absolute value inequalities?
Simplifying expressions with fractional exponents?
Completing the square?
Finding the equations of straight lines, using the two-point or the point-slope formula?
Finding the line perpendicular to a given line (i.e., finding the ***normal*** line?)
Finding the center and radius of a circle?
Finding the vertex of a parabola?
Graphing some simple curves; e.g. absolute value, straight lines, simple parabolas, circles, y = 1/x, exponential and log functions?
Simplifying log expressions?
Solving equations with exponential and logarithmic functions?
Incidentally, what is "e"?

SOME THEORETICAL TERMS TO KNOW
Do you know the meaning of these terms that you might encounter?

1. **Counterexample:** This is a single example that disproves an assertion.

2. **Prove, Explain, Show:** These terms mean that you must give arguments, usually from the theory, to justify WHY something is true. Do not simply restate that the assertion is true or that it seems valid in your mind.

3. **Axiom:** This is a basic assumption on which mathematics is built; it cannot be proved.

4. **Theorem:** This is an important result for the theory. Since a change in wording can have unexpected implications, be careful with your wording. A **Lemma** is a result that leads to a Theorem, and a **Corollary** is a result which follows from a Theorem.

Provide the Graph

It is important to be able to provide quick sketches of the graphs of functions such as these. Do this without plotting points, but show the y-intercept and/or x-intercept if they exist. Which graphs show the function rising (growth), which falling (decay)? What is the effect of multiplying "x" in the exponent by a positive coefficient ? What is the effect of multiplying the function by a positive coefficient? **SUMMARIZE IN YOUR OWN WORDS the graphs of these functions!**

$$a)\, y = 2^x \qquad\qquad f)\, y = \left(\frac{1}{2}\right)^x$$

$$b)\, y = e^x \qquad\qquad g)\, y = 3e^{-4x}$$

$$c)\, y = 3e^x \qquad\qquad h)\, y = \log_2 x$$

$$d)\, y = 2^{-x} \qquad\qquad i)\, y = \ln x$$

$$e)\, y = e^{-x} \qquad\qquad j)\, y = \log_{1/2} x$$

Hints: (a),(b) are growth functions going through (0,1). (c) is a growth function going a little faster; it goes through (0,3). (d) is the same as (f): these are decay functions going through (0,1) . (e) is also a decay function going through (0,1) . (g) is a decay function, going through (0, 3). (h) is the graph of (a) reflected across the 45-degree line; it goes through (1,0). (i) is the graph of (b) reflected across the 45-degree line; it also goes through (1,0). (j) is the graph of (f) reflected across the 45-degree line; as a result, it swoops downward; it goes through (1,0) as well.

MAKE THE BEST MATCH OF THE GRAPHS OF THE FUNCTIONS ABOVE WITH THE GRAPH S BELOW:

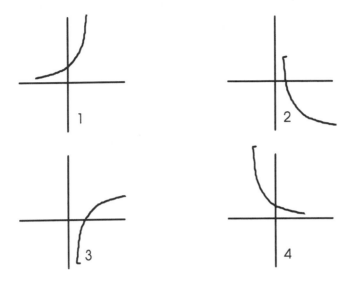

Answers: Match a,b,c to 1.   Match d,e,f,g to 4.   Match h,i to 3.   Match j to 2.

CHAPTER 2

"Methought I heard a voice cry, 'Sleep no more!' "....Shakespeare

The concept of "limit" is what calculus is about, and the history of the discovery and refinement of the concept are fascinating. In this chapter, the limit is introduced informally, then more precisely . With the concept of limit, it will be possible to explain what is meant by a continuous function.

For this chapter you will need a good grasp of the material in Chapter I. The new concepts introduced here will require special care in keeping your notation accurate.

SECTION 2.1

" All the modern inconveniences"...Mark Twain

LIMITS: Rule for finding limits, so far:

If $\lim_{x \to a} f(x)$ does not exist when we plug in x = a (we get $0/0$, $4/0$, ∞, etc.),

then DO ALGEBRA, such as cancellation, and keep trying to plug in x = a.

IMPORTANT PROPERTIES OF THE LIMIT $\lim_{x \to a} f(x)$

♦   **The limit should be the same as x approaches a from either the left or right.**

♦   **We do not care what actually happens at x = a when we find the limit.**

♦   **The limit tells us what "should" happen; it is what we predict will happen to f(x) as x approaches a.**

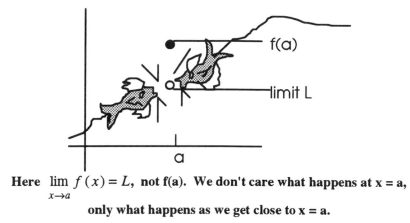

Here $\lim\limits_{x \to a} f(x) = L$, **not f(a). We don't care what happens at x = a,**

only what happens as we get close to x = a.

AN IMPORTANT GEOMETRICAL APPLICATION OF THE LIMIT.

The limit is defined with **algebra**, but it has some geometrical applications when combined with certain quotients of functions. We will see these quotients again and again in calculus.

THE SLOPE OF THE SECANT LINE through (a, f(a)) and (x, f(x)) is the quotient $\dfrac{f(x) - f(a)}{x - a}$. This quotient represents the "average" rate of change in f ; it does not require the concept of limit.

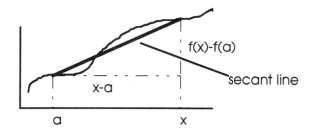

THE SLOPE OF THE TANGENT LINE to the graph of f at x = a (when it exists) is given by: $\lim\limits_{x \to a} \dfrac{f(x) - f(a)}{x - a}$. This is the "instantaneous" rate of change in f at x = a and cannot be defined without the concept of limits.

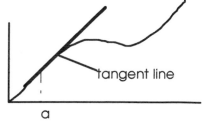

CAUTION: The slope of the tangent line is only one example of how a limit can be used. It is an **application** of the concept of limit which is extremely important; however, there are other uses for the limit, and this **should not be confused with the definition of limit**.

NOTE: Some functions do not have tangent lines. The function $f(x) = |x|$ has no tangent line at $x = 0$ (although we can draw tangent lines at other points). Many students feel that democratically, the tangent line at $x = 0$ should exist, and should be the x-axis. However, the definition of limit requires that the limit be the same when we approach $x = 0$ from either the right or left direction.

slope of tangent line here is -1   slope of tangent line here is +1

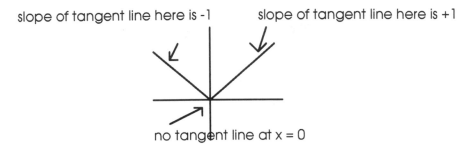

no tangent line at x = 0

The limits below are not obviously solved with algebra techniques.

IMPORTANT LIMITS TO MEMORIZE

$$\lim_{x \to 0} \frac{e^x - 1}{x} = 1$$

$$\lim_{x \to 0} \frac{\sin x}{x} = 1$$

EXAMPLE: Find the value of the limit: $\displaystyle\lim_{x \to 3} \frac{2x + 4}{5x - 1}$.

Answer: Since "plugging in" $x = 3$ gives a real number, we do so, and $\displaystyle\lim_{x \to 3} \frac{2x + 4}{5x - 1} =$

$\dfrac{2(3) + 4}{5(3) - 1} = \dfrac{10}{14} = \dfrac{5}{7}$.

EXAMPLE: Find the value of the limit $\displaystyle\lim_{x \to 3} \frac{x^2 - 5x + 6}{x - 3}$.

Answer: When "plugging in" leads to the answer "0/0", we must use algebra.

Factoring, $\displaystyle\lim_{x \to 3} \frac{x^2 - 5x + 6}{x - 3} = \lim_{x \to 3} \frac{(x - 3)(x - 2)}{x - 3} = \lim_{x \to 3} (x - 2) = 1$.

EXAMPLE: Find an expression for the quotient $\dfrac{f(x)-f(a)}{x-a}$ and guess the slope of the

tangent line to the graph of the function f at x = a, where $f(x) = x^2 + 3x + 1$ and a = 4.

Answer: We are specifically looking for $\dfrac{f(x)-f(a)}{x-a}$ and then, $\lim\limits_{x\to a}\dfrac{f(x)-f(a)}{x-a}$.

Now $\dfrac{f(x)-f(a)}{x-a} = \dfrac{x^2+3x+1-29}{x-4} = \dfrac{x^2+3x-28}{x-4} = \dfrac{(x+7)(x-4)}{x-4} = ?$ If x ≠ 4,

we can cancel and obtain $\dfrac{f(x)-f(a)}{x-a} = x + 7$. For the second part of the question:

$\lim\limits_{x\to a}\dfrac{f(x)-f(a)}{x-a} = \lim\limits_{x\to 4}\dfrac{(x+7)(x-4)}{x-4} = \lim\limits_{x\to 4}(x+7) = 11$.   Therefore, we anticipate

the slope of the tangent line to be 11, when x = 4.

SECTION 2.2

" The wisest man is he who does not fancy that he is so at all"...Despreaux

EXAMPLE: In the graph below, determine if f has a limit at the designated points, and
what the limit is.

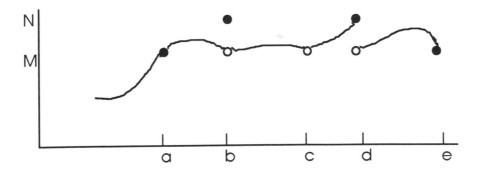

Answer: At x = a the limit is M, at x = b, the limit is M, at x = c, the limit is M, at x = d
there is no limit, and at x = e, again there is "officially" no limit because the function is not
defined to the right of e . We will eventually say that there is a "left-sided" limit here.

EXAMPLE: Use the results of this section to evaluate $\lim\limits_{x\to 2}(x+3)$.

Answer: We use the limit law : $\lim\limits_{x\to a}(bx+c) = ba+c$.   Therefore, $\lim\limits_{x\to 2}(x+3) = 2 + 3 =$

5. (**Note:** The limit laws are proved rigorously. We are solving this problem by citing laws

that may seem obvious to you, but this is so we remain aware of the rigor necessary when
dealing with limits.)

What's Wrong Here? Can you correct the errors in the following work?

1. $\lim\limits_{x \to 0} \dfrac{3x}{5x} = \dfrac{0}{0}$, therefore the limit does not exist.

2. $\lim\limits_{x \to 2} \dfrac{x^2 - 4}{x - 2} = x + 2 = 4$.

3. $\lim\limits_{x \to 2} \dfrac{x^2 - 4}{x - 2} = \lim\limits_{x \to 2} x + 2 = \lim\limits_{x \to 2} 4$

4. $\underline{\lim\limits_{x \to 2}} \ x + 2 = 4$

Answers: (1) A preliminary answer of "0/0" means "do more algebra", not "does not exist". (2) Keep the limit sign after 1st equality (3) Drop the limit sign after 2nd equality; answer is 4. (4) The limit sign is not underlined.

THE TANGENT LINE TO THE GRAPH OF A FUNCTION

Given a curve, $y = f(x)$, and a point on the curve $(a, f(a))$, it is possible under certain conditions to find a tangent line to the curve. The "smoothness" of the curve is important. The tangent line to the curve will have these properties:

- the slope of the tangent line will be $\lim\limits_{x \to a} \dfrac{f(x) - f(a)}{x - a}$.

- the tangent line will go through the point $(a, f(a))$.

We might say the SLOPE is the important thing here: we know the point. From a point and a slope, we can determine the tangent line.

> **The equation of the tangent line to the graph of f**
> **at the point (a, f(a)) is given by**
> $$y - f(a) = m(x - a) \text{ where } m = \lim\limits_{x \to a} \dfrac{f(x) - f(a)}{x - a}$$

EXAMPLE: Find the tangent line to the graph of f at $(2,19)$ where $f(x) = 4x^2 + 3$.

(1) *FIND THE SLOPE OF THE TANGENT LINE.*

Answer:

$$\lim_{x \to a} \frac{f(x) - f(a)}{x - a} = \lim_{x \to 2} \frac{4x^2 + 3 - (4(4) + 3)}{x - 2} = \lim_{x \to 2} \frac{4x^2 + 3 - 19}{x - 2} = \lim_{x \to 2} \frac{4x^2 - 16}{x - 2} = \lim_{x \to 2} \frac{4(x^2 - 4)}{x - 2} =$$

$$\lim_{x \to 2} \frac{4(x - 2)(x + 2)}{x - 2} = \lim_{x \to 2} 4(x + 2) = 4(2 + 2) = 16.$$

(2) *FIND THE EQUATION OF THE TANGENT LINE.*

Answer: The equation is given by $y - f(a) = m(x - a)$ where m =16, from above. So y - 19 = 16 (x - 2), or in slope-intercept form, y = 16 x - 13 .

VELOCITY

The constructs $\dfrac{f(x) - f(a)}{x - a}$ and $\lim_{x \to a} \dfrac{f(x) - f(a)}{x - a}$ introduced in Section 2.1 have real world applications.

IF.......

♦ **x represents time, t, and**
♦ **f(x) represents the position of an object at time t.**

 (In this particular case, the limiting value "a " will look more like a time variable if we call it t_0)

...THEN

$$\frac{f(x) - f(a)}{x - a} = \frac{f(t) - f(t_0)}{t - t_0} \quad \text{represents}$$

AVERAGE VELOCITY from t_0 to t , and

$$\lim_{x \to a} \frac{f(x) - f(a)}{x - a} = \lim_{t \to t_0} \frac{f(t) - f(t_0)}{t - t_0} \quad \text{represents}$$

(INSTANTANEOUS) VELOCITY at t_0 .

Note: sometimes instantaneous velocity is called simply, velocity.

EXAMPLE: Suppose the position of an object is given by $f(t) = -5t^2 + 3$. Find the velocity $v(t_0)$ when $t_0 = 4$.

Answer: This is a question of finding $\lim_{t \to t_0} \dfrac{f(t) - f(t_0)}{t - t_0}$. But

$$\lim_{t \to t_0} \frac{f(t) - f(t_0)}{t - t_0} = \lim_{t \to 4} \frac{5t^2 + 3 - (5(4)^2 + 3)}{t - 4} = \lim_{t \to 4} \frac{5t^2 - 80}{t - 4} = \lim_{t \to 4} \frac{5(t^2 - 16)}{t - 4}$$

$$= \lim_{t \to 4} \frac{5(t - 4)(t + 4)}{t - 4} = \lim_{t \to 4} 5(t + 4) = 40. \text{ So the velocity is } 40 \text{ (no units given)}.$$

EXAMPLE: A stone is thrown upwards from a height of 30 meters. If the initial velocity is 8 m/sec, determine the **velocity and speed** of the stone after 3 seconds.

Answer: The position of the object is given by the formula: $h(t) = -4.9t^2 + v_o t + h_o$ where v_o is the initial velocity and h_o is the height from which the stone is thrown. So $h(t) = -4.9t^2 + v_o t + h_o = -4.9t^2 + 8t + 30$. Work will show that $\lim_{t \to 3} \frac{h(t) - h(3)}{t - 3}$ = -21.4 , which we interpret as the stone moving *downwards* at a rate of 21.4 m./sec.

Speed is the absolute value of velocity. Here, the speed is 21.4 m./sec.

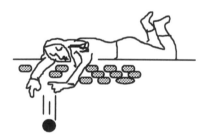

EXAMPLE: Suppose a ball is thrown downward or dropped from a height of 30 meters. Assume that after 1 sec. the velocity of the ball is -9.8 m/sec. What is the ball's initial velocity?

Answer: We are being given : $\lim_{t \to 1} \frac{h(t) - h(1)}{t - 1} = -9.8$, where h(t) = $-4.9t^2 + v_o t + 30$.

Now

$$v(1) = \lim_{t \to 1} \frac{h(t) - h(1)}{t - 1} = \lim_{t \to 1} \frac{-4.9t^2 + v_0 t + 30 - (-4.9 + v_0 + 30)}{t - 1} = \lim_{t \to 1} \frac{-4.9(t^2 - 1) + v_0(t - 1)}{t - 1}$$

$$= \lim_{t \to 1} \frac{-4.9(t - 1)(t + 1) + v_0(t - 1)}{t - 1} = \lim_{t \to 1} -4.9(t + 1) + v_0 = -9.8 + v_0.$$

But velocity at time t = 1 is given to be -9.8, which means v_o = 0, and the ball was not thrown downward, but dropped.

THE FORMAL DEFINITION OF LIMIT :

We say $\lim\limits_{x \to a} f(x) = L$ if:

given any $\varepsilon > 0$, there exists a $\delta > 0$ such that $0 < |x - a| < \delta \Rightarrow |f(x) - L| < \varepsilon$

Note: the symbol "$\Rightarrow$" is read "implies".

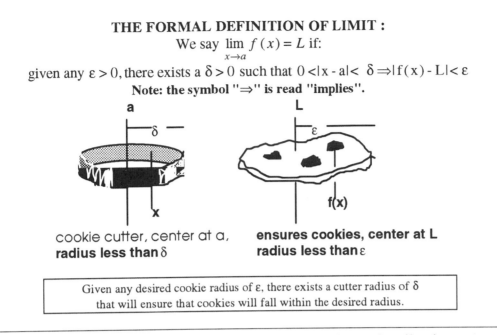

cookie cutter, center at a,
radius less than δ

**ensures cookies, center at L
radius less than ε**

Given any desired cookie radius of ε, there exists a cutter radius of δ
that will ensure that cookies will fall within the desired radius.

EXAMPLE: Use the *formal definition of the limit* (with deltas and epsilons) to prove

$$\lim_{x \to 4} (3x + 5) = 17.$$

Answer: We must show, given any $\varepsilon > 0$, there exists a $\delta > 0$ such that

$\|x - a\| < \delta \Rightarrow \| f(x) - L\| < \varepsilon$, or	(State what needs to be proved.)
$\|x - 4\| < \delta \Rightarrow \| 3x + 5 - 17\| < \varepsilon$, or	(Plug in a, f(x), L.)
$\|x - 4\| < \delta \Rightarrow \| 3x - 12\| < \varepsilon$, or	(Do algebra to get R.S. to resemble L.S.)
$\|x - 4\| < \delta \Rightarrow 3\|x - 4\| < \varepsilon$, or	
$\|x - 4\| < \delta \Rightarrow \|x - 4\| < \varepsilon/3$.	(When L.S. resembles R.S., conclude)
So take $\delta \le \varepsilon/3$.	(With this δ all statements above are true.)

Note: Here "R.S." and "L.S." denote "right side" and "left side" of the implication, respectively. Sometimes "R.H.S." and "L.H.S." are used for "right hand side" and "left hand side."

SECTION 2.3

"The lion is not so fierce as they paint him."...George Herbert

EXAMPLE: Find $\lim\limits_{x \to 27} \dfrac{x^{2/3} - 9}{x - 27}$.

Answer:

$$\lim_{x \to 27} \frac{x^{2/3} - 9}{x - 27} = \lim_{x \to 27} \frac{(x^{1/3})^2 - 3^2}{(x^{1/3} - 3)(x^{2/3} + 9x^{1/3} + 3^2)} = \lim_{x \to 27} \frac{(x^{1/3} - 3)(x^{1/3} + 3)}{(x^{1/3} - 3)(x^{2/3} + 9x^{1/3} + 3^2)}$$

$$= \lim_{x \to 27} \frac{(x^{1/3} + 3)}{(x^{2/3} + 9x^{1/3} + 3^2)} = \frac{6}{15} = \frac{2}{15}$$ where we have used the first of the two factoring rules :

$$\boxed{\begin{array}{l} a^3 - b^3 = (a - b)(a^2 + 3ab + b^2) \\ a^3 + b^3 = (a + b)(a^2 - 3ab + b^2) \end{array}}$$

EXAMPLE: Find $\lim_{x \to 0} x\left(1 - \dfrac{2}{x}\right)$.

Answer: We cannot simply plug in x = 0; this leads to an undefined answer. However

$$\lim_{x \to 0} x\left(1 - \frac{2}{x}\right) = \lim_{x \to 0} x - 2 = 0 - 2 = -2.$$

EXAMPLE Find $\lim_{x \to a} \dfrac{f(x) - f(a)}{x - a}$ where f(x) = $\sqrt{x}$.

Answer:

$$\lim_{x \to a} \frac{f(x) - f(a)}{x - a} = \lim_{x \to a} \frac{\sqrt{x} - \sqrt{a}}{x - a}$$

$$= \lim_{x \to a} \frac{\sqrt{x} - \sqrt{a}}{x - a}\left(\frac{\sqrt{x} + \sqrt{a}}{\sqrt{x} + \sqrt{a}}\right)$$

$$= \lim_{x \to a} \frac{x - a}{(x - a)(\sqrt{x} + \sqrt{a})}$$

$$= \lim_{x \to a} \frac{1}{\sqrt{x} + \sqrt{a}}$$

$$= \frac{1}{2\sqrt{a}}.$$

EXAMPLE Find $\displaystyle\lim_{x\to a}\frac{f(x)-f(a)}{x-a}$ where $f(x)=1/x$.

Answer:

$$\lim_{x\to a}\frac{f(x)-f(a)}{x-a}=\lim_{x\to a}\frac{1/x-1/a}{x-a}$$

$$=\lim_{x\to a}\frac{(a-x)/xa}{x-a}$$

$$=\lim_{x\to a}\frac{a-x}{(x-a)xa}$$

$$=\lim_{x\to a}\frac{-1}{xa}$$

$$=-\frac{1}{a^2}.$$

CONTINUITY is defined at a point.
We say f is continuous at x = a if
$$\lim_{x\to a}f(x)=f(a)$$

The equation above says 3 things :
(1) the right side must exist (this means, be a real number, not 0/0 or infinity, etc.)
(2) the left side must exist
(3) the left side must equal the right side (meaning the limiting value equals f (a)).

This says, *WHAT WE PREDICT WILL HAPPEN* (left side).......
.......*ACTUALLY DOES* (right side).

It is not a definition to say f is continuous if you can draw the graph without lifting your pencil (although this is certainly an idea behind the concept). Why?

EXAMPLE: Let $f(x)=\begin{cases}2x+3 & \text{for } x<1\\ 7 & \text{for } x=1\\ 10x-5 & \text{for } x>1\end{cases}$. How can we redefine the function f at

x = 1 to make it continuous there?

Answer:
f approaches 2(1) + 3 = 5 from the left of x = 1, where x < 1, and
f approaches 10(1) - 5 = 5 from the right of x = 1, where x > 1.
So to be continuous at a = 1, f should be defined so that f(1) = 5, not 7.

EXAMPLE: The following functions are not continuous at x = a . Determine which property in the definition of continuity fails.

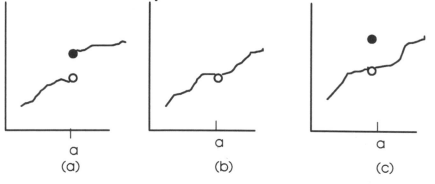

(a)   (b)   (c)

Answer: (a) $\lim\limits_{x \to a}\ f(x)$ does not exist,   (b) f (a) does not exist,   (c) $\lim\limits_{x \to a}\ f(x) \neq f(a)$

SECTION 2.4

" But in science, the credit goes to the man who convinces the world, not to the man to whom the idea first occurs."....Sir Francis Darwin

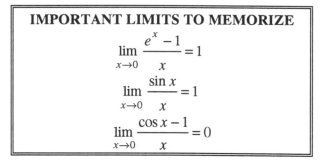

IMPORTANT LIMITS TO MEMORIZE

$$\lim_{x \to 0} \frac{e^x - 1}{x} = 1$$

$$\lim_{x \to 0} \frac{\sin x}{x} = 1$$

$$\lim_{x \to 0} \frac{\cos x - 1}{x} = 0$$

FINDING MORE LIMITS -- USING THE SUBSTITUTION RULE

EXAMPLE:

$\lim\limits_{x \to 0} \dfrac{\sin 2x}{x} = \lim\limits_{u \to 0} \dfrac{\sin u}{u / 2} = \lim\limits_{u \to 0} \dfrac{2 \sin u}{u} = 2.$ We substituted u = 2x : as x → 0, u→ 0.

EXAMPLE:

$\lim\limits_{x \to e} \dfrac{\ln(\ln x)}{\ln x} = \lim\limits_{u \to 1} \dfrac{\ln u}{u} = \dfrac{\ln 1}{1} = 0.$ Here we substituted u = ln x, and see that as

x → e, u → 1.

NOTE: Generally we do not write out the u-substitution, but rather, "think" it. Your teacher may want to see the work, however. It's wise to offer some documentation with all limits you evaluate, to give evidence of your reasoning.

THE SQUEEZING THEOREM

This theorem offers an important way of finding limits when traditional algebraic methods fail. Learn the statement of the theorem literally and conceptually from your text.

EXAMPLE: Using the Squeezing Theorem. Find $\lim\limits_{x \to 0} x \sin\dfrac{1}{x}$.

Answer:

We know that $-1 \le \sin(1/x) \le 1$. Multiplying by x, where x > 0, gives $-x \le x \sin(1/x) \le x$. Multiplying by x, where x < 0, gives $x \le x \sin(1/x) \le -x$. Combining these two inequalities using the definition of absolute value, we obtain $-|x| \le x \sin(1/x) \le |x|$. Now by the

Squeezing Theorem, $\lim\limits_{x \to 0} x \sin\dfrac{1}{x} = 0$.

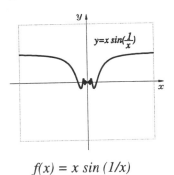

$$f(x) = x \sin (1/x)$$

$$\text{Using } \lim_{x \to 0} \frac{\sin x}{x} = 1$$

(In general one can obtain accurate answers to these limits using the approximation sin x ≈ x for small x. This method is not strictly legal, so we do not endorse it.)

EXAMPLE: Find $\lim\limits_{x \to 0} \dfrac{\sin 3x}{x}$.

Answer: Here we want a "3x" in the denominator to balance "sin 3x". $\lim\limits_{x \to 0} \dfrac{\sin 3x}{x} =$

$\lim\limits_{x \to 0} \dfrac{3}{3} \cdot \dfrac{\sin 3x}{x} = 3 \lim\limits_{x \to 0} \dfrac{\sin 3x}{3x} = \lim\limits_{u \to 0} \dfrac{\sin u}{u} = 3$.

EXAMPLE: Find $\lim\limits_{x \to 0} \dfrac{\sin^2 4x}{x^2}$.

Answer: Rewrite this as

$$\lim_{x\to 0}\frac{(\sin 4x)(\sin 4x)}{x^2}=\lim_{x\to 0}\frac{(\sin 4x)(\sin 4x)}{x^2}=\lim_{x\to 0}\frac{(\sin 4x)(\sin 4x)(16)}{4x\cdot 4x}=16\left(\lim_{u\to 0}\frac{\sin u}{u}\right)^2=16$$

where we let $u=4x$.

EXAMPLE: Find $\lim_{x\to 0}\dfrac{\tan x^2}{5x}$.

Answer: Rewrite with sin and cos. $\lim_{x\to 0}\dfrac{\tan x^2}{5x}=\lim_{x\to 0}\dfrac{\sin x^2}{5x\cos x^2}=$

$$\lim_{x\to 0}\frac{\sin x^2}{5x}\lim_{x\to 0}\frac{1}{\cos x^2}=\lim_{x\to 0}\frac{x\sin x^2}{5x^2}\lim_{x\to 0}\frac{x}{5}\lim_{x\to 0}\frac{\sin x^2}{x^2}\lim_{x\to 0}\frac{1}{\cos x^2}0\cdot 1\cdot 1=0.$$

$$\text{Using }\lim_{x\to 0}\frac{\cos x-1}{x}=0\ \left(\text{also}:\ \lim_{x\to 0}\frac{1-\cos x}{x}=0\right)$$

EXAMPLE: Find $\lim_{x\to 0}\dfrac{1-\cos x^2}{x^2}$.

Answer: Let $u=x^2$. Then $\lim_{x\to 0}\dfrac{1-\cos x^2}{x^2}=\lim_{u\to 0}\dfrac{1-\cos u}{u}=0.$

EXAMPLE: Find $\lim_{x\to 0}\dfrac{1-\cos x}{\sin x}$.

Answer: Remembering that $\sin x\approx x$ near $x=0$, we realize this limit is similar to

$\lim_{x\to 0}\dfrac{(1-\cos x)}{x}$, which we know. So, multiply above and below by x: $\lim_{x\to 0}\dfrac{1-\cos x}{\sin x}=$

$$\lim_{x\to 0}\frac{(1-\cos x)}{x}\cdot\frac{x}{\sin x}=\lim_{x\to 0}\frac{(1-\cos x)}{x}\lim_{x\to 0}\frac{x}{\sin x}=0\cdot 1=0.$$

$$\text{Using }\lim_{x\to 0}\frac{e^x-1}{x}=1$$

EXAMPLE: Find $\lim_{x\to 0}\dfrac{1-e^{2x}}{3x}$.

Answer: We use algebra to make this limit resemble the limit above. Multiply and divide

by 2 and factor a negative sign from the numerator. Then $\lim_{x\to 0}\dfrac{2(1-e^{2x})}{3(2x)}=$

$-\dfrac{2}{3}\lim_{u\to 0}\dfrac{e^u-1}{u}=-2/3,$ where we have substituted $u=2x$.

NOTE: While the laws about continuity of functions in this section may seem completely natural, they follow from laws about limits and illustrate the care we must take when dealing with the concept of limit. We can now say, without blinking, that the function f given by $f(x) = \dfrac{e^{x-5}(\sin 3x)}{\sqrt{x^2+1}}$, is continuous at any x. This is because of rules guaranteeing the continuity of the sum, difference, product, and composition of continuous functions (and the quotient, as long as we don't divide by 0!).

REVIEW OF CONCEPTS

♦ What is the definition of the limit? We say $\lim\limits_{x \to a} f(x) = L$ if.....

♦ One example of the limit that has a geometrical interpretation is the limit of a certain quotient. What is this quotient, and what does it represent geometrically?

♦ A "real-world" application of the limit of a quotient involves time, position, velocity, etc. What is this application, specifically?

♦ Some (at least 3) limits that are important to memorize are...

♦ The Squeezing Theorem states ....

♦ To find the tangent line to the graph of a function f at a point we....

♦ A function f is continuous at x = a if...

♦ Some laws about continuous functions are...

♦ It is important to use radians in Calculus because....

Correct the Error: The following is actual student work. Can you understand what the student thinking of and where the answer can be considered incomplete or wrong?

1. A function is continuous at x = a provided for all values of a, f(x) exists.
2. A function is continuous at x = a provided $\lim\limits_{x \to a} f(x) = \lim\limits_{x \to a} f(a)$.
3. There is no value where the function is undefined.
4. The function is continuous provided $\lim\limits_{x \to a} \dfrac{f(x) - f(a)}{x - a}$ exists.

Solve A Problem

Let f be given by $f(x) = \dfrac{1}{x-5}$. *Find the tangent line to the graph of f at the point x = 4.*

Divide a sheet of paper into two columns. In the left column, write out the steps that solve this problem. Simultaneously, in the right column, explain what you have done at each step. *Often we think we understand an idea until we try to put it in writing!*

SECTION 2.5

"Science demands an undivided allegiance from its followers. In your work and in your research there must always be passion."...Ivan Pavlov

ONE-SIDED LIMITS

There are three situations that can occur when taking limits of quotients:

- $\lim\limits_{x \to a} \dfrac{f(x)}{g(x)}$ gives a real number (possibly 0). Keep it.

- $\lim\limits_{x \to a} \dfrac{f(x)}{g(x)}$ looks like $\dfrac{0}{0}$. Use **algebra** or Squeezing Theorem or substitutions.

- $\lim\limits_{x \to a} \dfrac{f(x)}{g(x)}$ looks like $\dfrac{b}{0}$ where $b \neq 0$. Then there is a vertical asymptote, $x = b$.

Your answer will now be one of the following:
    (1) $+ \infty$ if both one-sided limits give $+ \infty$,
    (2) $- \infty$ if both one-sided limits give $- \infty$, or
    (3)   "Does not exist", if the two one-sided limits do not agree.

NOTE: Vertical asymptotes have the form " **x = b**". They occur in the "middle" part of the graphing plane, like *chimneys to infinity*.

These functions all have vertical asymptotes (the bold lines).

NOTE: Limits do not exist only if the two-sided limits do not agree, or if there is no hope of finding even a one-sided limit (rare, pathological cases.) Some teachers may say a limit "does not exist" if it is ∞ or $-\infty$. Be clear about this with your teacher!

EXAMPLE: Find $\lim\limits_{x \to 0} \dfrac{-x}{x^3}$.

Answer: Since this limit is of the form $\dfrac{0}{0}$, we use algebra (cancel) and obtain $\lim\limits_{x\to 0}\dfrac{-1}{x^2}$.

This limit is of the form $\dfrac{b}{0}$, so we know we have a vertical asymptote, $x = 0$. Now

$\lim\limits_{x\to 0^+}\dfrac{-1}{x^2} = -\infty$ (we think of the denominator as going to a "positivish" zero) and

$\lim\limits_{x\to 0^-}\dfrac{-1}{x^2} = -\infty$ (where the same thing happens). Since the one-sided limits agree,

$\lim\limits_{x\to 0}\dfrac{-x}{x^3} = -\infty.$

$$f(x) = \dfrac{-x}{x^3} = -\dfrac{1}{x^2}$$

USEFUL GRAPHS TO KNOW

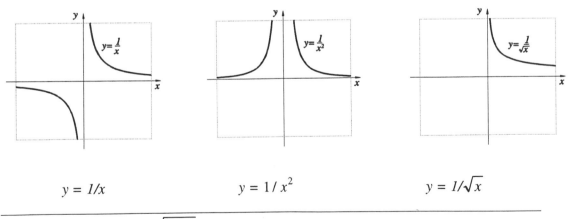

$y = 1/x$ $\qquad\qquad y = 1/x^2 \qquad\qquad y = 1/\sqrt{x}$

EXAMPLE: Find $\lim\limits_{x\to 3}\sqrt{1 - \dfrac{3}{x}}$.

Answer: We need to worry about square roots only when the function under the radical can be zero or negative, and this is one of those times. We need two-sided limits. To find these, we can ask, for example, what happens if x = 3.001 or 2.999?

$\lim\limits_{x\to 3^+}\sqrt{1 - \dfrac{3}{x}} = 0$. But $\lim\limits_{x\to 3^-}\sqrt{1 - \dfrac{3}{x}}$ does not exist. Therefore we have a right-sided, but not left-sided limit. The two-sided limit does not exist.

EXAMPLE: Find $\lim\limits_{x\to 1}\dfrac{|x-1|}{x-1}$.

Answer: Let u = x-1. Then $\lim\limits_{x\to 1}\dfrac{|x-1|}{x-1}=\lim\limits_{u\to 0}\dfrac{|u|}{u}$, which is of the 0/0 form. Now

$\lim\limits_{u\to 0^{+}}\dfrac{|u|}{u}=\lim\limits_{u\to 0^{+}}\dfrac{u}{u}=1$, and $\lim\limits_{u\to 0^{-}}\dfrac{|u|}{u}=\lim\limits_{u\to 0^{-}}\dfrac{-u}{u}=-1$, so the limit does not exist.

EXAMPLE: Find $\lim\limits_{x\to 1}\dfrac{|x-1|}{(x-1)^{2}}$.

Answer: Rewrite as: $\lim\limits_{u\to 0}\dfrac{|u|}{u^{2}}$. Terms are positive, and this limit is $+\infty$

EXAMPLE: Find $\lim\limits_{x\to 5}\dfrac{1}{x(x-5)}$.

Answer: Since this limit is of the form $\dfrac{b}{0}$, we know there will be a vertical asymptote : x = 5.

Now $\lim\limits_{x\to 5^{+}}\dfrac{1}{x(x-5)}=+\infty$ (where we think of the fraction as $\dfrac{1}{5(+0)}$, "+0" representing

"positivish 0"). And $\lim\limits_{x\to 5^{-}}\dfrac{1}{x(x-5)}=-\infty$ (where we think of the denominator as being

"negativish" 0.) If you like, you can plug in numbers like 5.001 and 4.999 which represent

numbers to the right and left of x = 5.. So $\lim\limits_{x\to 5}\dfrac{1}{x(x-5)}$ does not exist.

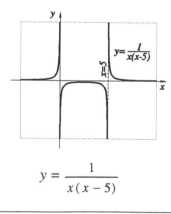

$$y = \dfrac{1}{x(x-5)}$$

TANGLING WITH ∞

We must be cautious when talking about ∞. Even though calculus is probably the first
course in which you have seen this entity, you may have some preconceived ideas.

♦ ∞ is not a number and does not obey the laws of arithmetic. Example: it seems reasonable to think that ∞ + 3 = ∞, but "canceling" ∞, we have **3 = 0**. Canceling ∞ is not legitimate.

♦ ∞ lawfully occurs only in a limit equation, as in $\lim\limits_{x\to 0} \dfrac{1}{x^2} = \infty$.

♦ The limit laws (dealing with sums, differences, products, quotients) do not hold when one of the limits is ∞ or -∞ . For example:

$$\lim\limits_{x\to 0}\; x^2 \cdot \frac{1}{x^2} = 1 \qquad\qquad\text{(like } 0 \cdot \infty = 1)$$

$$\lim\limits_{x\to 0}\; x^2 \cdot \frac{-2}{x^2} = -2 \qquad\qquad\text{(like } 0 \cdot \infty = -2)$$

$$\lim\limits_{x\to 0}\; x^3 \cdot \frac{1}{x^2} = 0 \qquad\qquad\text{(like } 0 \cdot \infty = 0)$$

$$\lim\limits_{x\to 0}\; x^2 \cdot \frac{1}{x^4} = \infty \qquad\qquad\text{(like } 0 \cdot \infty = \infty)$$

$$\text{Also: } \lim\limits_{x\to 0}\; \left(2 + \frac{1}{x^2}\right) - \frac{1}{x^2} = 2 \qquad\text{(like } \infty - \infty = 2)$$

$$\lim\limits_{x\to 0}\; \frac{2}{x^2} - \frac{1}{x^2} = \infty \qquad\qquad\text{(like } \infty - \infty = \infty)$$

Can you come up with limits that seem to say ∞ - ∞ = 3? ∞ - ∞ = 0? ∞ - ∞ = -∞?

EXAMPLE: Discuss the continuity of f, where $f(x) = \ln(x^2 - 1)$.

Answer: This function is continuous at x if $x^2 - 1 > 0$, that is, if | x | > 1. (It is not defined if x = ±1.) The function approaches -∞ as x approaches ±1. Vertical asymptotes are the lines x = ±1.

EXAMPLE: Discuss whether the function below is right or left continuous, continuous, or not continuous at the designated points.

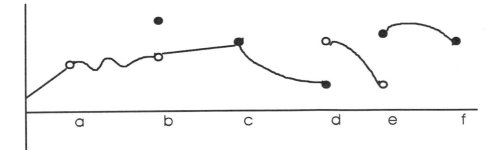

SECTION 2.6

"Education is... hanging around until you've caught on"....Robert Frost

CONTINUITY OF FUNCTIONS ON INTERVALS

It is easier to find where a function is not continuous. Watch out for:

1. division by 0.
2. square roots of negative numbers
3. some trigonometric functions (like tan x, which is a quotient of trig functions)
4. piecewise defined functions

EXAMPLE: Let f be given by $f(x) = \dfrac{1-\cos x}{x}$. Explain why f is continuous on

$(-\infty, 0), (0, \infty)$.

Answer: The function f is continuous since $\displaystyle\lim_{x \to a} \dfrac{1-\cos x}{x} = \dfrac{1-\cos a}{a}$ if $a \neq 0$.

EXAMPLE: Where is $f(x) = \sqrt{x^2 - 4}$ continuous?

Answer: Since f is continuous wherever f is defined, we require $x^2 - 4 \geq 0$ or $x^2 \geq 4$ or $|x| \geq 2$.

NOTE: THE INTERMEDIATE VALUE THEOREM is one of the most hard-working, behind-the-scenes theorem in all of calculus. The drawing below demonstrates the essential meaning: a continuous function on a closed interval [a,b] "assumes" (takes on) every y-value between f(a) and f(b).

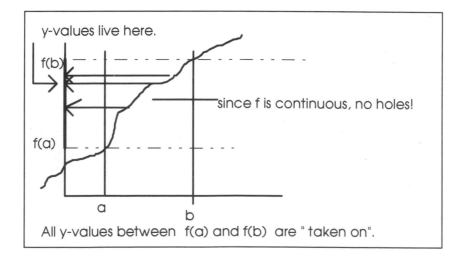

y-values live here.

f(b)

since f is continuous, no holes!

f(a)

a   b

All y-values between f(a) and f(b) are "taken on".

MORE PRECISELY (explaining what is meant by values being "taken on"):

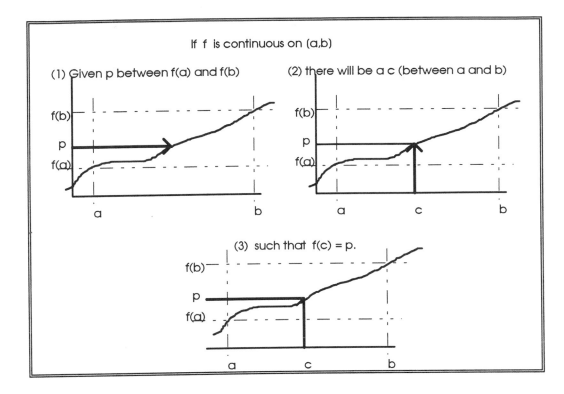

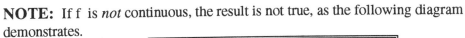

NOTE: If f is *not* continuous, the result is not true, as the following diagram demonstrates.

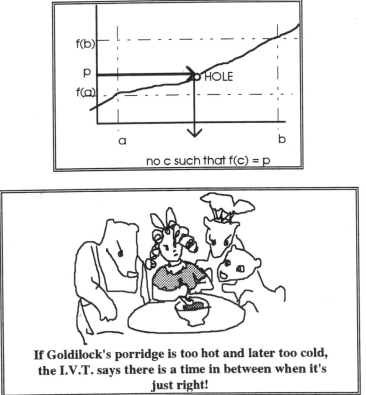

EXAMPLE: Using the I.V.T. to find where functions are equal. Show that there is a value of x in [0, π/2] for which sin x = 1- x.

We see from the graphs below that the functions y = sin x and y = 1- x intersect. But now we have a means to prove that this must happen. This is the typical method:

Consider the function f(x) = sin x - (1- x) . We will show f is zero somewhere on the given interval, and there, sin x = 1-x. Now f(0) = -1 < 0 and f(π/2) = -1+π/2 > 0 . Since f is continuous, the I.V.T. tells us that there must be a point x = c in the interval (0, π/2) such that f(c) = 0, and so sin c = 1- c.

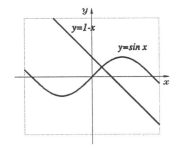

y = sin x and y = 1-x on the same graph.

NOTE: The Intermediate Value Theorem is an *Existence Theorem*: it does not tell us **where** this point occurs.

💣 DRAWBACK TO THE GRAPHING APPROACH

It might seem to you that a graph can show if limits exist, when a function is continuous, or whether the "c" promised by the Intermediate Value Theorem exists.   However, many teachers would not consider a graph an adequate explanation: they are looking for careful reasoning using the methods of this section.

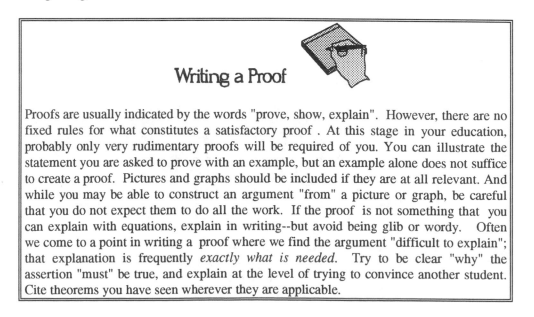

Writing a Proof

Proofs are usually indicated by the words "prove, show, explain". However, there are no fixed rules for what constitutes a satisfactory proof . At this stage in your education, probably only very rudimentary proofs will be required of you. You can illustrate the statement you are asked to prove with an example, but an example alone does not suffice to create a proof. Pictures and graphs should be included if they are at all relevant. And while you may be able to construct an argument "from" a picture or graph, be careful that you do not expect them to do all the work. If the proof is not something that you can explain with equations, explain in writing--but avoid being glib or wordy.   Often we come to a point in writing a proof where we find the argument "difficult to explain"; that explanation is frequently *exactly what is needed*.   Try to be clear "why" the assertion "must" be true, and explain at the level of trying to convince another student. Cite theorems you have seen wherever they are applicable.

EXAMPLE: PROVE that there is a value x = c in [0, $\pi/2$] where cos x = x.

☹**NOT GOOD:** The graph of y = cos x crosses the graph of y = x in this interval as we can see from the picture provided...

*This explanation , however believable, does not suffice as a proof. We want to know **why this result must happen.***

☺**BETTER:** Consider the function given by f (x) = cos x - x.   Notice f (0) = 1 > 0 and f($\pi/2$)= -1 < 0. Since f is continuous on [0, $\pi/2$], the Intermediate Value Theorem guarantees the existence of c in (0, $\pi/2$) such that f (c) = 0.   At this value, cos c - c = 0, or cos c = c.

This explanation anchors the result on a theorem which deals with an essential feature of continuous functions and explains how the hypotheses of the theorem are met (f is continuous on a closed interval).

AN APPLICATION OF THE INTERMEDIATE VALUE THEOREM:
THE BISECTION METHOD

✋**CAUTION:** The *Bisection Method* and the *Intermediate Value Theorem* should not be confused. The Bisection Method is a practical technique for finding roots of equations, using the theoretical result of the Intermediate Value Theorem.

THE ERROR TERM: The error in approximating the true value of f(c) (which we might not know) by some value p , is given by the absolute value of the difference:
$$\textbf{ERROR} = |\, \textbf{f(c) - p}\,|.$$

EXAMPLE: Use the Bisection Method to approximate $\sqrt{11}$ with an error less than 1/16.

Answer: Consider f(x) = $x^2 - 11$. We must find an interval (preferably short) on which f(x) assumes both "+" and " - " values. Since f is continuous everywhere, the I.V.T. promises there will be an x-value where f(x) = 0.   Thinking $\sqrt{11} < \sqrt{12} = 2\sqrt{3}$, where $\sqrt{3} \sim$ 1.7, we will use the interval [2(1.6),   2(1.8)] or [3.2, 3.6]. (Often the interval will be given.) Note f(3.2) < 0 and f(3.6) > 0, so this is a suitable interval. (If it seems like "cheating" to have an idea about $\sqrt{3}$, use the interval [3,4] instead.)

INTERVAL	LENGTH	MIDPOINT, x = c	f(c)
[3.2, 3.6]	.4	(3.6+3.2)/2= 3.4	.56 > 0. Use left half. Replace 3.6 with 3.4
[3.2, 3.4]	.2	(3.2+3.4)/2=3.3	-.11< 0 Use right half. Replace 3.2 with 3.3
[3.3, 3.4]	Interval has length .1 < 1/8. We know the value of c for which f(c) = 0 lies in this interval.	(3.3+3.4)/2= 3.35 **STOP!** The distance from 3.35 to the value we want is less than 1/16.	

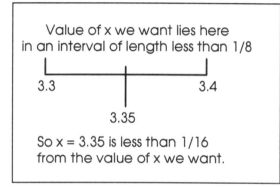

Thus our approximate solution to $f(x) = 0$, or our approximation to $\sqrt{11}$ is 3.35. While we do not have $f(c) = 0$ we are within 1/16 of the exact value of $\sqrt{11}$.

𝒮elf-assessing

Getting a grade *after* a test may not always help you to determine how your methods of studying are working. Try this type of exercise--perhaps it will help you *before* the next test.

BEFORE A TEST, ON A SHEET OF PAPER	AFTER A TEST, ON A SHEET OF PAPER
Write down what you feel are your strengths and weaknesses in this chapter; what you think you will do well on, what you think you will have trouble with.	After the test, write down where your strengths and weaknesses lay, and how the way you studied might have affected your performance. What can you do to improve your performance in the future?

CORRECT THE ERROR:

The problem here was to state the **Intermediate Value Theorem**. Compare the correct statement of the theorem with these statements. All of these statements have some truth component, but need to be made more carefully. Can you rewrite the statements? Many of these statements confuse the *function* with the *graph of the function*. Remember the function *takes on* y-values *at* x-values.

1. If points (a, f(a)) and (b,f(b)) are on a continuous function then all points in between these two points exist.
2. If a given continuous function has points greater than 0 and less than 0 then there exists some point on the function where it equals 0.
3. If a function is continuous on a given interval, [a,b], the function will cross every y-value in the interval.
4. If a function is continuous between two points [a,b], there is a point c where c is between those two points.
5. If a function is continuous and exists at two points, then the function takes on every value between those two points.

PROVIDE A GRAPH OF

a function that has a right sided limit but not a limit at x = a.	
a function that has no limit at x = a although f(a) exists.	
a function that is continuous from the right but not from the left at x = a.	
a function that does not satisfy the conditions of the Intermediate Value Theorem although the function is continuous.	
a function that does not satisfy the conditions of the Intermediate Value Theorem although the function is defined on a closed interval.	
a function that is continuous but has no tangent line at a point	
a function that is continuous everywhere except at x = a, and there it could be redefined so as to make it continuous.	
a function that has a limit at x = a , but is not continuous there.	
a function whose limit requires the Squeezing Theorem	

REVIEW OF CHAPTER

1. When finding limits, give some motivation for your answer--don't simply write down the answer. Be careful with your notation. It is difficult for a teacher to assign partial credit for a wrong limit if there is no written documentation.

2. You are probably not used to having to memorize definitions and theorems. Practice by writing them down, then check back with the text. Know why the hypotheses of the theorems are necessary.

3. Work problems in the review at the end of the chapter. Review all homework problems, all worked examples in the text, and your notes. Teachers write tests based on what they have lectured on, and they remember, even if we don't! While practicing problems, try to "typecast" problems. One category would be limits involving sin x and x. Scramble the order in which you solve these problems, so as not to fall into a rote mode. On a test, problems can be presented in any order, and "recognition" memory may fail you. Devise a sample test of problems and work your sample test with a group of others.

KEEPING A DOUBLE - ENTRY JOURNAL

This is a useful way to keep track of what you know and how to learn. Keep a notebook with two columns. A few times a week, keep a record of your study patterns, what subjects are causing problems, questions you need to ask, what you do and don't feel solid with. A week later, in the right hand column, compare how you feel with how you felt before. Look back at all of your comments before a quiz or test. You will have a record of your progress that will be encouraging and useful. It has been verified again and again that writing things down helps your retention. And what you have written will give you an invaluable reflection of *how* you learn . Some students learn in "jumps" from plateau to plateau, like the greatest integer function; other students seem to learn in a more linear (straight line) way.

CHAPTER 3

"If I have seen farther than Descartes, it is by standing on the shoulders of giants"...Isaac Newton

In this chapter, we study the particular limit of a quotient that was presented in Chapter Two, and that will be identified as the "derivative". We discover ways in which the derivative provides information about a function, and how to apply this information.

Careful attention must be paid to the new notation introduced. Lots of practice using this notation is necessary. Like anything new that is worth learning, this material may seem frustrating at first. Go carefully and read slowly!

WHAT IS PROPERLY SIMPLIFIED is often a question of taste. There are no clear cut rules. Sometimes it might be troublesome to get your answers to agree, in form, with the answers in the back of the text; nevertheless, this is probably a good exercise in algebra.

HINT FOR SIMPLIFYING:

Frequently, simplified answers are written in factored form.

Positive Integer powers: Factor out the term having lowest power:

$$x^3 + 3x^2 + 4x = x(x^2 + 3x + 4) \ .$$

Any other powers: Same rule applies.

$$(x+3)^{-1/3} + 2(x+3)^2 + 3(x+3)^{5/3} = (x+3)^{-1/3}(1 + 2(x+3)^{2+1/3} + 3(x+3)^{5/3+1/3})$$

$$= (x+3)^{-1/3}[1 + 2(x+3)^{7/3} + 3(x+3)^2]$$

SECTION 3.1

"Physical concepts are free creations of the human mind, and are not, however it may seem, uniquely determined by the external world."...Einstein

THE DERIVATIVE

The derivative of f at x = a is the familiar limit of a quotient we saw in Section 2.1 .

$$f'(a) = \lim_{x \to a} \frac{f(x) - f(a)}{x - a}$$ is the derivative of f with respect to x at x = a, if this limit

exists. We also write this limit as: $$f'(a) = \frac{df}{dx}\bigg|_{x=a}$$. This expression gives a **number**.
When we want the derivative expressed as a function of x, we write:

$$f'(x) = \lim_{t \to x} \frac{f(t) - f(x)}{t - x}$$. We also write this latter expression as $\frac{df}{dx}$. This expresses

the derivative **in terms of the variable x.** (When using these expressions, we must be careful that df/dx and f'(x) exist.)

GEOMETRICAL INTERPRETATION OF THE DERIVATIVE: The derivative of f with respect to x, evaluated at x = a, is **the slope of the tangent line to the graph of f at x = a.**

NOTE: Many students think the derivative gives the tangent line itself. It doesn't; the derivative at x = a is *a real number*.

NOTE: The quantity $$\frac{f(x) - f(a)}{x - a}$$, without the limit, is **the slope of the secant line**
joining (a, f(a)) and (x, f(x)). It represents the "average" slope of the function between these two points.

NOTE:
(1) f does not have a derivative at cusps or sharp points on its graph. Functions with derivatives are "smoother" than functions without derivatives.
(2) If f has a derivative at x = a, then f must be continuous at x = a. In order for the limit $$f'(a) = \lim_{x \to a} \frac{f(x) - f(a)}{x - a}$$ to be defined as the denominator approaches 0, the numerator must. (This essentially, provides the continuity of f at x = a.)
(3) The fact that f is continuous is not a guarantee that f has a derivative. For example, the function f(x) = |x| is continuous everywhere, but has not derivative at x = 0.
(4) If a function has a derivative at a point, it is said to be "differentiable" there.
(5) The process of finding the derivative is called "differentiation". We "differentiate f at the point x=a when we find $$f'(a) = \lim_{x \to a} \frac{f(x) - f(a)}{x - a}$$

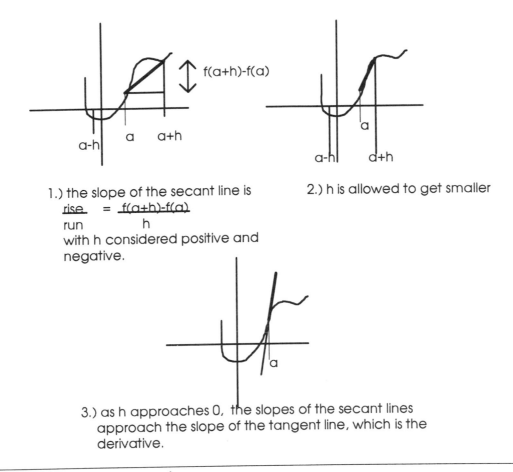

1.) the slope of the secant line is

$$\frac{rise}{run} = \frac{f(a+h)-f(a)}{h}$$

with h considered positive and negative.

2.) h is allowed to get smaller

3.) as h approaches 0, the slopes of the secant lines approach the slope of the tangent line, which is the derivative.

EXAMPLE: Let $f(x) = \dfrac{1}{x}$. Find the derivative of f at $x = 4$.

Answer: $\boxed{f'(a) = \lim_{x \to a} \dfrac{f(x) - f(a)}{x - a}} = \lim_{x \to 4} \dfrac{\dfrac{1}{x} - \dfrac{1}{4}}{x - 4}$

$= \lim_{x \to 4} \dfrac{4 - x}{4x(x - 4)} = -\dfrac{1}{16}$. We can write this as $f'(4) = -\dfrac{1}{16}$ or $\dfrac{df}{dx}\bigg|_{x=4} = -\dfrac{1}{16}$.

EXAMPLE: Let $f(x) = \dfrac{1}{x}$. Find the derivative of f as a function of x.

Answer: $\boxed{f'(x) = \lim_{t \to x} \dfrac{f(t) - f(x)}{t - x}} = \lim_{t \to x} \dfrac{\dfrac{1}{t} - \dfrac{1}{x}}{t - x} = \lim_{t \to x} \dfrac{-(t - x)}{tx(t - x)} = -\dfrac{1}{x^2}$.

Note: the answer comes out in terms of "x" because we let "x" take the place of "a" in the formula in the previous exercise. Then we were forced to replace the "x" in that exercise by "t" ! But the method is essentially the same in these two exercises.

EXAMPLE: If $f(x) = \dfrac{1}{x}$, find $f'(-3)$.

Answer: From the previous example, we have created a **formula:** $f'(x) = -\dfrac{1}{x^2}$. We can plug in x = -3, so $f'(-3) = -\dfrac{1}{9}$. We can also write: $\left.\dfrac{df}{dx}\right|_{x=-3} = -\dfrac{1}{9}$.

EXAMPLE: At what points below does the derivative of the graphed function fail to exist?

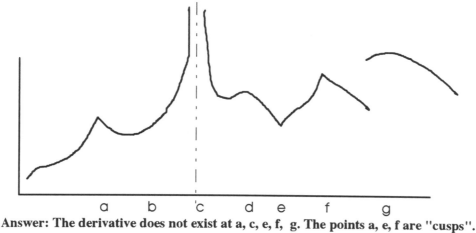

Answer: The derivative does not exist at a, c, e, f, g. The points a, e, f are "cusps".

NOTE: We saw the equation of the tangent line in a previous chapter. We add the word "derivative" to this equation now.

THE EQUATION OF THE TANGENT LINE TO THE GRAPH OF f AT (x_o, y_o)
is given by $\boxed{y - y_o = m(x - x_o)}$ where the slope, m, **is the derivative of f** at (x_o, y_o).

NOTE: The normal line at (x_o, y_o) has slope equal the negative reciprocal of the tangent line's slope. So the slope of the normal line is $-\dfrac{1}{m}$, and the equation of the normal line is given by $y - y_o = -\dfrac{1}{m}(x - x_o)$.

EXAMPLE: Let f be given by $f(x) = \dfrac{1}{x}$. Find the **equation of the tangent line** at (-3, -1/3) to the graph of the function f .

Answer: We have just seen that the slope of the tangent line is the derivative when x = -3, is f'(-3) = $-\dfrac{1}{9}$. The tangent line shares a point in common with the curve: $(-3, -\dfrac{1}{3})$.

Thus the equation of the tangent line is $y - y_o = m(x - x_o)$ or $\boxed{y + \dfrac{1}{3} = -\dfrac{1}{9}(x + 3),}$ or

simplified, $y = -\dfrac{1}{9}x - \dfrac{2}{3}$.

EXAMPLE: Find **the equation of the normal line** to the graph of the function f above.

Answer: The slope of the normal line is 9. The equation of the normal line is

$\boxed{y + \dfrac{1}{3} = 9\ (x + 3)}$ or simplified, y = 9 x +80/3.

EXAMPLE: Let $f(x) = 5$. Find the derivative of f as a function of x. (Note: f is a constant function).

Answer: $\boxed{f'(x) = \lim_{t \to x} \dfrac{f(t) - f(x)}{t - x}} = \lim_{t \to x} \dfrac{5 - 5}{t - x} = \lim_{t \to x} 0 = 0$. The limit describes

the behavior of $\dfrac{5 - 5}{t - x}$ when t is near x , not equal to x, and the denominator is never

actually 0.

NOTE: The derivative of any constant function is 0.

EXAMPLE: Estimate the slope of the secant line between the x- and y-intercepts of the function graphed below.   Estimate the derivative at the x-intercept. Estimate the derivative at x = 2.

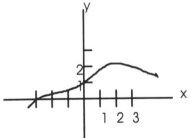

Answer: The slope of the secant line between (-3,0) and (0,1) is m = rise/run , approximately 1/3. The slope of the tangent line at (0,1) we might estimate as 1. At x = 2, we might estimate it to be 0.

COMMON ERRORS

EXAMPLE: If $f(x) = \dfrac{1}{x}$, find f '(5).

INCORRECT: Since $f(5) = \dfrac{1}{5}$, we have $f'(5) = 0$. (Don't plug in the value x = a until the derivative is taken.)

CORRECT: $f'(x) = -\dfrac{1}{x^2}$, so $f'(5) = -\dfrac{1}{25}$.

EXAMPLE: If $f(x) = \dfrac{1}{x}$, find f'(x)

INCORRECT: $f(x) = \dfrac{1}{x} = \dfrac{-1}{x^2}$. (The function and the derivative have been equated. These are two different functions.)

CORRECT: Since $f(x) = \dfrac{1}{x} = x^{-1}$, we have $f'(x) = x^{-2} = -\dfrac{1}{x^2}$.

EXAMPLE: For the function given by $f(x) = \begin{cases} 3-x & \text{for } x \geq 3 \\ 2+x & \text{for } x < 3 \end{cases}$, find f'(5).

Answer: For the point x = 5 we use the definition of f that applies for x ≥3; namely, f(x) = 3 - x. Thus f ' (5) = -1.

EXAMPLE: Let f(x) = | x | - x . Determine whether f has a derivative at x = 0.

Answer: $\boxed{f'(a) = \lim_{x \to a} \dfrac{f(x) - f(a)}{x - a}} = \lim_{x \to 0} \dfrac{|x| - x - 0}{x - 0} = \lim_{x \to 0} \dfrac{|x|}{x} - 1.$ This limit

does not exist because $\lim_{x \to 0} \dfrac{|x|}{x} = \begin{cases} 1 \text{ if x approaches 0 from the right} \\ -1 \text{ if x approaches 0 from the left} \end{cases}$.

The function does *not* have a derivative at x = 0. That is, $\lim_{x \to 0^+} \dfrac{|x|}{x} \neq \lim_{x \to 0^-} \dfrac{|x|}{x}$.

NOTE: f(x) = |x| DOES have a derivative if x ≠ 0. This derivative is:

$$f'(x) = f'(x) = \begin{cases} 1 & \text{if } x > 0 \\ -1 & \text{if } x < 0 \end{cases}$$

EXAMPLE: Let $f(x) = \begin{cases} x^2 - 3 & \text{for } x \geq 3 \\ 2x & \text{for } x < 3 \end{cases}$. Does f have a derivative at x = 3?

Answer: f is defined piecewise, and f is differentiable "on pieces". We see that f is even continuous at x = 3, because both definitions of f agree when x = 3 . Now, from the definition of f on the interval x ≥ 3, where $f(x) = x^2 - 3$, we have

$\lim\limits_{x \to 3^+} \dfrac{f(x) - f(3)}{x - 3} = 2x$. And on the interval $x < 3$, where $f(x) = 2x$,

$\lim\limits_{x \to 3^-} \dfrac{f(x) - f(3)}{x - 3} = \lim\limits_{x \to 3^-} \dfrac{2x - (3^2 - 3)}{x - 3} = 2$. Since the right and left hand limits in the

definition of the derivative do not agree, f has no derivative at $x = 3$.

SECTION 3.2

"The independent scientist who is worth the slightest consideration as a scientist has a consecration which comes entirely from within himself: a vocation which demands the possibility of supreme self-sacrifice."....Norbert Wiener

---

FORMULAS TO MEMORIZE:

$$\frac{d}{dx} x^n = nx^{n-1}$$

$$\frac{d}{dx} \sin x = \cos x$$

$$\frac{d}{dx} \cos x = -\sin x$$

NOTE: At this point we know the formula $\dfrac{d}{dx} x^n = nx^{n-1}$ ONLY for **n** a positive integer. As it turns out, this formula will hold for any real-valued **n**. But we will need to wait for this result.

EXAMPLE: If $f(x) = e^{2x}$, find the derivative of f as a function of x: .

Answer:

$$\boxed{f'(x) = \lim_{h \to 0} \frac{f(x+h) - f(x)}{h}} = \lim_{h \to 0} \frac{e^{2(x+h)} - e^{2x}}{h} = \lim_{h \to 0} \frac{e^{2x}e^{2h} - e^{2x}}{h} =$$

$$\lim_{h \to 0} \frac{e^{2x}(e^{2h} - 1)}{h} = \lim_{h \to 0} \frac{2e^{2x}(e^{2h} - 1)}{2h} = \lim_{u \to 0} (2e^{2x})\frac{e^u - 1}{u} = 2e^{2x} \text{, where u = 2h.}$$

NOTE: We have described the derivative as the slope of the tangent line to the graph of a function. We can go one step further and *graph the derivative*, or graph this slope. At first, this may seem confusing. The graph of the derivative is just a record, at every point, of the slope of the tangent line to the function .

EXAMPLE: Show that the graph of $y = \cos x$ gives the derivative of $y = \sin x$.

Hint: The beginning of an explanation follows. (Try to make it more general.)

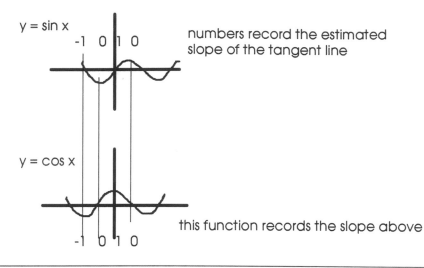

numbers record the estimated
slope of the tangent line

this function records the slope above

EXAMPLE: Let $f(x) = x^2 + \sqrt{x}$. Show f is differentiable on $(0, \infty)$.

Answer: We need only show $\boxed{f'(x) = \lim_{h \to 0} \dfrac{f(x+h) - f(x)}{h}}$ exists for $x > 0$. But

$$\lim_{h \to 0} \frac{f(x+h) - f(x)}{h} = \lim_{h \to 0} \frac{(x+h)^2 + \sqrt{x+h} - (x^2 + \sqrt{x})}{h}.$$

We might worry that $\sqrt{x+h}$ is undefined. But if $x > 0$, (as it is when $x \, \varepsilon \, (0,\infty)$), there is no problem: h may be positive or negative as long as it is small -- and h is considered to be small. Therefore $x + h > 0$. Now the limit above is:

$$\lim_{h \to 0} \frac{x^2 + 2xh + h^2 + \sqrt{x+h} - x^2 - \sqrt{x}}{h} = \lim_{h \to 0} \frac{+ 2xh + h^2 + \sqrt{x+h} - \sqrt{x}}{h} =$$

$$\lim_{h \to 0} \frac{2xh + h^2 + \sqrt{x+h} - \sqrt{x}}{h} = \lim_{h \to 0} \frac{2xh + h^2}{h} + \frac{\sqrt{x+h} - \sqrt{x}}{h} =$$

$$\lim_{h \to 0} (2x + h) + \lim_{h \to 0} \frac{(\sqrt{x+h} - \sqrt{x})}{h} \frac{(\sqrt{x+h} + \sqrt{x})}{(\sqrt{x+h} + \sqrt{x})} =$$

$$2x + \lim_{h \to 0} \frac{x+h - x}{h(\sqrt{x+h} + \sqrt{x})} = 2x + 1 + \lim_{h \to 0} \frac{h}{h(\sqrt{x+h} + \sqrt{x})} = 2x + \frac{1}{2\sqrt{x}}.$$

As long as $x > 0$, all steps above are valid. So f is differentiable when $x > 0$.

NOTE: The derivative dy/dx expresses the **"rate of change of y with respect to x"**. So for example, dA/dr is "the rate of change of A with respect to r." Think about climbing a mountain: then the "instantaneous slope", or the immediate rate of change in height with respect to lateral distance (if there is such a concept!) is what we mean by the derivative.

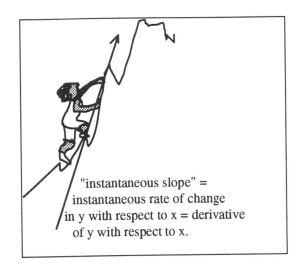

"instantaneous slope" = instantaneous rate of change in y with respect to x = derivative of y with respect to x.

EXAMPLE:

1.Find the rate of change of the area of a circle with respect to its radius.

Answer: This is given by dA/dr or A ' (r), where $A = \pi r^2$. From the formulas for finding the derivative, dA/dr = $2\pi r$.

2. Find the rate of change of the volume of a cylinder with respect to its radius.

Answer: Since the volume of a cylinder is $V = \pi r^2 h$ (we think of h as constant since it is not changing) dV/dr = $2\pi r$h.

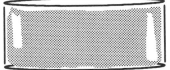

The volume of the cylinder changes as the radius changes. This rate of change is given by dV/dr.

SECTION 3.3

" Well, if I called the wrong number, why did you answer the phone?"... James Thurber

DIFFERENTIATION RULES

$$POWER\ RULE:\ \frac{d}{dx}x^n = nx^{n-1},\ if\ n\ is\ an\ integer.$$

EXAMPLE: Find the derivative of $f(x) = \dfrac{6}{x^8}$.

Answer: Rewrite as $f(x) = 6x^{-8}$. By our new techniques, $\dfrac{df}{dx} = 6(-8)x^{-9} = -48x^{-9}$.

NOTE: " $\mathbf{D_x}$" stands for "derivative with respect to x". We use this notation in this section for its convenience in helping you to memorize formulas (hopefully). Practice saying the formulas aloud. You can read $\mathbf{D_x}$ as "D sub x" or more simply as "$\mathbf{D}$".

$$PRODUCT\ RULE: \quad \frac{df}{dx} = (D_x FIRST)(SECOND) + (FIRST)(D_x SECOND)$$

(It helps if you practice saying these rules aloud! We mean by "FIRST", the "first factor", and by "SECOND", the "second factor")

EXAMPLE: Differentiate $f(x) = (x^2 + 3x + 5)(\sin x)$

Answer: In this case: $\dfrac{df}{dx} = (2x + 3)(\sin x) + (x^2 + 3x + 5)(\cos x)$. (Simplify.)

$$TRIPLE\ PRODUCT: \quad \frac{df}{dx} = (D_x FIRST)(SECOND)(THIRD) +$$
$$(FIRST)(D_x SECOND)(THIRD) + (FIRST)(SECOND)(D_x THIRD)$$

EXAMPLE: Differentiate $f(x) = (x^2 + 2x)(\sin x)(\cos x)$

Answer: In this case the unsimplified answer is:

$\dfrac{df}{dx} = (2x + 2)(\sin x)(\cos x) + (x^2 + 2x)(\cos x)(\cos x) + (x^2 + 2x)(\cos x)(-\sin x)$.

(Simplify.)

$$QUOTIENT\ RULE:$$
$$\frac{df}{dx} = \frac{(BOTTOM)(D_x TOP) - (TOP)(D_x BOTTOM)}{(BOTTOM)^2}$$

Practice saying this aloud ! Many students confuse this formula. The position of the minus signs makes a difference.

EXAMPLE: Differentiate $f(x) = \dfrac{\sin x}{x^2 + 4x + 6}$

Answer: Our derivative is: $\dfrac{df}{dx} = \dfrac{(x^2 + 4x + 6)(\cos) - (\sin x)(2x + 4)}{(x^2 + 4x + 6)^2}$. (Simplify!)

NOTE: Because of the quotient rule, we can now find derivatives of all trigonometric functions. Memorize this list. Be able to find these derivatives *with speed!*

FUNCTION	DERIVATIVE
sin x	cos x
cos x	−sin x
tan x	$\sec^2 x$
sec x	sec x tan x
csc x	-csc x cot x
cot x	- $\csc^2 x$

Note that all " CO - " functions have negative derivatives.

Quick self-quiz:

Function	$\cos x$	$\tan x$	$\sin x$	$\csc x$	$\cot x$	$\cot x$
Derivative?						
Function	$\sin x$	$\cos x$	$\csc x$	$\cot x$	$\tan x$	$\sec x$
Derivative?						

EXAMPLE: Suppose $f(x) = ax^2 + bx$, and the line $y = x + 4$ is the tangent line to the graph of f at the point (1,5). Find the function f.

Answer: Since (1,5) is a point on the graph of the function, $5 = a(1^2) + b(1)$, or a + b = 5. We also know that the slope of the tangent line, given as $y = x + 4$, is m =1. Since f '(x) = 2a x + b, and f '(1) = 1 (given), we have f '(1) = 2a + b = 1. Solving these two equations for a and b, we find that a = -4 and b = 9 so $f(x) = -4x^2 + 9x$.

COMMON ERRORS

INCORRECT: If $y = x^3 + 3x + 2$, then $\dfrac{dy}{dx}(x^3 + 3x + 2) = 3x^2 + 3$.

CORRECT: If $y = x^3 + 3x + 2$, then $\dfrac{d}{dx}(x^3 + 3x + 2) = 3x^2 + 3$.

INCORRECT: If y = sec x then dy/dx = sec tan x.
CORRECT: If y = sec x then dy/dx = sec x tan x.

SECTION 3.4

" We should all be concerned about the Future because we will have to spend the rest of our lives there."...Charles Francis Kettering

---

TWO NEW DERIVATIVES:

If $f(x) = \ln x$, then $f'(x) = \dfrac{1}{x}$

If $f(x) = e^x$, then $f'(x) = e^x$

NOTE: It will be useful from now to think of the rules for derivatives in terms of "*actions*". For example, although these rules will soon be modified, it is useful to think:

♦ the derivative of $\ln \square$ is "One over $\square$" or "$1/\square$".

♦ the derivative of $\cos \square$ is $-\sin \square$.

♦ the derivative of $e^{\square}$ is $e^{\square}$.

THE CHAIN RULE

When a function is a composite function (as with $f \circ g$), we must use the chain rule to calculate derivatives.

THOUGHT PROBLEM: If
MARY runs 5 times as fast as JIM and
JIM runs 4 times as fast as BOB,
how much faster does MARY run than BOB?

We multiply rates: Mary runs $5 \cdot 4$, or 20 times faster than Bob. But derivatives represent rates. In fact, when we find the derivative of the composition of functions, we multiply derivatives. But be careful how this is done!

$$\frac{d(f \circ g)}{dx} = \frac{df}{dg} \cdot \frac{dg}{dx} = \frac{df}{dg} \cdot \frac{dg}{dx}$$

$\uparrow$ MARY to BOB $\uparrow$ MARY to JIM $\uparrow$ JIM to BOB

Often this expression is written with **u**:

$$\frac{d(f(u(x)))}{dx} = \frac{df}{du} \cdot \frac{du}{dx}$$

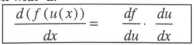

EXAMPLE: If $f(x) = (5x+3)^3$, find the derivative of f with respect to x.

Answer: This function is the composite: $x \xrightarrow{5(\ )+3} 5x+3 \xrightarrow{(\ )^3} (5x+3)^3$.

(1) the derivative of the $5x+3$ function, or the $5(\ )+3$ function is **5**.
(2) the derivative of the cubing function $(\ )^3$ is $3(\ )^2$.
(3) Consequently, we MULTIPLY these derivatives in reverse order.

$$f'(x) = 3(5x+3)^2(5) \text{ or simplifying, } f'(x) = 15(5x+3)^2$$
$$\uparrow \qquad \uparrow$$

deriv of cubing   deriv of 5 ( )

NOTE: $\cos^4 x$ is the same as $(\cos x)^4$. But $\cos x^4$ is the same as $\cos(x^4)$.

EXAMPLE: If $f(x) = \cos^5(x^3+1)$, find f '(x) .

Answer: This function is the composite:
$$x \xrightarrow{(\ )^3+1} x^3+1 \xrightarrow{\cos(\ )} \cos(x^3+1) \xrightarrow{(\ )^5} \cos^5(x^3+1)$$
$$\uparrow \qquad\qquad \uparrow \qquad\qquad \uparrow$$

this deriv is 3 ( )   this deriv is - sin( )   this deriv is 5 ( )^4

So working backwards,

$$f'(x) = 5(\cos(x^3+1))^4(-\sin(x^3+1)) \quad (3x^2) = \quad -15x^2(\cos^4(x^3+1))(\sin(x^3+1)).$$
$$\uparrow \qquad\qquad \uparrow \qquad\qquad \uparrow$$

deriv of ( )^5 deriv of cos( )   deriv of ( )^3 + 1

COMBINING CHAIN RULE AND PRODUCTS AND QUOTIENTS

A good rule of thumb is: look for products and quotients before using the chain rule.

EXAMPLE: Find f '(x) where $f(x) = e^{2x}\cos(x^2+1)$.
Answer: In a diagram, the product looks "split" -- because of a product.

$$x \longrightarrow \begin{cases} \xrightarrow{2(\ )} 2x \xrightarrow{e^{(\ )}} e^{2x} \\ \xrightarrow{(\ )^2+1} x^2+1 \xrightarrow{\cos(\ )} \cos(x^2+1) \end{cases}$$

$$\frac{df}{dx} = (D_x(e^{2x}))\,(\cos(x^2+1)) \quad + \quad e^{2x} \quad (D_x(\cos(x^2+1)))=$$

$$[e^{2x}][2] \qquad [\cos(x^2+1)] \quad + \quad [e^{2x}]\,[-\sin(x^2+1)]\,[2x] \;=$$

↑　　　　　　　↑　　　　　　↑　　　　　　　↑

chain rule, deriv e^{2x}　　hold fixed　　　　hold fixed　　chain rule, deriv $\cos(x^2+1)$

$$2e^{2x}[\cos(x^2+1)-x\,\sin(x^2+1)].$$

EXAMPLE: Here the product is "inside" the chain rule. If $f(x)=\ln(x\sin 3x)$, find $f\,'(x)$.

Answer:

(1) The "outer" function is ln, and the derivative of **ln (u)** is $\dfrac{1}{(\mathbf{u})}$ (times du / dx)

(2) The next "outer" function is a PRODUCT: x sin3x

(3) The next "outer" function is 3 ( )

Thus the derivative is

$$f'(x)=\left(\frac{1}{x\sin 3x}\right)(\;\sin 3x \quad + \quad x(\cos 3x)(3)\;)=\frac{\sin 3x+3x\cos 3x}{x\sin 3x}=\frac{1}{x}+3\cot 3x$$

↑　　　　　　　　↑

deriv of ln　　　　deriv of product

NOTE: With the chain rule, any quotient can now be represented with products and powers. You may prefer this method to the quotient rule. (See below.)

EXAMPLE: Find the derivative of $f(x)=\dfrac{\cos x}{x^3+1}$.

Answer: We rewrite this as $f(x)=(\cos x)(x^3+1)^{-1}$ and now use the PRODUCT RULE, rather than the quotient rule.

$$f\,'(x)=(-\sin x)(x^3+1)^{-1}\;+\;(\cos x)(-(x^3+1)^{-2}(3x^2)).$$

This takes some effort to simplify.

SECTION 3.5

"A round man cannot be expected to fit in a square hole right away. He must have time to modify his shape."...Mark Twain

HIGHER DERIVATIVES

THE SECOND DERIVATIVE AS ACCELERATION: Given y = h (t), where h gives **position** as a function of **time**, h' (t) represents **velocity** at time t, and h"(t) represents **acceleration**. Velocity is the rate of change of position with respect to time. Acceleration is the rate of change of velocity with respect to time.

EXAMPLE: Find the distance, velocity and acceleration at time $t = \pi$ of a particle whose position is given by   f (t) = 2 sin t + cos t + t.
Answer:

Compute derivatives without evaluating ...   *When asked to evaluate , do it last!*

$f(t) = 2\sin t + \cos t + t \;\rightarrow$	$Position = f(\pi) = 2\sin \pi + \cos \pi + \pi = -1 + \pi$
$f'(t) = 2\cos t - \sin t + 1 \;\rightarrow$	$Velocity = f'(\pi) = 2\cos \pi - \sin \pi + 1 = -1$
$f''(t) = -2\sin t - \cos t \;\rightarrow$	$Acceleration = f''(\pi) = -2\sin \pi - \cos \pi = 1$

EXAMPLE: Give an example of a function that has a first derivative but not a second derivative, at some value of x.

Answer: The function given by $f(x) = x^{4/3}$ has a first derivative but not a second derivative at x = 0.

EXAMPLE: Find $\dfrac{d^2 y}{dx^2}$ for $y = (x^3 - \sec x)^4$.

Answer: First,

$\dfrac{dy}{dx} = 4(x^3 - \sec x)^3 (3x^2 - \sec x \tan x).$ Then

$d^2y/dx^2 = [12(x^3 - \sec x)^2 (3x^2 - \sec x \tan x)] \ (3x^2 - \sec x \tan x) +$

$\qquad\qquad \textbf{[D}_\textbf{x}\textbf{(first factor)]} \qquad\qquad\qquad\qquad \textbf{second factor}$

$\qquad\qquad\qquad + \ (4(x^3 - \sec x)^3)[6x - (\sec x(\sec^2 x) + \tan x(\sec x \tan x))]$

$\qquad\qquad\qquad\qquad \textbf{second factor} \qquad\quad \textbf{[D}_\textbf{x}\textbf{(second factor)- - requires product rule]}$

which can be simplifed as :

$\dfrac{d^2y}{dx^2} = 4(x^3 - \sec x)^2 [3(3x^2 - \sec x \tan x)^2 \ + \ (x^3 - \sec x)(6x - \sec x(\sec^2 x + \tan^2 x))]$

EXAMPLE: Find the 27th derivative of $f(x) = \sin 3x$.

Answer: The derivatives of sine and cosine follow a pattern: the derivatives repeat with a cycle length of 4.

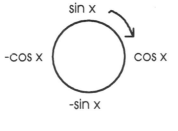

The derivatives are:

$$y = \sin x$$
$$y' = \cos x$$
$$y'' = -\sin x$$
$$y^{(3)} = -\cos x$$
$$y^{(4)} = \sin x \ \text{(back to the start!)}$$

Since $27 = 6(4) + 3$, we go through the cycle 6 times, and it will be as if we are taking the *3rd derivative* of sine, which by the chart above, is cosine. However, we are taking the derivative of $y = \sin \mathbf{3x,}$ so there will be a factor of $\mathbf{3}$ every time we take a derivative. The answer is $y^{(27)}(x) = 3^{27} \cos 3x$.

EXAMPLE: Using initial conditions. If $f''(x) = -\sin x + x$, with $f'(0) = 1$ and $f(0) = 2$, find $f(x)$.

Answer:

$f''(x) = -\sin x + x$. Working backwards,

$f'(x) = \cos x + x^2/2 + C.$ Now the fact that

$f'(0) = 1$ which is given , implies that

$$\begin{cases} 1 \ = \ \cos 0 \ + \ 0 \ + \ C, \text{ and} \\ C \ = \ 0 \quad \text{Thus} \end{cases}$$

$f'(x) \ = \ \cos x \ + \ x^2 \,/\, 2.$ Now, working backwards again,

$f(x) \ = \ \sin x \ + \ x^3 \,/\, 6 + C.$ But

$f(0) \ = \ 2$ implies that

$$\begin{cases} 2 \ = \ 0 \ + \ 0 \ + \ C, \text{ and} \\ C \ = \ 2. \quad \text{Thus} \end{cases}$$

$f(x) \ = \ \sin x \ + \ x^3 \,/\, 6 \ + 2.$

SECTION 3.6

" Youth is wholly experimental."...Robert Louis Stevenson

IMPLICIT DIFFERENTIATION

There are some equations that cannot be solved directly for y, but they nevertheless tell us how y is influenced by x. These equations are "implicitly" defined. It is "implied" that y is a function of x.

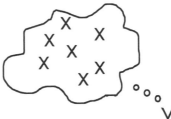

EXPLICIT FORM: In explicit form, y is an equation in terms of x, with x - terms only on the right side. E.g., $y = \sin x + x + 3.$

IMPLICIT FORM: In implicit form, the equation is not solved for y, but x and y are related by the equation. In fact, we may not be able to find y in terms of x.
E.g., in the equation $xy = y \sin y \ + \ x \cos y \ + \ x,$ we *cannot* find y explicitly.

EXAMPLE: Find y' if x + y = cos y
Answer:

- For now, we will rewrite as x + $\boxed{y}$ = cos$\boxed{y}$. The boxes remind us that y is a
  function of x: think of the box as having x's "inside" it.

- Differentiate, left to right. Note that $\dfrac{d(x)}{dx} = 1$ and $\dfrac{d(y)}{dx}$ can be written as y' .

$$1 \quad + \quad y' \qquad = \qquad\qquad (-\sin y) \qquad (y')$$

$$\uparrow \qquad \uparrow \qquad\qquad\qquad \uparrow \qquad\qquad \uparrow$$

deriv of x   deriv of y   deriv of cos ( )   deriv of y with respect to x

USE   CHAIN   RULE   HERE

- Simplify: Gather all y' terms to the left side, and all other terms to the right:
  $y' + y'\sin y = -1$

- Factor the y' from the left side: $y'(1 + \sin y) = -1$.

- Solve for y' if desired: $y' = \dfrac{-1}{1 + \sin y}$.

- Note that by dividing above, y cannot take on some values it formerly could.

EXAMPLE: Given the relation $x^2 + y^2 = 1$, find dy/dx at the points (1,0) and (0,1) .
(This is the equation of a circle centered at the origin of radius 1.)

Answer: Differentiating $x^2 + y^2 = 1$ gives $2x + 2y\ y' = 0$. Then $y' = -\dfrac{x}{y}$ and at (1,0),

y' does not exist. Indeed, the tangent line to the circle at the point (1,0) is vertical. Now at
the point (0,1),   y' = 0,   which corresponds to our knowledge that the tangent line at this
point on the circle is horizontal.

The derivative at (x_0, y_0) is $\dfrac{dy}{dx} = -\dfrac{y_0}{x_0}$, from which we see that the slope of the ray from

the origin to the point (x_0, y_0) is perpendicular to the tangent line at (x_0, y_0).

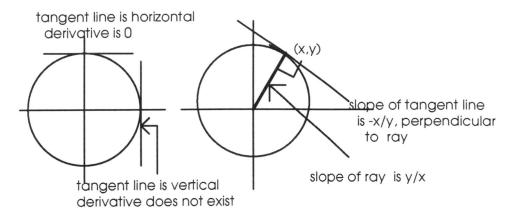

tangent line is horizontal
derivative is 0

(x,y)

slope of tangent line
is -x/y, perpendicular
to ray

tangent line is vertical
derivative does not exist

slope of ray is y/x

NOTE: Some students like to differentiate all variables (x as well as y) with respect to x in implicit differentiation. (See , in the following example, if you prefer this method.)

EXAMPLE: Find dy/dx for $\sin y = \cos x + y + 3x$.

Answer:

$$(\cos y)\boxed{\frac{dy}{dx}} = (-\sin x)\boxed{\frac{dx}{dx}} + \boxed{\frac{dy}{dx}} + 3\boxed{\frac{dx}{dx}}. \text{ Then, using that } \boxed{\frac{dx}{dx}} = 1,$$

$$(\cos y)\frac{dy}{dx} = (-\sin x) + \frac{dy}{dx} + 3. \text{ (Simplify again...)}$$

EXAMPLE: Find y" for the implicitly given function $xe^y = y$.

Answer:

Taking first derivative, we have $xe^y y' + e^y = y'$.

Solve for y', then differentiate again to find y''.

$$y' = \frac{-e^y}{xe^y - 1}, \quad \text{so} \quad y'' = \frac{(xe^y - 1)(-e^y y') - (-e^y)(xe^y y' + e^y)}{(xe^y - 1)^2}. \text{ Now plug in } y' \text{ and}$$

simplify.

EXAMPLE: Given the astroid, $x^{2/3} + y^{2/3} = 2$, find the equation of the tangent line at

the point (1,1). ✦ (**⬅This is an astroid.**)

Answer:

$\frac{2}{3}x^{-1/3} + \frac{2}{3}y^{-1/3}y' = 0$, and at (1,1) , $\frac{2}{3}(1) + \frac{2}{3}(1)y' = 0$, or $y' = -1$. Thus the tangent

line at (1,1) is : $y - 1 = (-1)(x-1)$ or $y = -x + 2$.

NOTE: It is interesting to see the curves $x^n + y^n = 1$, on your graphing calculator.

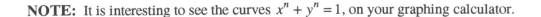

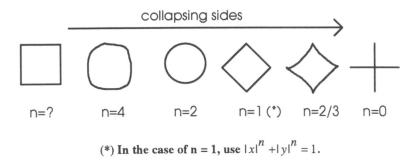

n=?	n=4	n=2	n=1 (*)	n=2/3	n=0

(*) **In the case of n = 1, use** $|x|^n + |y|^n = 1$.

EXAMPLE: Differentiate $x^2 + yx = 3 + y$ with respect to t, where x and y are functions of t.

Answer: Now our answer must contain dy/dt and dx/dt terms, because both it is to be understood that both x and y are functions of t.

$$2x\frac{dx}{dt} + \frac{dy}{dt}x + y\frac{dx}{dt} = 0 + \frac{dy}{dt} \text{ or } (2x+y)\frac{dx}{dt} + (x-1)\frac{dy}{dt} = 0.$$

IF YOU DON'T DO WELL ON TESTS

Everyone is entitled to a "bad day", but if not doing well on tests is a chronic problem, you may need to look closely at the way you study. Chances are, you are learning *passively* rather than *actively*. This type of learning stresses recognizing rather than constructing and is probably encouraged by traditional approaches to teaching and learning. In mathematics we read and are lectured to in a "foreign language", but tests ask us to *speak* this language fluently. In order to make learning more of an active process, try writing summaries and explanations for yourself. Rewrite your notes: teachers write tests based on the material they cover, and they remember. And above all, keep up with the material--cramming never works!

Your Phone Rings ☎

Friend: "I don't get this implicit stuff. What's the difference between an implicit and an explicit function? And what's the good of this anyway?

You say:----------------------

Friend: "How do you take the derivative of an implicit function? Please go slowly; I get confused. And I never know how to simplify. What do you do?"

You say:----------------------

Friend: "I get really mixed up when we have to take the second derivative of an implicit function. What should I do?"

You say:----------------------

SECTION 3.7

" 'Then you should say what you mean,' the March Hare went on. 'I do,' Alice hastily replied, 'at least --at least I mean what I say--and that's the same thing, you know!' "...Lewis Carroll

RELATED RATES PROBLEMS

The text suggests a method of approach to these problems that is very useful. At first you may feel stumped by these problems, but take heart: there are only so many types of related rates problems. A brief outline of some of the familiar types is provided here.

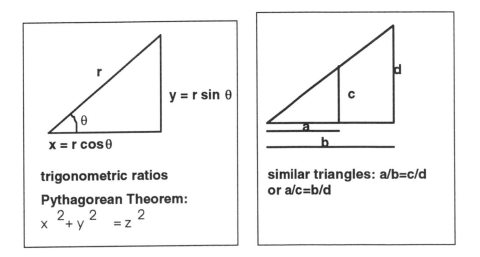

trigonometric ratios

Pythagorean Theorem:
$$x^2 + y^2 = z^2$$

similar triangles: a/b=c/d
or a/c=b/d

FORMULAS TO KNOW

Volume of a Sphere	$\frac{4}{3}\pi r^3$ or $\frac{2}{3}$ the volume of the encompassing cylinder
Volume of a Cylinder	$\pi r^2 h$ (all prismatic solids have volume = base $\cdot$ height)
Volume of a Cone	$\frac{1}{3}\pi r^2 h$ or $\frac{1}{3}$ the volume of the encompassing cylinder
Surface Area of a Sphere	$4\pi r^2$ or the area of 4 "great circles" *(circles whose radius is that of the sphere)*
Surface Area of a Cylinder	$2\pi rh$ (imagine this area stretched out as a rectangle)

NOTE: Hint with trigonometry. Often we are taught that $\cos\theta$ and $\sin\theta$ are the ratios, x/r and y/r, respectively, in a triangle. It would be much better if we were taught that $x = r\cos\theta$ and $y = r\sin\theta$. Think of these terms as giving different "projections" of a ray from the origin.

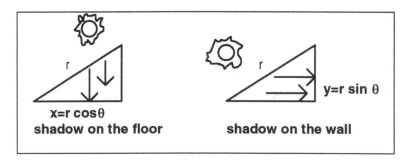

x=r cosθ
shadow on the floor

y=r sin θ
shadow on the wall

THE EXPANDING OR SHRINKING OBJECT

EXAMPLE: The radius of a cylinder is expanding at the rate of 3 ft. /sec., while the height stays the same at 6 ft. How fast is the volume of the cylinder changing when the radius is 5 ft.?

Answer:

(1) Always draw a picture. Your picture should not show the result at the final moment, but should illustrate the event in progress.

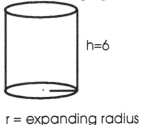

h=6

r = expanding radius

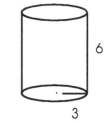

6

3

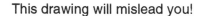

This drawing shows the event in progress -- good!   This drawing will mislead you!

(2) Look at the last line, generally, for the clue: What rate are we asked for? Look for words like: how fast, changing, increasing, decreasing, etc.

In this problem, we are asked for rate of change of volume.

(3) Give names to variables . Decide what we know and what we want to know. We let V = volume, r = radius, height = 6, and from (2), we want dV/dt. We are given that d r/ dt = + 3 (dr/dt is positive since r is increasing).

(4) Find the formulas that will be useful to you. We are attempting to RELATE RATES so we need to relate variables.
$$V = \pi r^2 h = \pi r^2 (6) = 6\pi r^2, \text{ since h is the constant 6.}$$

(5) Differentiate implicitly with respect to time. (All *related rates* problems require differentiation with respect to *time*.)
$$\frac{dV}{dt} = 12\pi r \frac{dr}{dt}$$

(6) Now we are ready to plug in the values given in the problem that describe the situation at the final moment.
In this problem, r = 5 at the "final moment" and
$$\left.\frac{dV}{dt}\right|_{r=5} = 12\pi r \left.\frac{dr}{dt}\right|_{r=5} = 12\pi(5)(3) = 180\pi \text{ cu. ft. / sec.}$$

🖐 CAUTION! Don't "plug in" first, and differentiate later. This method always gives an erroneous answer of 0, so be a little suspicious if this is your answer!

NOTE: "Rate of change" of a quantity is positive if the quantity is enlarging; negative if the quantity is diminishing.

THE PUSHED STONE

EXAMPLE: A boy is on a ramp that rises a height of 5 ft. over a horizontal distance of $5\sqrt{3}$ ft.. (See diagram) If he pushes a stone up the ramp so that lateral distance, x, changes at a rate of 6 ft./sec., how fast is the stone rising?

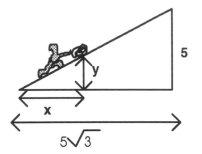

Answer: We are given dx/dt = +6, and seek dy/dt where y is the height of the stone.

We search for a relation between y and x. Let's use similar triangles:
$\dfrac{y}{x} = \dfrac{5}{5\sqrt{3}} = \dfrac{1}{\sqrt{3}}$ or $y = \dfrac{1}{\sqrt{3}}x$. Now differentiating , $\dfrac{dy}{dt} = \dfrac{1}{\sqrt{3}}\dfrac{dx}{dt}$. Since $\dfrac{dx}{dt} = 6$,

$\dfrac{dy}{dt} = \dfrac{6}{\sqrt{3}}\, ft./\sec.$

ANGLES IN A RECTANGULAR REGION : THE BASEBALL GAME

EXAMPLE: A baseball diamond is a square with sides 90 ft. long. A baseball player is advancing from 2nd to 3rd base at a rate of 20 ft. /sec. Let θ be the angle between the third base line and the line from the catcher at home-plate to the runner. How fast is θ changing when the runner is $\dfrac{90}{\sqrt{3}}$ ft. from 3rd base?

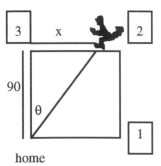

Answer:
What we know: Letting x = distance of runner from 3rd base, we know dx/dt = -20 (negative because this distance is decreasing).

What we seek: We are looking for $d\theta/dt$ when $x = \dfrac{90}{\sqrt{3}}$.

The relationship: We must relate x to θ : here $\tan\theta = x/90$. (1)

Differentiate: $(\sec^2\theta)\dfrac{d\theta}{dt} = \dfrac{1}{90}\dfrac{dx}{dt}$ (2)

Use what problem gives, for the "final moment": We still need θ at the moment when $x = \dfrac{90}{\sqrt{3}}$. From (1), $\tan\theta = x/90$, so $\tan\theta = \dfrac{1}{\sqrt{3}}$ and $\theta = \pi/6$. Thus $\sec\theta = 2/\sqrt{3}$.
From (2), $(4/3)\,d\theta/dt = -20/90$, and $d\theta/dt = -(1/6)$ rad/sec.

DISTANCES IN A RECTANGULAR REGION : THE STREET INTERSECTION

EXAMPLE: Two streets are perpendicular to each other. A stationary truck is 3 miles from the intersection on one street. A moving car is approaching the intersection at 40mph on the other street. How fast is the distance between the car and truck changing when the moving car is 4 miles from the intersection?

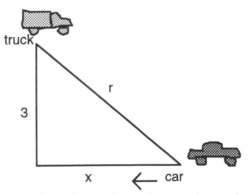

Answer: Let x be the distance from the car to the intersection and r, the distance between the car and truck. Then
$r^2 = x^2 + 3^2 = x^2 + 9$

$2r\dfrac{dr}{dt} = 2x\dfrac{dx}{dt}$ or $r\dfrac{dr}{dt} = x\dfrac{dx}{dt}$. (1)

Now $\dfrac{dx}{dt} = -40$. And when $x = 4$, we have $r = 5$ (a "3-4-5" triangle). So from (1),

$5\dfrac{dr}{dt} = 4(-40) = -160$, and $\dfrac{dr}{dt} = -\dfrac{160}{5} = -32$mph. (The distance between car and truck is decreasing .)

EXAMPLE: Variation: Suppose the truck is not stationary but is moving away from the intersection at 60 mph and the car is moving toward the intersection at 40 mph. How is the distance between the car and truck changing when the car is 4 miles and the truck 3 miles from the intersection?

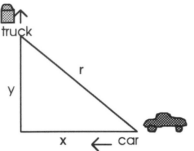

Answer: The work is essentially the same as above, but now y is the variable distance of the truck from the intersection. So

$$r^2 = x^2 + y^2 \text{ and } 2r\frac{dr}{dt} = 2x\frac{dx}{dt} + 2y\frac{dy}{dt} \text{ or } r\frac{dr}{dt} = x\frac{dx}{dt} + y\frac{dy}{dt}. \quad (1)$$

Now $\dfrac{dy}{dt} = 60$ and $\dfrac{dx}{dt} = -40$. When x = 4 and y = 3, r = 5.

Plug this information into (1) to find dr / dt = 4 mph.

HINT: Actually study--don't just work-- the problems in the text and try to categorize them into types. Learn a problem of each type. If **angle** is asked for, use **trigonometric relations**. If not, the **Pythagorean Theorem and similar triangles** should suffice.

THE SLIPPING LADDER
These problems generally rely on the Pythagorean Theorem.

EXAMPLE: A ladder 15 ft. long rests against a wall. When the bottom of the ladder is x ft. from the wall, it is being pushed **toward** the wall at a rate of 2x ft. per sec. How fast is the top of the ladder moving up the wall when the bottom of the ladder is 5 ft. from the wall?

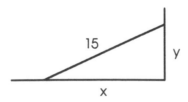

Answer: Let x and y be as in the picture; x the lateral distance of the ladder from the wall, and y the ladder's height. We have:

$x^2 + y^2 = 15^2$, and $\dfrac{dx}{dt} = -2x$, so $2x\dfrac{dx}{dt} + 2y\dfrac{dy}{dt} = 0$, or $2x(-2x) + 2y\dfrac{dy}{dt} = 0$.

(The dx/dt term is *negative* as x is diminishing.) We seek dy/dt. But first we must find y when x =5. From the triangle, x = 5 implies $y = 10\sqrt{2}$. So

$-4x^2 + 2y\dfrac{dy}{dt} = 0$ gives $-4(5^2) + 2(10\sqrt{2})\dfrac{dy}{dt} = 0$, and $\dfrac{dy}{dt} = \dfrac{100}{20\sqrt{2}} = \dfrac{5}{\sqrt{2}}\, ft./\sec..$

The ladder is moving up the wall at this rate.

THE SHADOW KNOWS!
These problems generally rely on similar triangles.

EXAMPLE: A boy 4 ft. tall is walking away from a lamp-post 10 feet high at a rate of 5 ft./sec. At what rate is the length of his shadow on the ground changing?

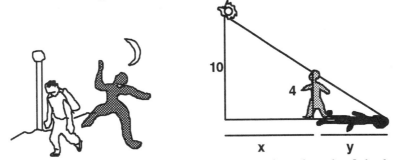

Answer: Letting x = distance of boy from lamp-post and y = length of shadow,

$\dfrac{4}{10} = \dfrac{y}{x+y}$ or $4(x+y) = 10y$, or $4x = 6y$. Differentiating, $4\dfrac{dx}{dt} = 6\dfrac{dy}{dt}$ and

$4(5) = 6\dfrac{dy}{dt}$ so $\dfrac{dy}{dt} = \dfrac{10}{3}\, ft./\sec$. The length of the shadow increases as we walk away

from the lamp post at a rate of $10/3\, ft./\sec.$

I ᴺᵛᴱᴺᵀ ᴀ Pʀᴏʙʟᴇᴍ

Variations on the pushed stone problem above might ask for dx/dt, dy/dt, or dr/dt, where r is the distance of the stone along the slant-height of the triangle. Create some variations on this problem. (This is the best of all exercises for understanding--create your own problems.

THE FILLING (OR UNLOADING) CONE/TROUGH

EXAMPLE: A right circular cone with base radius 8 ft. is unloading sand. The height of the cone is 24 ft. The radius is changing at the rate of 5 ft./ min. At what rate is the volume of sand changing when the radius of the sand in the container is 4 ft.? (No angles are asked for. Let's try similar triangles.)

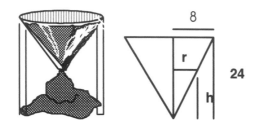

Answer:

$V = \dfrac{1}{3}\pi r^2 h$ is the volume of a cone of radius r and height h

The radius of the cone of sand is, r a variable. The radius of the conical container is 8.

The height of the cone of sand is, h a variable. The height of the conical container is 24.

Now dr/dt = -5 is given, and we are seeking dV dt when r = 4.

Differentiating the volume equation (product rule): $\dfrac{dV}{dt} = \dfrac{2}{3}\pi r \dfrac{dr}{dt} h + \dfrac{1}{3}\pi r^2 \dfrac{dh}{dt}$

By similar triangles $\dfrac{r}{h} = \dfrac{8}{24}$ so $3 r = h$ (1)

Differentiating this equation $3\dfrac{dr}{dt} = \dfrac{dh}{dt}$ Now substituting

$$\dfrac{dV}{dt} = \dfrac{2}{3}\pi r \dfrac{dr}{dt} h + \dfrac{1}{3}\pi r^2 (3\dfrac{dr}{dt}). \qquad (2)$$

When r = 4, we can find h = 12 from (1) and knowing $\dfrac{dr}{dt} = -5$ we can find $\dfrac{dV}{dt}$

from (2). $\dfrac{dV}{dt} = \dfrac{2}{3}\pi(4)(-5)(12) + \dfrac{1}{3}\pi(16)(3)(-5) = -240\pi$ cu. ft./min.

The volume of sand is decreasing at this rate.

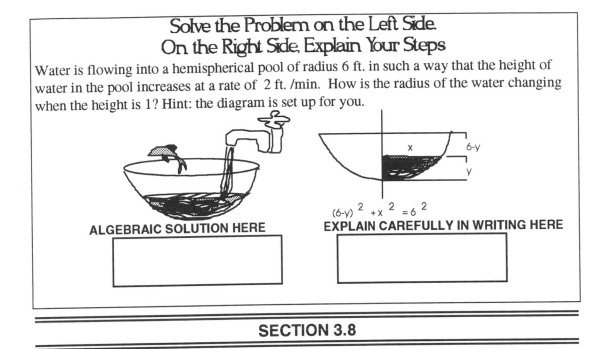

Solve the Problem on the Left Side. On the Right Side, Explain Your Steps

Water is flowing into a hemispherical pool of radius 6 ft. in such a way that the height of water in the pool increases at a rate of 2 ft./min. How is the radius of the water changing when the height is 1? Hint: the diagram is set up for you.

$(6-y)^2 + x^2 = 6^2$

ALGEBRAIC SOLUTION HERE

EXPLAIN CAREFULLY IN WRITING HERE

SECTION 3.8

"Truth is the most valuable thing we have. Let us economize it."...Mark Twain

THE DIFFERENTIAL

There are two important concepts presented in this section. They must be understood geometrically. Look carefully at the graphs that accompany the definitions.

(1) THE DIFFERENTIAL

THE DIFFERENTIAL

$$df = f'(a)\ h,$$

or when a is not given, and we want the most general formula, we write:

$$df = f'(x)\ dx$$

Notice that in the second equation, we write $x = a$, and $h = x - a = dx$

(2) THE LINEAR APPROXIMATION. "Linear" means "straight line.

THE LINEAR APPROXIMATION

$$f(a+h) \approx f(a) + f'(a)\ h$$

◆ Given the linear approximation : f (a+h) ≈ f(a) + f ' (a) h ,
if we substitute x = a + h, and h = x - a we have:

$$f (x) ≈ f (a) + f ' (a) (x-a) .$$

◆ Compare this approximation above with the *equality*,

$$f (x) = f (a) + f ' (a) (x-a) .$$

◆ This expression above is the equation of the **tangent line** to the graph of the function f
at x = a. We are thus saying in the approximation, that values of f(x) are approximated
by values of y from the tangent line.

◆ Note that the term given the differential is hidden in the linear approximation.

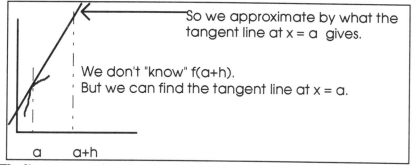

So we approximate by what the
tangent line at x = a gives.

We don't "know" f(a+h).
But we can find the tangent line at x = a.

a   a+h

The linear approximation is what the tangent line would predict for f(x).

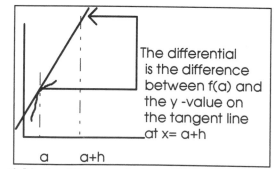

The differential
is the difference
between f(a) and
the y -value on
the tangent line
at x= a+h

a   a+h

The differential is the difference between f(a) and what the tangent line
predicts for f at a+h.

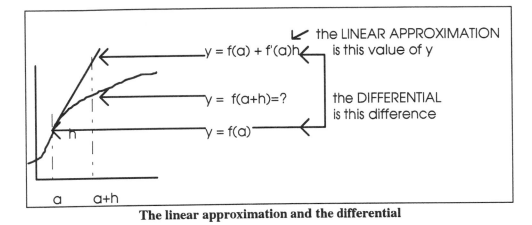

The linear approximation and the differential

EXAMPLE: In the two pictures below, label the differential and the linear approximation to f(x) at x = a.

Answer:

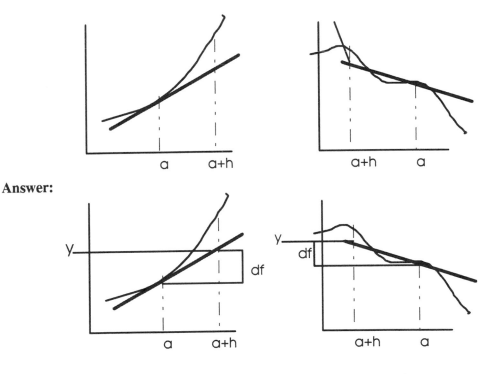

y gives the linear approximation to f at a+h.
This is the y-value the tangent line predicts
when we don't know the real f(a+h).

EXAMPLE: Approximate $\sqrt[4]{15}$, using differentials: .

Answer: We use $f(a+h) \approx f(a) + f'(a)(h)$

with $f(x) = \sqrt[4]{x} = x^{1/4}$, $a = 16$, (since we know $\sqrt[4]{16} = 2$) , and $h = x-a = 15-16 = -1$.

Then $f'(x) = \dfrac{1}{4}x^{-3/4}$, and $f'(a) = f'(16) = \dfrac{1}{4}(16)^{-3/4} = \dfrac{1}{4}2^{-3} = \dfrac{1}{32}$. Therefore,

$f(a+h) \approx f(a) + f'(a)(h)$, or $\sqrt[4]{15} \approx \sqrt[4]{16} + \dfrac{1}{32}(-1)$, or $\sqrt[4]{15} \approx 2 - \dfrac{1}{32} = \dfrac{63}{32}$.

EXAMPLE: Let $f(x) = \cos(\sin x)$. Find df as a function of x.

Answer: $f'(x) = -[\sin(\sin x)][\cos x]$. Since $df = f'(x)\ dx$,

$df = -[\sin(\sin x)][\cos x]dx$.

EXAMPLE: Approximate the volume of the shell of a sphere with inner radius 5 and outer radius 5.01.

Answer: We are being asked for dV, given that $dr = 0.01$. Volume is $V = \dfrac{4}{3}\pi r^3$, and

$dV = \dfrac{4}{3}\pi(3r^2)dr$. So when $r = 5$, $dV = \dfrac{4}{3}\pi(3)(5^2)(0.01) = 4\pi(.25) = 3.14159$. Thus

the volume of the spherical "shell" we are asked for is 3.14159 cubic units.

Chapter Review

1.   Can you differentiate explicitly and implicitly? Can you find the equation of the tangent line and the normal line to a function at a point? What is the Quotient Rule?

2.   Have you studied examples of related rates problems of various types listed here? What special tools do you have to use, in working a related rates problem? What steps do you go through to solve a related rates problem?

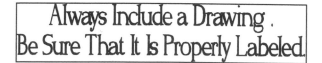

Always Include a Drawing .
Be Sure That It Is Properly Labeled.

3.   What IS the differential? Give a graphical illustration showing what the differential represents. How do you use the differential to find approximations? Give an example. What is the relationship between the linear approximation of a function and the tangent line? What does "linear" approximation mean?

4.   Don't forget the Newton Raphson method in this chapter! Can you illustrate it graphically? Can you use this method to find approximations? How does this method compare with the Bisection Method, studied in the previous chapter? Which method uses more information about the function? In what situations might the Newton-Raphson method not produce results?

5.   In this chapter, we have seen some "pay-offs" of the derivative. Explain in what ways the derivative helps us understand a function better.

Styles of Problems

The exercises you work as homework stress many features of mathematical understanding .

1.   **State** a definition or theorem.
2.   **Interpret** geometrically from a graph.
3.   **Describe** a function or graph (to show why certain properties hold).
4.   **Derive** a formula or **Prove/Explain/Show** why something is true.
5.   **Find** an interpretation (the equation of the tangent line, etc.).
6.   **Use** a construct (the differential, etc.) **in order to** (approximate..)
7.   **Find where** something happens (the tangent line intersects...).
8.   **In what situations** can we use a result?
9.   **Give an example** of something (a function that is continuous but not differentiable...)
10. **Compare your results** (compare your answer with the Bisection method with the Newton-Rhapson method.

Add your own examples to this list!

FROM THE STUDENT:

I couldn't set up the equation. (Try to stay organized; remember similar triangles, the Pythagorean theorem, or trigonometric relations! Remember the "categories" of types of problems. The problem you see must be similar to something you have already done.)

I didn't know how to draw the picture. (Many students don't draw a picture, and don't label it. But a picture will help you get started, and covers a multitude of problems, such as forgetting to identify variables --"let x = .." Imagine the pictures you've seen, and just begin drawing; the ideas will come.)

I couldn't remember the formula. (Memorize! Try to link these formulas to something unusual or interesting that you will remember.)

The problem wasn't like the ones we worked. (There are a great many related rates problems possible. But the problem given is probably close enough to something you have seen for you to be able to tackle it. This is why it is a good idea to try to think of these problems as falling into categories, and to have some general strategies for attack.)

FROM THE TEACHER:

There is no picture.
The picture isn't labeled.
The differentiation is wrong, which makes the whole problem wrong.
The formula (for the volume of a sphere, etc.) isn't right, so the whole problem is wrong.
There is so little organization, there seems to be nothing to give partial credit for!!

CHAPTER 4

"Nothing great was ever achieved without enthusiasm"...Ralph Waldo Emerson

In this chapter we examine applications of the derivative -- particularly as the derivative applies to optimization (finding maximum and minimum values) and graphing. These are among the most widely used applications of calculus. The chapter begins with some theorems that are worthwhile memorizing. The Mean Value Theorem is particularly important.

Brush up on the review in Chapter 1 of this guide; and know your trigonometry and geometry formulas!

SECTION 4.1

"Nothing quite new is perfect"...Cicero

MAXIMUM AND MINIMUM VALUES

MAXIMUM-MINIMUM THEOREM : (See your text for the authors' exact wording). The *idea* behind this theorem is that a continuous function on a closed interval takes on a maximum and minimum value on that interval.

As with any theorem in mathematics, we need to understand why the hypotheses are necessary.

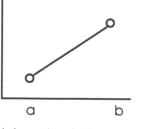

Interval not closed,
f has no max or min on (a,b)

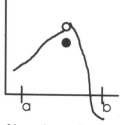

f is not continuous,
f has no max on (a,b)

NOTE: A **critical number** is a value of x at which f ' is 0 or does not exist, and which is in the domain of f. We do not have a guarantee that a max or min will occur at a critical number--that is, even if f' is 0 or fails to exist at a critical number, the function may not have a max or min there. The function given by $f(x) = x^3$, for example, has derivative $f'(x) = 3x^2$ which equals 0 at $x = 0$, but there is no max or min at $x = 0$.

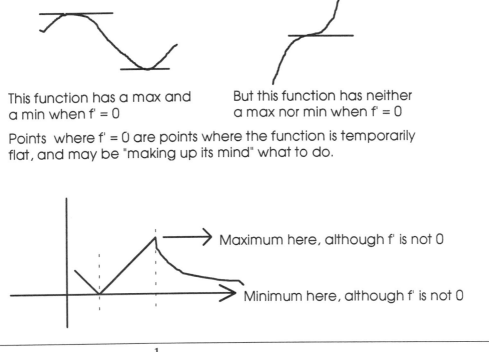

This function has a max and a min when f' = 0

But this function has neither a max nor min when f' = 0

Points where f' = 0 are points where the function is temporarily flat, and may be "making up its mind" what to do.

Maximum here, although f' is not 0

Minimum here, although f' is not 0

EXAMPLE: Let $f(x) = x + \dfrac{1}{x^2}$. Find the critical numbers (if any) of f.

Answer:

$f'(x) = 1 - \dfrac{2}{x^3}$. Now f ' fails to exist if x = 0 but this value of x is not in the domain of f .

And f ' = 0 if $1 = \dfrac{2}{x^3}$ or x = $\sqrt[3]{2}$. In other words, the critical number is c = $\sqrt[3]{2}$.

At this value of x , f might have a max or min , or possibly - -neither.

CAUTION! **The max or min is a y-value.** It is taken on at an x-value. Critical numbers are x-values.

The maximum of f is M taken on when x = a

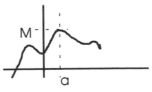

NOTE ON SOLVING PROBLEMS IN THIS SECTION

When asked, in this section, to find **extreme values** of a function, we have very little to go on other than the graph. However, we will soon develop techniques for examining where the function is increasing or decreasing.

EXAMPLE: Given $f(x) = 4x^3 + 1$, find all extreme values of f . (A graphing calculator helps here.)

Answer:

Since f is differentiable, we seek critical numbers x = c where $f'(c)$ is 0 or fails to exist.

$f'(x) = 12x^2 = 0$ if x = 0. Thus the point $(0, f(0)) = (0,1)$ is the point we want to examine. We don't know if a y-value of "1" yields a max, min, or "nothing".

Plugging in points near x = 1 gives values of y slightly greater and less than 5. We conclude that this function does **not** have an extreme value.

Here is another way to argue the point. You will see this method developed in a later section, but it seems intuitive now. Since f' ≥ 0 for all x, the slope of the tangent line is everywhere non-negative, and the function is non-decreasing. So f does not have a max or min

---

EXAMPLE: Find the extreme values of $f(x) = x^2 e^x$.

Answer:

Since $f'(x) = 2xe^x + x^2 e^x$ exists everywhere, we check where f'(x) = 0; i.e. , where

$xe^x(2 + x) = 0$. This occurs if x = 0 or x = -2 (e^x is never 0.) Now check the y-values:

critical number c	f (c)
c = 0	f(0) = 0
c = -2	$f(-2) = 4e^{-2}$

The fact that f(0) < f(-2) does not in itself mean that (0,0) gives a minimum and $(-2, 4e^{-2})$ gives a maximum. (See Errors, below.) This function tends to ∞ as x does, so **there is no maximum**. Is there a minimum? Because f(x) ≥ 0, we can say that **(0,0) is a minimum.** Notice that the point $(-2, 4e^{-2})$ amounts to nothing. (*Find the graph on a graphing calculator to confirm this.*)

Common Error

♦   When setting a function equal to 0, **DO NOT CANCEL** immediately--you may lose roots. Factor instead.

INCORRECT: $x^2 - x = 0$ or $x^2 = x$, so (cancelling the "x"), x = 1.

CORRECT: $x^2 - x = 0$ or x(x - 1) = 0 so x = 0 or 1. Above we canceled
even when x might have been 0, which was not legitimate.

MAXIMUM AND MINIMUM PROBLEMS

Like related rates problems, these problems fall into categories which are worth knowing.
Some are introduced here, some in the next section.

cow attempting to learn to swim

THE FARMER AND THE FENCE

EXAMPLE: Maximize the area. A farmer is building a rectangular grazing field for
cows and goats. One side of the field does not require fencing, because a stream provides a
natural boundary . The farmer intends to use 200 yds. of fencing to fence in 3 sides of a
pasture with a partitioning fence in the middle. Cows and goats will both have access to the
stream. What is the largest total grazing area possible?

Answer:

STEP 1. Draw a picture .

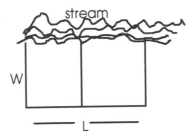

STEP 2. Determine what we are trying to maximize or minimize.
Here we are trying to maximize area. Let L = length of pasture, W = width, A = pasture
area. Then A = L W.

STEP 3. Express this equation in terms of a single variable.
The relationship between length and width is given by: $200 = 3W + L$, so $L = 200 - 3W$. Then $A = (200 - 3W)(W) = 200W - 3W^2$.

STEP 4. Differentiate the equation in the previous step, and set the derivative to zero.
Find when $dA / dW = 0$. But $dA /dW = 200 - 6W = 0$ when $W = 100/3$.

STEP 5. Be sure that a maximum or minimum really makes sense in the problem. Some
functions do not have maxima and minima.
Our area function has a maximum because the equation for area, $A = 200W - 3W^2$, is a
parabola opening downward. The minimum area makes no sense -- this would be 0,
occurring when L = 200 and W = 0 or when W = 200/3 and L = 0, and these are infeasible
values for W and L.

STEP 6. Check back: what exactly does the question asks for? In fact, this question asks
for the **area** of the pasture. Since $L = 200 - 3W = 200 - 3(100/3) = 100$, we have
$A = L W = (100)(100/3) = 10,000/3$ square yards.

MAXIMIZING THE AREA INSCRIBED IN A FIGURE

NOTE: The shape giving the maximum area for a given perimeter is a circle, and the shape giving the maximum volume for a given surface area is a sphere. (Nature demonstrates this, with water ripples and soap bubbles. Surface tension attempts to keep boundaries to a minimum, and the water or air inside the boundary tries to expand.)

In many problems you will see analogies to this. For example, the greatest area for a rectangle of given perimeter is achieved when the rectangle is a square.

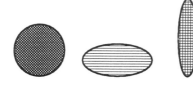 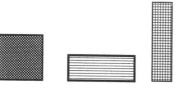

NOTE : As we work more of these types of problems, we begin to combine steps. These steps do not represent number for number, the steps we went through in the previous problem.

EXAMPLE: If a rectangle is inscribed inside a circle of radius r, what is the maximum area it may have ? (Note the clue above. Unfortunately, this is not a proof or explanation.)

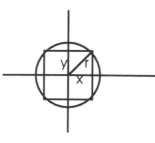

Answer:
STEP 1. Draw the picture.

STEP 2. We are attempting to maximize the area of a rectangle. Find dimensions for the rectangle, 2x and 2y, labeled above. Set up the equation for area: $A = (2x)(2y) = 4xy$. Express A in terms of one variable. Since $y = \sqrt{r^2 - x^2}$ on a circle, $A = 4x\sqrt{r^2 - x^2}$.

STEP 3. Set derivative to 0.
$$\frac{dA}{dx} = 4\sqrt{r^2 - x^2} + \frac{4x}{2\sqrt{r^2 - x^2}}(-2x) = 4\sqrt{r^2 - x^2} - \frac{4x^2}{\sqrt{r^2 - x^2}} = 0 \quad . \text{ Factoring,}$$

$$\frac{dA}{dx} = \frac{4}{\sqrt{r^2 - x^2}}((r^2 - x^2) - x^2) = \frac{4}{\sqrt{r^2 - x^2}}(r + \sqrt{2}x)(r - \sqrt{2}x). \text{ The critical numbers are}$$

$x = r, \ x = \pm r/\sqrt{2}$. The negative answers do not make sense geometrically in the way we set up the problem. So our only possible candidate is $x = r/\sqrt{2}$, in which case, $y = r/\sqrt{2}$ also. Note: we are seeing our optimal "square" again!

We would have obtained the same answer if we had set up the equation for A in terms of y . We can conclude (in many ways) that $x = y = r/\sqrt{2}$. (In other words , the square is inscribed with one corner on the 45^0 - line.)

STEP 4. Check that a maximum area makes sense. We are maximizing the continuous function A over the region x ε [0,r]. A graph shows that it has a maximum on this area. (The Maximum-Minimum Theorem to be proved will guarantee this result.) The minimum area is 0, occurring at the endpoints of the interval, when x = 0, or x = r.

STEP 5. Check back : what does the original problem asks for? Here it asks for area, which is $A = 4xy = 4 (r/\sqrt{2})^2 = 2r^2$. (**What does this say geometrically?**)

EXAMPLE: (**Hint**) To find the maximum volume for a cylinder inscribed in a cone, use that the volume of the cylinder, $V = \pi r^2 h$, is the equation to differentiate. But we need the relationship between r and h. This is given in terms of R and H , the radius and height of the cone, which are constants. Using similar triangles,: r/R = H-h/H, so r = R(H-h)/H.

Then $V = \pi r^2 h = \pi \left(\dfrac{R(H-h)}{H} \right)^2 h$ with R and H constants.

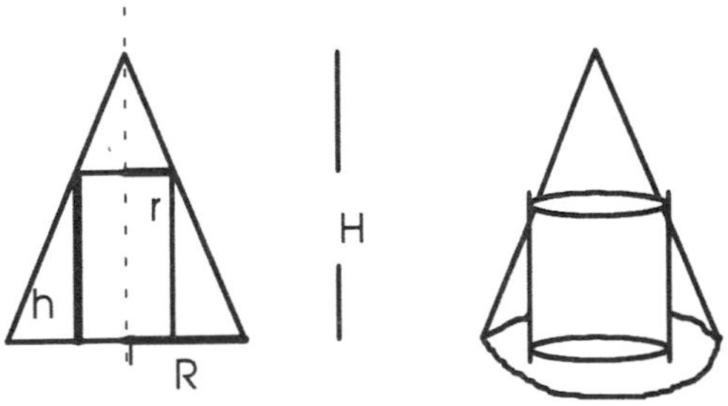

SUMS AND PRODUCTS OF NUMBERS

EXAMPLE: Find two numbers whose sum is 10 and whose product is maximal.

Answer: Let x and y be the numbers. Then x + y = 10. We wish to maximize the product, P, given by $P = xy = x (10\text{-}x) = 10x - x^2$. Now dP / dx = 10 - 2x = 0 when x = 5 and y = 5. (The product is 25.) We know this is maximal because P describes a downward opening parabola and the boundary points, x = 0 and x = 10, give a product P = 0. (**Do you see why this is a variation of the "area " problem?**)

PERCENTAGE OF ON-CAMPUS JOB OFFERS, 1988

HARD CALCULUS FIELDS	PERCENTAGE OF OFFERS	AVG. ANNUAL SALARY
Engineering	50%	28,872
Hard Sciences	42%	29,208
SOFT CALCULUS FIELDS	45%	23,160
Economics	3%	24,612
Business Management	40%	23,112
Ag/Bio/Health Science	2%	22,164
NO CALCULUS FIELDS	5%	20,772
Social Sciences	2%	21,768
Humanities	1%	20,760
Communications	2%	19,476

Data from Lucy W. Sells, Source: CPC Salary Survey: A Study of Beginning Offers 1987-1988, July 1988.

SOLVING A WORD PROBLEM

**Solve the following word problem without looking back in the space below.
YES! RIGHT NOW!
Then check with the grading system on the next page to find your score.**

The problem to solve: A farmer wants to fence in a pasture 1 sq. km. in area with barbed wire and wood fencing. He wants 3 sides of the pasture to be fenced with barbed-wire, and one side with wood. Wood costs 7 times as much as barbed wire. What dimensions should the pasture be if the farmer is to pay the least money? To solve this problem, let x be the length of wood side, and y , the length of the barbed-wire side. (Please differentiate with respect to x.)

QUESTION	YES ✌	NO 👎
Did you draw a picture? y x		
Label the picture?		
Write down on paper : x y = 1?		
Write down the cost equation? C = 7x + x + 2y = 8x + 2/x		
Differentiate correctly to find the critical numbers? With respect to x, this would be dC/dx = 8-2/x^2 = 0.		
Solve correctly? x = 1/2 km. , y = 2 km.		
Check that this answer provides the minimum cost? (To do this, note that cost increases to ∝ as x approaches 0, or as y approaches 0. In this way, we know that our answer provides a minimum C.)		
Give both x and y in the final answer?		

Grading your problem. From 25 points, subtract 3 for every "no". Multiply the result by 4. This is your score based on 100 points.

SECTION 4.2

"I'm very good at integral and differential calculus; I know the scientific names of beings animalculus"....Gilbert & Sullivan

The theorems you see should be accompanied by pictures, in your mind. Can you draw the picture that accompanies Rolle's Theorem?

> **ROLLE'S THEOREM :** If f is continuous on [a,b] and differentiable on (a,b), and f(a) = f (b) = 0, then there exists a point c in (a,b) such that f ' (c) = 0.   (See text.)

The function must be differentiable on the open interval. For example, if
f (x) = - | x | +5 is defined on [-5,5] then f(-5)= f(5) = 0, but there is no point c in (-5,5) for which f '(c) = 0. Since this function is not differentiable in the interval (-5,5), Rolle's Theorem does not apply.

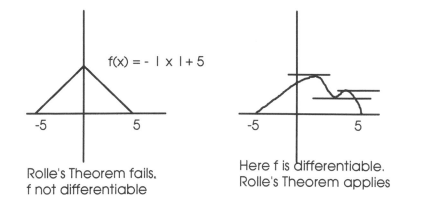

f(x) = - | x | + 5

Rolle's Theorem fails.
f not differentiable

Here f is differentiable.
Rolle's Theorem applies

THE MEAN VALUE THEOREM states that if f is continuous on [a,b] and
differentiable on (a,b), then there exists a point c in (a,b) where

$$f'(c) = \frac{f(b) - f(a)}{b - a}$$

NOTE: Understanding the theorem: The theorem allows us to say that if we travel from
A to B and average 55 mph, then there is a moment in time when our speedometer reads 55
exactly. That is, there is a moment when the instantaneous velocity is 55 mph .

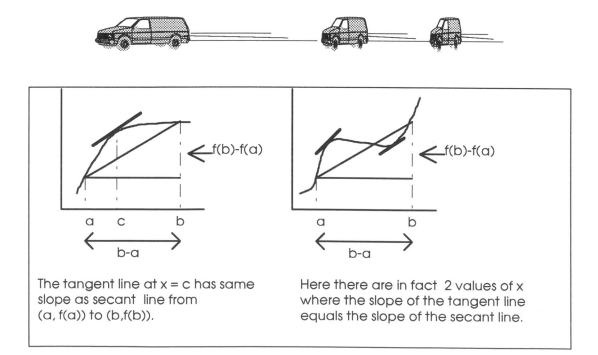

The tangent line at x = c has same
slope as secant line from
(a, f(a)) to (b,f(b)).

Here there are in fact 2 values of x
where the slope of the tangent line
equals the slope of the secant line.

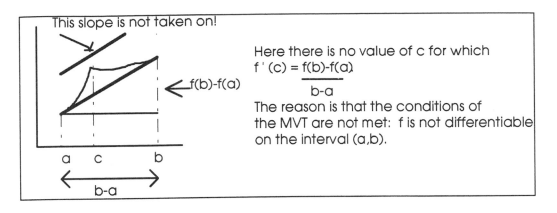

THE MEAN VALUE THEOREM AND THE INTERMEDIATE VALUE THEOREM

The Mean Value Theorem (M.V.T.) and the Intermediate Value Theorem (I.V.T.) have different hypotheses. The Intermediate Value Theorem tells us what we can expect if the function is merely continuous on [a,b]. The Mean Value Theorem tells us what we can expect if the function is also differentiable on (a,b). It is a sharper result about the behavior of the function. Notice the language of theorems is usually that a function be **"continuous on the closed interval"**, and/or **"differentiable on the open interval".** The Mean Value Theorem tells us something about the immediate behavior of a function (at some point interior to an interval) from information about its behavior at the endpoints of the interval. In some ways, this result should be surprising. The trouble with both the M.V.T. and the I.V.T. is that they do not produce the point, x = c, at which the good results happen. These theorems are "existence" theorems--they merely say that the point, x = c, exists. In some rare cases, we may be able to find "c" (see the Example following). But these two theorems derive their great importance for their usefulness in the development of the theory of calculus

---

EXAMPLE: Find a number c on the interval (0,4) at which the line tangent to the graph of $f(x) = -4 + \sqrt{x}$ is parallel to the line joining (0, f(0)) and (4, f(4)) .

Answer: We let a = 0, b = 4. Then the slope of the secant joining these points is:
$$\frac{f(b) - f(a)}{b - a} = \frac{(-4 + \sqrt{4}) - (-4 + \sqrt{0})}{4 - 0} = 1/2.$$ We seek c in (0,4) such that

$f'(c) = \dfrac{1}{2\sqrt{x}} = 1/2.$ Now f ' (c) $= \dfrac{1}{2\sqrt{c}}$, so we set $\dfrac{1}{2\sqrt{c}} = \dfrac{1}{2}$ to determine that c = 1.

We check that c = 1 lies in the given interval (0, 4), and it does.

EXAMPLE: Harder Problem. Find upper and lower bounds for $f(x) = 29^{1/3}$.
Answer:
When f is a function on an interval [a,b] such that the conditions of the Mean Value Theorem are met, then there exists a c in (a,b) such that [f(b)-f(a)] / (b-a) = f '(c).

Let $M = \max\{|f'(x)|: x \in (a,b)\}$ (Check this maximum exists --it does.) Then $-M \le f'(x) \le$ M for $x \in (a,b)$, and in particular, $- M \le f'(c) \le M$. Therefore, $-M(b-a) \le f'(x) \le M(b-a)$. Now let f be the function $f(x) = x^{1/3}$. We let "a" be a point where we know something about f: namely, a = 27. We let b = 29.5, because we want to bound f(29) and we want 29 $\in$ (a,b). (In this case, however, it will not hurt to use b = 29, because our function is defined for x > 29).   Now $f'(x) = (1/3)x^{-2/3}$. We find M such that $|(1/3)x^{-2/3}| \le M$ for all x in [27, 29.5]. One such M is found by letting x = 27, with M = 1/27. Using the inequality above,

$$f(27) - (1/27)(29.5\text{-}27) \le f(29) \le f(27) + (1/27)(29.5\text{-}27), \text{ or}$$
$$3\text{-}(2.5)/27 \le 29^{1/3} \le 3 + (2.5)/27, \text{ or } 2.907 \le 29^{1/3} \le 3.093.$$

The bounds on f(29) could be made sharper by tightening the interval (a,b).

SECTION 4.3

"Let early education be a source of amusement; you will then be better able to find out the natural bent."...Plato

EXAMPLE: Find all antiderivatives of $f(x) = -\cos x + x^2 + 4x^5$.

Answer: Antiderivatives are given by functions

$F(x) = -\sin x + \dfrac{x^3}{3} + 4 \cdot \dfrac{x^6}{6} + C,$ where C represents any constant. (Simplify this!)

Check your answer by showing that $F'(x) = f(x)$.

EXAMPLE: Using initial conditions. Find f(x) satisfying:

$f'(x) = -\dfrac{1}{2}x^3; \quad f(1) = 0.$

Answer:

$$f(x) = -\frac{1}{2}\left(\frac{x^4}{4}\right) + C$$

$$f(x) = -\frac{x^4}{8} + C. \quad \text{Now}$$

$$f(1) = -1/8 + C$$

$$0 = -1/8 + C, \text{ and}$$

$$C = 1/8. \text{ So:}$$

$$f(x) = -\frac{x^4}{8} + \frac{1}{8}.$$

CHECK WORK $f'(x) = -(4/8)x^3$ and $f(1) = 0$. **Check!**

EXAMPLE: Using initial conditions. Find the function f satisfying f''(x) = cos x, f'(π) = 0, f(0) = 1.

Answer:

$$f''(x) = \cos x \text{ implies}$$
$$f'(x) = \sin x + C$$
$$f'(\pi) = \sin \pi + C, \text{ or } 0 = 0 + C, \text{ or } C = 0, \text{ so}$$
$$f'(x) = \sin x. \text{ This implies}$$
$$f(x) = -\cos x + C$$
$$f(0) = -\cos 0 + C, \text{ or } 1 = -1 + C, \text{ or } C = 2, \text{ so}$$
$$f(x) = -\cos x + 2.$$

CHECK WORK: $f(0) = 1$. $f'(\pi) = 0$. $f''(x) = \cos x$. **Check!**

NOTE: When finding intervals on which f is increasing and decreasing, we will find values of x for which f'(x) = 0 and for which f'(x) does not exist. We will investigate these values of x *even when x is not in the domain of f(x)*, because changes in the behavior of a function can occur at such points. Sometimes these values of x are called "critical numbers", because they are points where the situation may change. This would represent a slight difference in the interpretation the word had in max-min problems, where x was a critical number only if f(x) was defined.

EXAMPLE: Let $f(x) = x^5 + x^3 + 2x - 1$. Find the intervals on which f is increasing or decreasing:

Answer: We check the sign of the derivative, $f'(x) = 5x^4 + 3x^2 + 2$. This is always positive, so f is increasing everywhere.

EXAMPLE: If $f(x) = x^5 + x^3 - 2x - 1$, find the intervals on which f is increasing and decreasing.

Answer: First find values x where f'(x) = 0 (f' will always exist if f is a polynomial).
$f'(x) = 5x^4 + 3x^2 - 2 = (5x^2 - 2)(x^2 + 1) = 0$ when $5x^2 = 2$, or $x = \pm\sqrt{2/5}$. Now draw a chart, showing where f'(x) = 0. Using test points (or factors --see Chapter 1 of this Guide) find where f' is positive and negative.

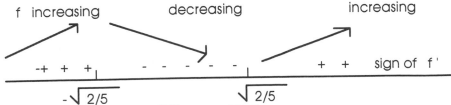

Thus, f is increasing on $(-\infty, -\sqrt{2/5})$ and $(\sqrt{2/5}, \infty)$ and decreasing elsewhere.

Specifically, f is decreasing on $(-\sqrt{2/5}, \sqrt{2/5})$.

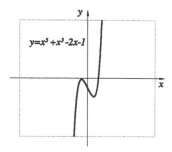

$$f(x) = x^5 + x^3 - 2x - 1$$

EXAMPLE: If f is given by $f(x) = \dfrac{1}{5 - x}$, find where f is increasing and decreasing.

Answer:

Since $f'(x) = \dfrac{1}{(5 - x)^2} > 0$ on its domain of definition , this function is always

decreasing. Here $f'(x)$ does not exist at x = 5.

sign of f '

5

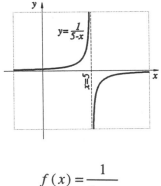

$$f(x) = \dfrac{1}{5 - x}$$

EXAMPLE: If $f(x) = \sin 2x + x$, find where f is increasing and decreasing.

Answer:

$f'(x) = 2\cos 2x + 1$. Setting this to 0, $\cos 2x = -1/2$, Restricting $2x$ to lie in $[0, 2\pi]$, or equivalently, x in $[0, \pi]$, we have $2x = 2\pi/3$ or $4\pi/3$. Then $x = \pi/3$ or $2\pi/3$. Now we chart the sign of $f'(x)$.

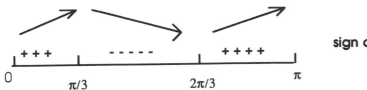

sign of f ' (x) = 2cos 2x + 1

So far, we have that f is increasing on $(0, \dfrac{\pi}{3})$ and $(\dfrac{2\pi}{3}, \pi)$ and decreasing on

$(\pi/3, 2\pi/3)$. Since f' is a periodic function with period π, we add $(n\pi, n\pi)$ to our

intervals. So f is increasing on $(n\pi, \dfrac{\pi}{3} + n\pi)$ and $(\dfrac{2\pi}{3} + n\pi, \pi + n\pi)$,

and decreasing on $(\dfrac{\pi}{3} + n\pi, \dfrac{2\pi}{3} + n\pi)$, where n $= 0, \pm 1, \pm 2, \ldots$.

EXAMPLE: **The sign of f ' need not alternate from interval to interval.** If
$f(x) = \dfrac{x^4}{4} - \dfrac{x^3}{3} - \dfrac{x^2}{2} + x$, find where f is increasing and decreasing:

Answer:

☞ **For advice on factoring, see the NOTE following.**

We set $f'(x) = x^3 - x^2 - x + 1 = 0$. Then $(x-1)^2(x+1) = 0$, or $x = \pm 1$. Now the

chart of signs for $f'(x)$ is :

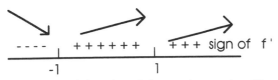

and we see that f is increasing on (-1, ∞) and decreasing on (-∞,1).

NOTE: Finding the roots of a cubic polynomial. When finding the roots of a cubic polynomial, formulas exist, but chances are, the problem you see is "rigged". Guess a number that makes the polynomial 0; we usually guess x = 0, x = 1 or -1, or x = 2 or -2. In the problem above, we wanted to factor g(x) = $x^3 - x^2 - x + 1$. We saw that g(1) = 0. Then, since x = 1 makes the polynomial 0, it will follow that (x-1) is a factor. The other factor, which will be quadratic, can be found by dividing polynomials. (See Chapter 1 of this Guide for a review.)

Write the Theorem
Can you find a result? (There may be more than one answer.)

1. If f is continuous on a closed interval then ------------

2. If f is continuous on the closed interval [a,b] and differentiable on (a,b) then --------------

3. A theorem that guarantees a maximum for a function states --------------------

4. A theorem that guarantees a point x = c where the tangent line to the graph is parallel to the x-axis is -----------------and states -----------------.

5. The theorem that corresponds to the picture below states:

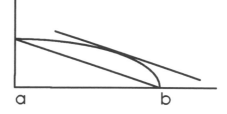

IF THE TESTS AREN'T LIKE THE HOMEWORK

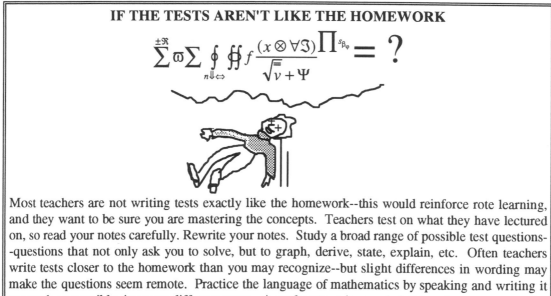

Most teachers are not writing tests exactly like the homework--this would reinforce rote learning, and they want to be sure you are mastering the concepts. Teachers test on what they have lectured on, so read your notes carefully. Rewrite your notes. Study a broad range of possible test questions--questions that not only ask you to solve, but to graph, derive, state, explain, etc. Often teachers write tests closer to the homework than you may recognize--but slight differences in wording may make the questions seem remote. Practice the language of mathematics by speaking and writing it as much as possible, in many different ways. A study group is very helpful . As you go over the homework, try to understand why these types of questions are being asked--think "around" the questions, rather than simply falling into the rote mode of doing what is asked. If you do the "long" problems, you will get a good conceptual grasp of the topics.

Before it's too late...........!

♦ *The concepts I have the most difficulty understanding are:*

♦ *The most important components of the theory in my mind, so far are:*

♦ *The problems I have the most trouble with are:*

♦ *What I need to do :*

SECTION 4.4

"Be not careless in deeds, nor confused in words, nor rambling in thought." Marcus Aurelius

EXPONENTIAL GROWTH AND DECAY FUNCTIONS

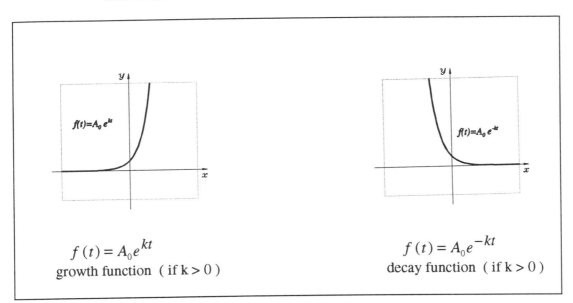

$$f(t) = A_0 e^{kt}$$
growth function (if $k > 0$)

$$f(t) = A_0 e^{-kt}$$
decay function (if $k > 0$)

EXAMPLE: **Procedure for solving a problem with an exponential growth function.**

These questions show a range of types of questions one can be asked about exponential growth.

A substance from outer space has a mass that follows an exponential growth equation. At the outset of the observation of this substance, 10 g. of the substance are present. Five minutes later, 15 g. are present. (a) **How much is present** 20 mins. after the initial observation? (b) **At what time** will the substance have mass 23 g.? (c) **How much was present** 1 min. **before** the observation began?

Answer: Before we can answer any of these questions, we must find the constants in the exponential equation, which must be memorized.

$$\boxed{A = A_o e^{kt}}$$

⊠ We must find k and A_0.

Since the amount present at the beginning of the observation is 10g., we have $A_0 = 10$.

(Note: We used that when $t = 0$, $A = A_0 = 10$.)

⊠ A_o always gives the amount present when t = 0.

We have then, $A = 10e^{kt}$.   When t = 5, A = 15, so

$15 = 10 e^{5k}$.

⊠   We solve for k. First we divide, then take ln ( ) of both sides.

(Note: You could take ln ( ) first and obtain ln 15 = ln 10 + 5k.)

$\dfrac{15}{10} = \dfrac{3}{2} = e^{5k}$ so $\ln(3/2) = 5k$. So k = (1/5) ln(3/2) = .08109. (Find on your calculator.)

So our formula for A becomes:

$$A = 10e^{.08109\, t}.\quad \textit{Now we can answer the questions!}$$

(a) At time t = 20, $A = 10 e^{.08109\,(20)} = 50.62194$ g.

(b) We want t, where $23 = 10 e^{.08109\, t}$. Divide before taking ln ( ).

$23/10 = e^{.08109\, t}$ and ln (23/10) = .08109 t, so t = 10.27142 min.

(c) We set t = -1 Negative values of t are used for times before the start of the experiment .

$A = 10 e^{.08109\,(-1)} = 9.22111$ g.

SECTION 4.5

"As a boy, Newton was only passively interested in his schoolwork until he suddenly woke up at the age of adolescence."...E.T. Bell

NOTE: We can tell if a polynomial function is increasing or decreasing without bound, by examining the term with the largest exponent.

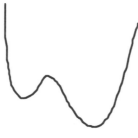   

Leading term has even exponent, with positive coefficient--curve bends up at both ends. An example of this type of function is:

$y = 3x^4 +$ (lower order terms)

Leading term has odd exponent, with positive coefficient-- curve goes from low y-values to high. An example of this type of function is:

$y = 3x^5 +$ (lower order terms)

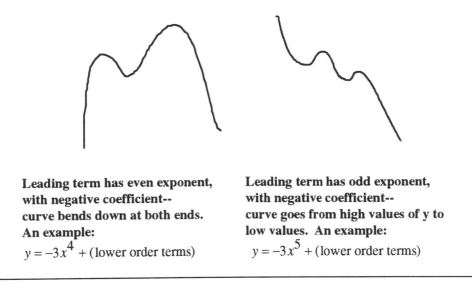

Leading term has even exponent, with negative coefficient-- curve bends down at both ends. An example:

$$y = -3x^4 + \text{(lower order terms)}$$

Leading term has odd exponent, with negative coefficient-- curve goes from high values of y to low values. An example:

$$y = -3x^5 + \text{(lower order terms)}$$

FIRST DERIVATIVE TEST: This test determines where f has relative max/mins

1. Find for what x, $f'(x) = 0$ or $f'(x)$ does not exist (These x-values are called critical numbers).
2. Draw a chart, showing the sign of f' and use this to find where f is rising or falling.
3. The chart will show where f has relative minima and maxima. Compare all of the y-values of the relative max/mins together with values of f at the end-points (if there are end-points) to determine the absolute maximum and minimum of f, **if they exist.**

EXAMPLE: Use the first derivative test to find and classify relative max/mins of the function $f(x) = x^4 - 8x^2 + 5$.

Answer: We find $f'(x) = 4x^3 - 16x^2$ and note $f'(x) = 0$ when $4x^2(x-4) = 0$, or x = 0 or 4. The sign of $f'(x)$ depends on the factor (x - 4) alone.

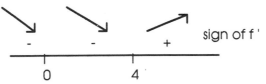

Relative max/mins: We see from the chart that f has a relative minimum at (4, f(4)) = (4, 133). There are no relative maxima. The point (0, f(0)) amounts to "nothing"

EXAMPLE: Find the relative maxima and minima of $f(x) = \sin x + x$.

Answer:

$f'(x) = \cos x + 1$. Restricting x to $[0, 2\pi]$, we find $\cos x + 1 = 0$ for $x = \pi$. Thus $f'(x) = 0$ for $x = \pi + 2n\pi$, n = 0, ±1, ±2, ... But these are not relative maxima or relative minima , by the sign chart below.

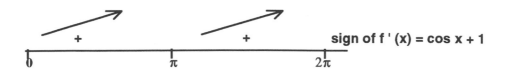

The function is everywhere increasing with no relative max/mins and no absolute max/min. This graph of y = sin x + x winds up the line y = x like ivy twining up a tree trunk.

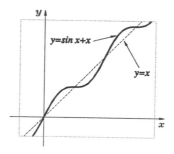

$$f(x) = \sin x + x$$

Hint in graphing this function: When sin x = 0, the curve is on the line y = x. When sin x = 1, the curve is "1 above" the line, and when sin x = -1, the curve is "1 below" the line.

EXAMPLE: If $f(x) = (x^2 - 1)^4$, find all relative maxima and minima of f.

Answer: We find the critical numbers by setting the derivative to 0. (In this example, the derivative exists everywhere.) $f'(x) = 4(x^2 - 1)(2x) = 0$ if x = 0 or x = ±1. We may write f '(x) = 8x (x - 1)(x + 1). Now it is easier to find the sign chart of f ' by the method of factors explained in Chapter 1 of this Guide (or you can use test points instead).

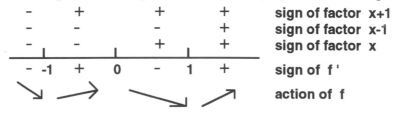

The function is increasing on (-1,0) and (1, ∞) and decreasing on (-∞ , -1) and (0,1). It is useful to make a chart of y-values:

Critical Number x = c	$f(c) = (c^2 - 1)^4$
$x = -1$	$f(-1) = 0$ (rel min)
$x = 0$	$f(0) = 1$ (rel max)
$x = 1$	$f(1) = 0$ (rel min)
Function increasing without bound?	Yes; looks like $f(x) = x^8$
Function decreasing without bound?	No

Conclusion: The function has a relative maximum at $(0, 1)$ and relative minima at $(-1, 0)$ and $(1, 0)$.

THE SECOND DERIVATIVE TEST

1. As before, find the critical numbers, values of x for which   f '(x) is 0 or does not exist.
2. At each critical number, determine the sign of f ''(x). If f ''(x) is positive, we have a relative min. If f ''(x) is negative, we have a relative max. If f ''(x) is zero, the test fails. Use the first derivative test instead or the method in the Exercise below.

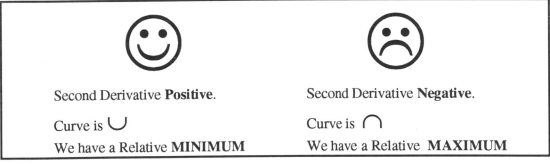

Second Derivative **Positive**.

Curve is $\cup$

We have a Relative **MINIMUM**

Second Derivative **Negative**.

Curve is $\cap$

We have a Relative **MAXIMUM**

EXAMPLE: Use the **Second Derivative Test** to find all relative maxima and minima of $f(x) = x^4 - 2x^2 + 5$.

Answer: First we find the critical numbers from the first derivative.
$f'(x) = 4x^3 - 4x = 0$ when $4x(x^2 - 1) = 0$, or x = 1, -1, or 0. We chart the sign of the second derivative, $f''(x) = 12x^2 - 4$, at each critical number:

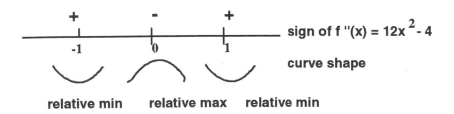

Critical number x = c	$f(c) = c^4 - 2c^2 + 5$
x = -1	f(-1) = 4 (rel min)
x = 0	f(0) = 5 (rel max)
x = 1	f(1) = 4 (rel min)
Function increasing without bound?	Yes
Function decreasing without bound?	No

Conclusions: A relative max is found at (0,5) and relative mins at (-1,4) and (1,4). These points are **candidates** for providing the absolute max and min of f. If x were to lie in a closed interval, we would need to compare the y-values of these candidates with the y-values of f at the endpoints of the interval. In fact, this function is a polynomial of 4th degree; it will have an absolute minimum, but no absolute maximum. The points ($\pm$1,4) are the absolute minima.

EXAMPLE: Use the **Second Derivative Test** to find all relative maxima and minima of the function $f(x) = 4x^5 - 5x^4$.

Answer: We begin by finding the critical numbers. Since this function is a polynomial, the derivative will exist everywhere.

$f'(x) = 20x^4 - 20x^3 = 20x^3(x-1) = 0$ when $x = 0$ and $x = 1$. We check the sign of

the second derivative , $f''(x) = 80x^3 - 60x^2$.

The test fails when x = 0, because f "(0) = 0. We could go back to the first derivative test. In this case, as a short cut, we have tested the sign of f " slightly to the right and left of x = 0. The fact that   f " < 0 near 0 tells us f is concave downward at x = 0.

Our chart now gives:

Critical Number x = c	$f(c) = -5c^4 + 4c^5$
$x = 0$	$f(0) = 0$ (nothing)
$x = 1$	$f(1) = -1$ (rel min)
Function increasing without bound?	Yes; looks like $f(x) = 4x^5$
Function decreasing without bound?	Yes

Conclusion: The function has a relative min at (1,-1) and a relative max at (0,0).

NOTE: When f "(c) = 0, this is an indication that the function is fairly flat at x = c.   The functions below have second derivative equal to 0 at x = 0.

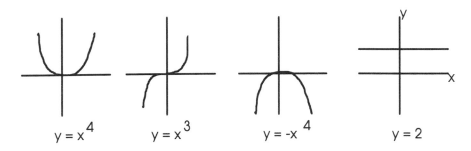

$$y = x^4 \qquad\qquad y = x^3 \qquad\qquad y = -x^4 \qquad\qquad y = 2$$

SECTION 4.6

"Where there are problems, there is life."...Alexander Zinoviev

EXTREME VALUES ON A CLOSED INTERVAL

NOTE: In the previous section, we examined functions that are continuous on the entire real line. These functions may not attain an absolute maximum and minimum. However:

> **If f is continuous on a closed interval, f will have an absolute max and an absolute min on that interval. To find it:**
>
> 1. Find the largest y-value of the relative max's and compare with f(endpoints).This gives the absolute maximum.
>
> 2. Find the smallest y-value of the relative min's and compare with f(endpoints).This gives the absolute minimum.

CAUTION! The relative and absolute maxima and minima are y-values .

The Maximum-Minimum Theorem guarantees that a continuous function on a closed interval takes on a max and min.

CLASSIFICATION OF EXTREMA
Classify the extrema of the function graphed below.

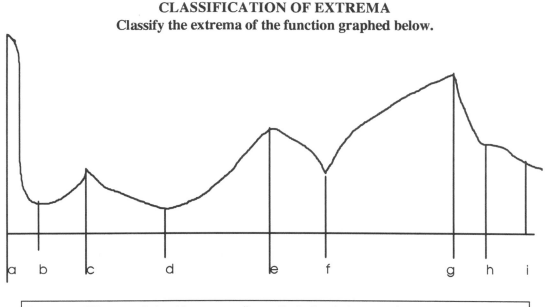

a   b   c   d   e   f   g   h   i

> **Answer:** When we "classify extrema" we find all relative max/mins and the absolute max/min, if they exist. In the diagram above, relative maxima are to be found when x = c, e, g. We exclude x = a because it is an endpoint. Relative minima are found at x = b, d, and f. (If we allowed endpoints, a relative minimum would exist at x = i.) The point x = h yields nothing, although the derivative is 0 there.

Conclusion: The absolute minimum of this function occurs at x = d in a near tie with x = b (same minimum, though) and the absolute maximum occurs at the endpoint x = a.

EXAMPLE: [This example was solved in the previous section as an example of the Second Derivative Test.] Let $f(x) = x^4 - 2x^2 + 5$ Find all extrema on the interval [-2, 4].

Answer: $f'(x) = 4x^3 - 4x = 0$ when $4x(x^2 - 1) = 0$, or x = 1, -1, or 0.

With endpoints the chart now looks like this:

Critical number x = c	$f(c) = c^4 - 2c^2 + 5$
x = -1	f(-1) = 4 (rel min)
x = 0	f(0) = 5 (rel max)
x = 1	f(1) = 4 (rel min)
Endpoint x = -2	f(-2) = 13
Endpoint x = 4	f(4) = 229

Conclusion: A relative max is found at (0,5) and the endpoints, and, from our previous work, relative mins at (-1,4) and (1,4). These are the candidates for the absolute max and min together with the values of f(x) at the endpoints. The absolute max is 229 and occurs when x = 4; the absolute min is 4 occurring at x = 1 and x = -1.

NOTE: The absolute maximum is selected from the relative maxima and the value of f at the endpoints; the absolute minimum from the relative minima and the value of f at the endpoints.

Common Errors

♦ Suppose f is differentiable on (a,b), f ' (a) = 0 , f ' (b) = 0, and f (a) > f (b). It does not follow that f (a) will be the maximum and f (b) the minimum.

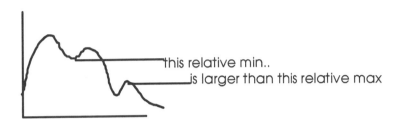

this relative min..
is larger than this relative max

Fill in the blanks with good explanations. It makes a good review to elaborate as much as possible .

Some major theory has been developed in this chapter. An important theorem is the Max-Min Theorem that says when f is ----------on a -----------then -------------.   Here is an example of how the theorem can fail if the conditions are not met.---------If the function is differentiable at a relative max or min, we can expect ------------. Rolle's Theorem states ----Geometrically, Rolle's Theorem says -------An example of a function for which Rolle's Theorem does not work is------- The INTERMEDIATE Value Theorem states ----------. We saw this in a former chapter. The MEAN Value Theorem states that if f is -------------then -----The geometrical significance of this theorem is that ---------. An example of the usefulness of this theorem is -----------. This theorem does not say where the point x = c lies. However, if asked to find this c, as in the homework problems we do the following ------. In finding maxima and minima of a function, we have these tools----------. Solving word problems comes up here. So far we have seen problems from a few categories; such as -----. Critical numbers are ---------. We use them to -----------. Extreme values are ------- . The First Derivative Test tells us ------. The Second Derivative Test tells us -----. We can determine the concavity of the curve by ---------.

MORE WORD PROBLEMS

These problems are similar to those we saw in Section 4.3. They might be considered to be a continuation. These problems have obvious, or implicit "endpoints" in the statement of the problem. Look back at the word problems from the previous sections, and see how much more easily they are solved now.

GREATEST VOLUME (OR AREA) FOR GIVEN SURFACE AREA (OR PERIMETER) Know your formulas!

EXAMPLE: Maximizing the area of a window. A Norman window has the shape of a rectangle with a semi-circle at the top. The perimeter of the window is 14 feet. What dimensions maximize the area?

Outline of a solution:

STEP 1. Draw the picture; label it:

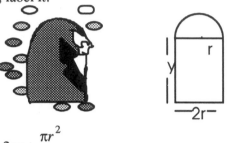

STEP 2. The area is $A = 2ry + \dfrac{\pi r^2}{2}$.

STEP 3. We seek a relation between r and y, to express the area in terms of a single unknown. The perimeter equation is :

$$14 = 2r + 2y + \pi r, \text{ or } 2y = 14 - r(2 + \pi), \text{ or } y = 7 - r(1 + \frac{\pi}{2}) .$$

Thus $A = 2r (7 - r (1 + \pi/2)) + \pi r^2 / 2 = 14r - 2r^2\left(1 + \dfrac{3\pi}{4}\right).$

STEP 4. Set dA/dr = 0. Find critical numbers, values of r.

STEP 5. Check this (these) value(s) of r in A" (r); we want A"(r) to be negative for a maximum. Find A associated with this r.

STEP 6. Find what happens at endpoints, if there are any. Compare these values of A with that found in Step 5. The endpoints give infeasible solutions (A = 0; they give the minimum A.)

STEP 7. Find y and 2r, because the problem asks for dimensions.

(The solution to this problem is: $r = y = \dfrac{14}{\pi + 4} \approx 1.96.$)

MINIMIZING THE COST OF A BOX

EXAMPLE: A manufacturer is making closed rectangular boxes with square bases, that must contain 100 cu. cm. If the material for the top costs twice as much per cu. cm. as the material used in making the rest of the box, what dimensions should the box be to minimize the cost?

Outline of a solution:

STEP 1. Draw the picture. Label it.

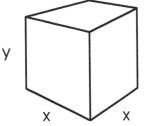

STEP 2. We are trying to minimize cost. So the equation for cost is
$C = x^2 + 4xy + 2x^2$ (where the unit of cost for the base and sides is 1, and the unit of
cost for the top is 2.) In other words, $C = 3x^2 + 4xy$.

STEP 3. From the volume equation, solve for y in terms of x:
$V = x^2 y$ or $100 = x^2 y$ or $y = 100 / x^2$.   Replace in the cost equation in STEP 2:
$C = 3x^2 + \dfrac{400}{x}$.

STEP 4. Set dC/dx = 0. We want x to satisfy C'' (x) > 0.

STEP 5. From the equation for y in (3), find y. Now we have the dimensions.

STEP 6. Are there endpoints? How large and small can x be? If x = 0 , Volume = 0, and if
x grows infinitely large, volume also tends to 0. Thus our solution must give the maximum
volume.

THE PAGE WITH MARGINS

EXAMPLE: Find the dimensions that will minimize the area of a rectangular page that
is to have 1 inch margins all around and 36 square inches of printed material.
  Outline of a solution:
STEP 1. Draw the picture. Let x and y be dimensions of the page.

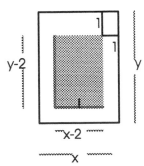

STEP 2. We want to minimize   area, A = x y , subject to (x - 2) (y - 2) = 36 (the
equation giving the area of the printed matter). Use this equation to solve for y in terms of
x, and write an expression for A in terms of x.

(Since (x-2)(y-2) = 36, we have $xy - 2x - 2y + 4 = 36$, or $xy - 2y = 32 + 2x$, or $y(x-2) = 32 + 2x$. From this we can solve for y.)

STEP 3. Find where dA/dx = 0. Find the corresponding y from the equation in (STEP 2). The problem asks for the dimensions of the page, and we have found them.

STEP 4. For this x, we could check that A" (x) > 0, corresponding to a minimum. But A, as a quadratic in x, gets large in an unbounded way. Therefore, our value must provide the minimum of A.

MINIMIZING DISTANCE FROM A POINT TO A CURVE

EXERCISE: Outline of a solution. Find the point(s) on the curve $y = x^2 + 2x$ the closest to the point (-1,0) .

Answer:
STEP 1. Draw the picture. (Use a graphing calculator?)

STEP 2. The distance from (-1,0) to a point on the curve is given by
$$D = \sqrt{(x-(-1))^2 + (y-0)^2} = \sqrt{(x+1)^2 + y^2} = \sqrt{(x+1)^2 + (x^2+2x)^2}.$$
Now minimizing D gives the same result as minimizing D^2 and
$D^2 = (x+1)^2 + (x^2+2x)^2$. (This is easier to differentiate.)
We set $\dfrac{d(D^2)}{dx} = 0$, or $2(x+1) + 2(x^2+2x)(2x+2) = 0$, or
$(x+1) + 2(x^2+2x)(x+1) = 0$, or $(x+1)(1+2x+2x^2) = 0$. This gives values of x = -1, and (using the quadratic formula) other roots which are complex.

STEP 3. Find the value of y on the parabola from the equation for the parabola, given above: from the equation, $y = x^2 + 2x$, our point on the curve is (-1, 3) .

STEP 4. To prove this gives the minimum distance, we argue as follows. D^2 is a function whose graph is a 4th degree polynomial, opening upwards, so the critical number (the only x where the derivative vanishes) must supply the minimum.

LAND OR WATER or WHERE TO CROSS OVER

EXAMPLE: Joe rows a boat on water more slowly than he walks on land. Assume he rows at 4 m. per min and walks on the bushy trail at 12 m. per min. He is attempting to reach a point across a river that is 50 m. wide and 40 m. downstream. He can go on land, then cut across the stream diagonally. What path will minimize his time?

Answer:
STEP 1. Draw a picture.

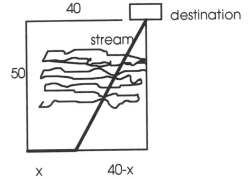

STEP 2. We wish to minimize TIME:

T = (Time on Land) + (Time on Water)

but to make the mathematics simpler, we will set the rate of travel on water to be 1 unit and
rate on land , 12/4=3 units. T now reports a time which is 4 times too slow, but this
maneuver will not affect the minimal path. Now with these units,

$T = (x/3) + \sqrt{(40-x)^2 + 50^2}$. Then

$\dfrac{dT}{dx} = \dfrac{1}{3} + \dfrac{(-2(40-x))}{2\sqrt{(40-x)^2 + 50^2}} = 0$ when $\dfrac{1}{3} = \dfrac{(40-x)}{\sqrt{(40-x)^2 + 50^2}}$ or

$\sqrt{(40-x)^2 + 50^2} = 3(40-x)$. Squaring both sides , $(40-x)^2 + 50^2 = 9(40-x)^2$. Then ,

$8(40-x)^2 = 50^2$ and $(40-x) = 50/\sqrt{8} = 25/\sqrt{2}$. So x = $40-25/\sqrt{2}$.

STEP 3. This finds x, the distance on land. To "find the path" , as the question asks, you
might want to find the distance traveled on water as well, suggested in the picture.

STEP 4. There are feasible endpoints implied here. Doing the entire trip on water, x = 0.
The other endpoint, x = 40, represents Joe doing 40 m on foot, and then crossing over the
river. We summarize this information:

x =	Actual time , $T = (1/4)[\, x/3 + \sqrt{(40-x)^2 + 50^2}\,]$	Classify
Endpoint x = 0	116.001 min	Longest Time
Endpoint x = 40	15.833 min .	(nothing)
$40 - 25/\sqrt{2} = 22.32$	15.118 min .	**Shortest Time**

Here we recall that our function T was 4 times too slow when we set up the problem.

We conclude Joe should travel $40 - (25/\sqrt{2})$ m. on land.

SECTION 4.7

"But for my part, it was Greek to me."...Shakespeare

CONCAVITY AND POINTS OF INFLECTION

The first derivative gives the rate of change of the function. If positive, the function is increasing, and if negative, the function is decreasing.

The second derivative gives the rate of change in the derivative. If positive, the slope of the tangent line is increasing, and the function is **concave upward.** If negative, the slope of the tangent line is decreasing, and the function is **concave downward.** Whereas the first derivative tells us about rising and falling, the second derivative describes the curvature of the curve.

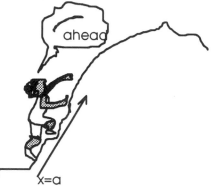

Knowledge of every derivative increases our understanding of the function. The second derivative gives information about the curve from farther away than the first derivative.

NOTE: Functions can be increasing or decreasing with different concavity.

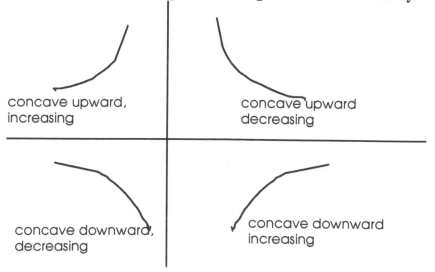

NOTE: A point *on the graph* of a function where the concavity changes is called an inflection point. We locate it by finding where f " (x) is zero or does not exist and checking that the concavity changes there.

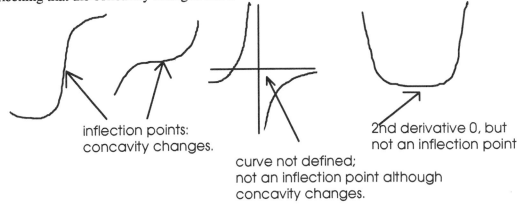

inflection points: concavity changes.

curve not defined; not an inflection point although concavity changes.

2nd derivative 0, but not an inflection point

EXAMPLE: Find where the function $f(x) = e^{-x^2}$ is concave upward or downward, increasing or decreasing. Find inflection points, if there are any.

Answer: We check both the first and second derivatives.

First Derivative: $f'(x) = -2xe^{-x^2} = 0$ when $x = 0$.

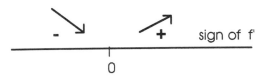

sign of f'

We see f is increasing on $(0, \infty)$, decreasing on $(-\infty, 0)$. The function has a relative minimum at $(0,1)$.

Second Derivative:
$f''(x) = -2e^{-x^2} + 4x^2e^{-x^2} = 0$ when $4x^2 - 2 = 0$, or when $x = \pm 1/\sqrt{2}$.
(Note that the sign of f'' depends on the sign of $4x^2 - 2$, or $2x^2 - 1$.)

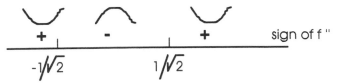

sign of f "

The function is concave upward on $(-\infty, -1/\sqrt{2})$ and $(1/\sqrt{2}, \infty)$ and concave downward on $(-1/\sqrt{2}, 1/\sqrt{2})$. The points $(\pm 1/\sqrt{2}, e^{-1/2})$ are the inflection points. The point $(0,1)$ is a relative maximum

EXAMPLE: If f (x) = 1/x, find the inflection points, if any.

Answer: $f'(x) = -1/x^2$ and $f''(x) = -2/x^3$. The function changes concavity at x = 0, as f " changes sign when x changes from negative to positive. But f is not defined at x = 0; we do not have an inflection point.

ABOUT POLYNOMIALS

The extent of "bumpiness" (the number of hills and valleys) in the graph of a polynomial is limited by the degree of the polynomial.

EXAMPLE: If f is a 4th degree polynomial with positive leading coefficient , graph some of its possible shapes.

Answer: A polynomial of degree 4, $f(x) = a_4 x^4 + a_3 x^3 + a_2 x^2 + a_1 x + a_0$,   can have at most 3 relative max or mins (where the derivative vanishes) because the derivative is a 3rd degree polynomial, which can have at most 3 distinct real roots.   Not every x for which the derivative vanishes will necessarily provide a relative max or min however; there may be "degeneracies".   Here are some possible shapes the polynomial a 4th degree polynomial can have if the leading coefficient is positive.

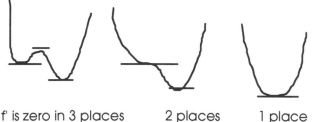

f' is zero in 3 places   2 places   1 place

What Would You Do If

❶   Asked to classify the extrema of a given function? (Go through all the steps). Did you use the first derivative test, the second? Did you attempt to find the overall max/min? Did you compare the y-values at endpoints as well?

❷ Asked to find the inflection points of a given function? (Go through all the steps). Did you remember that if x = a is to be an inflection point, the concavity of the function's graph must change at x = a?

❸   While checking whether x = c gives a relative max or min (or nothing), the Second Derivative Test produced f " (c) = 0 ? (Did you go back to the First Derivative Test? Did you check the concavity on either side of x = c?)

❹   Asked to work a problem with an exponential growth function? (What is the general equation? Would you be able to find the constants particular to the information of the problem? Would you be able to find quantities such as amount, and time?)

Draw an Example

Yes! These are all possible. Draw an example of a function which:

1. Is increasing everywhere and concave downward.
2. Is decreasing everywhere and concave upward.
3. Has no relative maximum or minimum at x = c, but the derivative is 0 there, nonetheless.
4. Has a relative maxima at a point where the derivative is not 0. (Hint: the derivative would not exist.)
5. Has a relative minimum which is not the absolute minimum.
6. Has an absolute maximum at an endpoint.
7. Has an absolute maximum that is the same as the absolute minimum, on a closed interval.
8. Is a 4th degree polynomial with maximal number of "bumps" allowable.
9. Is a 5th degree polynomial with maximal number of "bumps" allowable.
10. Is a 4th degree polynomial with the minimal number of "bumps" allowable.
11. Is concave downward on one side of the point x = c, and concave upward on the other side.
12. Is concave downward on one side of the point x = c, and concave upward on the other side, but for which the second derivative does not exist at x = c.
13. Has a relative minimum that is larger than a relative maximum.
14. Has a maximum at a point where the derivative is positive. (Hint: look for endpoints.)
15. Is a polynomial which has no real roots; that is, never touches the x-axis.
16. Has an absolute maximum for all x on the real line.

SECTION 4.8

"A youth who likes to study will in the end succeed...There are 6 Essentials in painting: The first is called spirit, the second rhythm, the third thought, the fourth scenery, and the fifth the brush, and the last is the ink"....Ching Hao

LIMITS AT INFINITY (as $x \to \pm\infty$)

EXAMPLE: Find $\displaystyle\lim_{x\to\infty} \frac{x^2 - x}{2x^2 + 1}$.

Answer: Divide above and below by the highest power of x in the rational function:

namely, x^2. Then $\displaystyle\lim_{x\to\infty} \frac{x^2 - x}{2x^2 + 1} = \lim_{x\to\infty} \frac{1 - 1/x}{2 + 1/x^2} = \frac{1}{2}$.

NOTE: A "rational function" is a quotient of polynomials.

NOTE: Notice that $\displaystyle\lim_{x\to\infty} \frac{1}{x} = \lim_{x\to-\infty} \frac{1}{x} = 0$.

EXAMPLE: Find $\lim\limits_{x \to \infty} \dfrac{\sin x}{x}$. Now $-1 \le \sin x \le 1$, so

$-\dfrac{1}{x} \le \dfrac{\sin x}{x} \le \dfrac{1}{x}$ when x > 0 (as it is, when large). Thus $\dfrac{\sin x}{x}$ gets "squeezed to 0" and

$\lim\limits_{x \to \infty} \dfrac{\sin x}{x} = 0$. Thus y = 0 is a horizontal asymptote.

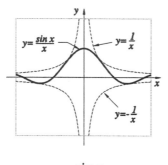

$$f(x) = \frac{\sin x}{x}$$

Hint in graphing: Graph the enveloping lines y = ±1/x. The graph of sin x wobbles between these lines, and when x →0, (sin x)/x →1.

NOTE: A function may cross its horizontal asymptote, as does the function above.

EXAMPLE: Find $\lim\limits_{x \to \infty} e^{-x^3}$.

Answer: As x gets large, the exponent gets large and negative, so this limit is 0. The function $y = e^{-x^3}$ has the horizontal asymptote, y = 0.

> **A HORIZONTAL ASYMPTOTE, y = b, occurs where**
> $$\lim\limits_{x \to \infty} f(x) = b \quad \text{or} \quad \lim\limits_{x \to -\infty} f(x) = b$$
> **where b is a real number.**

NOTE: In finding horizontal asymptotes, ask what happens as $x \to \infty$ **and** $x \to -\infty$ unless the function is not defined for large positive or large negative x.

EXAMPLE: If $f(x) = e^x \ln x$, find the horizontal asymptotes, if any.

Answer: Since ln x is defined only if x > 0, we find only $\lim\limits_{x \to \infty} e^x \ln x$, and this limit is ∞. There is no horizontal asymptote.

NOTE: When finding limits of rational functions, the polynomials behave like their leading term as x approaches ∞ or $-\infty$. In fact the limit above could be found like this:

$$\lim_{x \to \infty} \frac{x^2 - x}{2x^2 + 1} = \lim_{x \to \infty} \frac{x^2}{2x^2} = \frac{1}{2},$$ where we ignore every term but the leading term.

EXAMPLE: Given $f(x) = \frac{\sqrt{x+2}}{x^2 + 3}$, find the horizontal asymptotes, if any.

Answer: (Following our note above, this function behaves like $f(x) = \frac{\sqrt{x}}{x^2} = \frac{1}{x^{3/2}}$ for x

large, and the limit of this is 0. We anticipate this limit.) But we now find the limit more formally:

$$\lim_{x \to \infty} \frac{\sqrt{x+2}}{x^2 + 3} = \lim_{x \to \infty} \frac{\sqrt{x+2}}{x^2 + 3} \cdot \frac{\sqrt{x+2}}{\sqrt{x+2}} = \lim_{x \to \infty} \frac{x+2}{(x^2+3)(\sqrt{x+2})} \quad \text{(Rationalize denominator)}$$

$$= \lim_{x \to \infty} \frac{x/x^2 + 2/x^2}{(x^2/x^2 + 3/x^2)(\sqrt{x+2})} = \lim_{x \to \infty} \frac{x/x^2 + 2/x^2}{(x^2/x^2 + 3/x^2)(\sqrt{x+2})} = 0 \quad \text{(where we divide}$$

above and below by leading term) The numerator approaches 0 and the denominator , ∞.

Note that we divide ONE factor by x^2 in the denominator .

EXAMPLE: Finding the limit in the "$\infty - \infty$ " case. Find the limit:

$$\lim_{x \to \infty} x - \sqrt{2x^2 + 1} .$$

Answer: We make this limit look like a quotient.

$$\lim_{x \to \infty} x - \sqrt{2x^2 + 1} = \lim_{x \to \infty} x - \sqrt{2x^2 + 1} \cdot \left(\frac{x + \sqrt{2x^2 + 1}}{x + \sqrt{2x^2 + 1}} \right) \quad \text{(multiply above / below by conjugate)}$$

$$= \lim_{x \to \infty} \frac{x^2 - (2x^2 + 1)}{x + \sqrt{2x^2 + 1}} = \lim_{x \to \infty} \frac{-x^2 - 1}{x + \sqrt{2x^2 + 1}} = \lim_{x \to \infty} \frac{-x^2/x^2 - 1/x^2}{x/x^2 + \sqrt{(2x^2 + 1)/x^4}} =$$

$$\lim_{x \to \infty} \frac{-1 - 1/x^2}{1/x + \sqrt{(2x^2 + 1)/x^4}}.$$ The numerator approaches -1, and the denominator is

approaching 0 through positive values (x is large, so the denominator is positive).

EXAMPLE: Find the horizontal asymptotes, if any, for $f(x) = \frac{e^x + e^{-x}}{e^x + 3e^{-x}}$.

Answer: $\lim_{x \to +\infty} \frac{e^x + e^{-x}}{e^x + 3e^{-x}} = 1$ and $\lim_{x \to -\infty} \frac{e^x + e^{-x}}{e^x + 3e^{-x}} = 1$. (Since $e^{-x} = \frac{1}{e^x}$, these terms go

to 0.) Thus y = 1 is the horizontal asymptote.

EXAMPLE: Let $f(x) = \sqrt{\dfrac{1-x}{4-x}}$. Find the horizontal asymptotes. Then sketch the graph of the function.

Answer:

STEP 1. First find the domain, i.e., values of x where $\dfrac{1-x}{4-x} \geq 0$.

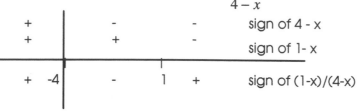

The vertical line means x cannot equal -4.

The domain of f consists of x in (-∞, -4) or [1,∞).

STEP 2. Find vertical asymptotes: x = - 4 is a vertical asymptote. Since f (x) > 0, the function will approach +∞ at this asymptote.

STEP 3.   Find intercepts: (1,0) is the x-intercept, and (0,1/2) is the y-intercept.

STEP 4. Find horizontal asymptotes: $\lim\limits_{x \to \infty} \sqrt{\dfrac{1-x}{4-x}} = 1$ and $\lim\limits_{x \to -\infty} \sqrt{\dfrac{1-x}{4-x}} = 1$. So y = 1 is a vertical asymptote as x → ∞ or x → -∞. Note that f(x) ≥ 0.

STEP 5.   Sketch the graph:

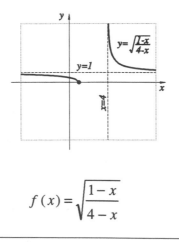

$$f(x) = \sqrt{\dfrac{1-x}{4-x}}$$

NOTE: A rational function f is a quotient of polynomials: $f(x) = \dfrac{p(x)}{q(x)}$.

In finding limits of rational functions, always divide if degree of q(x) ≤ degree p(x). Three things can happen.

1. If the degree of the numerator is less than the degree of the denominator,
   $\lim\limits_{x \to \pm\infty} f(x) = 0$ and $y = 0$ is a horizontal asymptote.

   EXAMPLE: $f(x) = \dfrac{x^2 + 20x - 5}{3x^3 - 400x^2}$. Degree of numerator is less than degree of
   denominator. Horizontal asymptote: $y = 0$.

2. If the degree of the numerator is greater than the degree of the denominator, then
   $\lim\limits_{x \to \infty} f(x) = \pm\infty$ and $\lim\limits_{x \to -\infty} f(x) = \pm\infty$. In this case, there is no horizontal asymptote.

   EXAMPLE: $f(x) = \dfrac{x^4 + 20x - 5}{3x^3 - 400x^2}$. Degree of numerator exceeds degree of
   denominator. No horizontal asymptote. (But by dividing, you may find other asymptotes,
   that are neither horizontal nor vertical.)

3. If the degree of the numerator equals the degree of the denominator, then

   $\lim\limits_{x \to \pm\infty} f(x) = \dfrac{a}{b}$ where a and b are the leading coefficients. The horizontal asymptote

   is $y = \dfrac{a}{b}$.

   EXAMPLE: $f(x) = \dfrac{x^3 + 20x - 5}{3x^3 - 400x^2}$. Degrees of numerator and denominator equal.

   Horizontal asymptote $y = \dfrac{1}{3}$.

PROOF : f APPROACHES A REAL LIMIT AS $x \to \infty$

We cannot talk about x getting "close to ∝". To avoid this, we speak of x getting larger
than any positive real number N we choose. In order for f(x) to have a real limit L ,
$$\lim\limits_{x \to \infty} f(x) = L ,$$
we mean that the y-values , f(x), get squeezed close to L, i.e. that they   get "arbitrarily"
close to L whenever x gets sufficiently large.

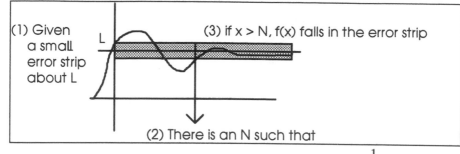

EXAMPLE: Use the formal definition of the limit to prove $\lim\limits_{x\to\infty}\dfrac{1}{x^2}=0$

Answer:
We must show: given any $\varepsilon > 0$, there exists an $N > 0$ such that

$x > N \Rightarrow |\ f(x) - L\ | < \varepsilon,$ or

$x > N \Rightarrow |\ \dfrac{1}{x^2} - 0\ | < \varepsilon$, or   (Replace terms in the definition)

$x > N \Rightarrow \dfrac{1}{x^2} < \varepsilon$, or   (Clear absolute values)

$x > N \Rightarrow \dfrac{1}{\varepsilon} < x^2$, or   (Get into the form "x >")

$x > N \Rightarrow x > \dfrac{1}{\sqrt{\varepsilon}}$ (Now we can conclude)

Take $N \geq \dfrac{1}{\sqrt{\varepsilon}}$.   (End of proof. This N makes all

statements above true.)

PROOF: f APPROACHES ∞ AS $x \to \infty$

$$\lim_{x\to\infty} f(x) = \infty$$

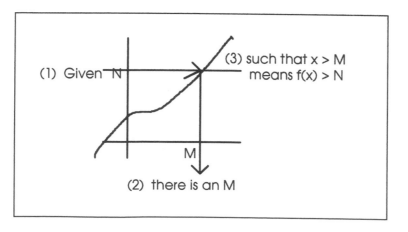

We can make the greyhound go as fast as we want (f(x) > N) by making the stuffed bunny's speed fast (x > M). When x > M, we will have f(x) > N. We want to know: can we find M? Generally it will be in a relation with N.

EXAMPLE: Use the formal definition of the limit to prove $\lim\limits_{x\to\infty} x^2 + 1 = \infty$

Answer:
We must show: given any $N > 0$, there exists an $M > 0$ such that

$x > M \implies f(x) > N$, or

$x > M \implies x^2 + 1 > N$, or   (Substitute terms into definition)

$x > M \implies x^2 > N - 1$, or   (Get into the form " x > ")

$x > M \implies x > \sqrt{N - 1}$.   (We are ready to conclude)

Take $M \geq \sqrt{N - 1}$.   (This M makes the above statements true.)

SECTION 4.9

"The concept of the function is the backbone of the whole of mathematics"....Rozsa Peter

GRAPHING

In this section, we combine all our knowledge about functions to produce graphs. Here are some tips to use when graphing.

PROPERTY	DEFINING PROPERTIES	THINGS TO WATCH FOR	EXAMPLES
Polynomial	Real coefficients, positive integer exponents.	Odd or even degree? Use derivatives. No asymptotes.	$f(x) = x^4 - 3x^2$
Rational function	Quotient of polynomials : $f(x) = \dfrac{p(x)}{q(x)}$	Vertical, horizontal asymptotes may govern behavior	$f(x) = \dfrac{x^3 - 3x}{x^3 + 2x}$.
Periodic function	Function repeats	Trig functions: have period, amplitude, frequency	$f(x) = 5\sin 2x$
Even function	$f(x) = f(-x)$	Same y for $\pm x$. Symmetric across y-axis	$f(x) = e^x + e^{-x}$ $f(x) = x^4 + \cos x$ $f(x) = \ln|x|$
Odd function	$f(-x) = -f(x)$	y changes sign with $\pm x$. Symmetric under a rotation by 180°	$f(x) = e^x - e^{-x}$ $f(x) = x^3 + \sin x$
Combinations of above	Products of odd and even functions	Odd times odd is even; odd times even is odd	$f(x) = x\sin x$ is even. $f(x) = \dfrac{\cos x}{x}$ is odd.
Damped or amplified function	Often, periodic function multiplied by 1/x or x or other functions	Periodic function oscillates inside frame of other function	$f(x) = \dfrac{\sin x}{x}$ is damped; $f(x) = x\sin x$ is amplified.
Functions which are added	$f(x) = g(x) + h(x)$	Graph can be found by adding y-values	$f(x) = x + \sin x$
Compositions of functions	$f \circ g(x)$	May be able to predict things from behavior of 2 functions	$\ln(\ )$ slows growth ; $e^{(\ )}$ speeds growth; $|\ |$ makes f ≥ 0

EXAMPLE: A polynomial. Graph $f(x) = x^4 - 2x^2$.

Answer:

♦   This is a polynomial of even degree, and $f \to \infty$ as $x \to \pm\infty$ with at most, 3 "bumps" possible.

♦   This polynomial is even i.e. symmetric with respect to y-axis. We need only graph this function for $x \geq 0$.

♦   Take derivative: $f'(x) = 4x^3 - 4x = 4x(x-1)(x+1)$; $f'(x) = 0$ if x = 0, 1, or -1.

Check the sign of the first derivative:

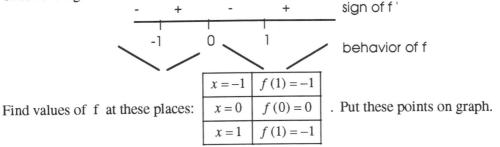

Find values of f at these places:

$x = -1$	$f(1) = -1$
$x = 0$	$f(0) = 0$
$x = 1$	$f(1) = -1$

. Put these points on graph.

Relative minima occur at (-1,-1) and (1,-1); relative max at (0,0). Increasing on (1,∞) and (-1,0) and decreasing on (-∞,-1) and (0,1).

♦ Second derivative concavity information $f''(x) = 12x^2 - 4 = 0$ when x $= \pm 1/\sqrt{3}$.

Find f at these places.

$x = 1/\sqrt{3}$	$f(x) = -5/9$
$x = -1/\sqrt{3}$	(same as above)

. Put these points on the graph.

The function is concave upward on (-∞, $1/\sqrt{3}$) and ($1/\sqrt{3}$, ∞) and concave downward on (-$1/\sqrt{3}$, $1/\sqrt{3}$).

♦ Find the intercepts: (0,0) is the only intercept.

♦ Graph the function.

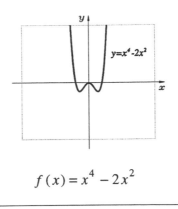

$$f(x) = x^4 - 2x^2$$

EXAMPLE: A rational function. Graph $f(x) = \dfrac{(x-1)^2}{x+1}$

Answer:

♦ Look for symmetry about obvious axes; none.

♦ It will be useful to have a chart of the sign of f:

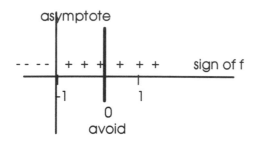

♦ Find the asymptotes: a vertical asymptote occurs where we divide by 0, or $x = -1$. There are no horizontal asymptotes since the degree of the numerator is greater than that of the denominator. Look for behavior at outer ranges of graph. But $\displaystyle\lim_{x\to\infty} \frac{(x-1)^2}{x+1} = \infty$ and $\displaystyle\lim_{x\to-\infty} \frac{(x-1)^2}{x+1} = -\infty$. (Sign chart above helps with this.)

♦ Determine the sign of the function on either side of the vertical asymptote: the sign is determined by $(x + 1)$, since the numerator is always positive. (The graph below illustrates what we know so far). We now seek relative max and mins .

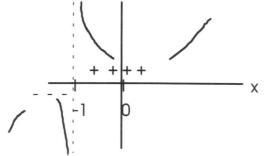

♦ Take the derivative:

$$f'(x) = \frac{(x+1)(2)(x-1)-(x-1)^2}{(x+1)^2} = \frac{(x-1)((2x+2)-(x-1))}{(x+1)^2} = \frac{(x-1)(x+3)}{(x+1)^2}$$

which is 0 if $x = 1$ or -3. The point $x = -1$ is also a critical number; however concavity will not change at $x = -1$ because the term $(x +1)$ is squared. Although the graph of f is fairly obvious now, we will determine where f is increasing, decreasing, etc., by making a sign chart of f'.

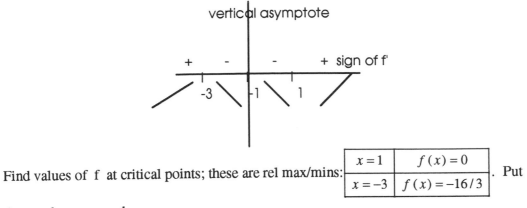

Find values of f at critical points; these are rel max/mins:

$x = 1$	$f(x) = 0$
$x = -3$	$f(x) = -16/3$

. Put

these values on graph.

◆ We should check one last thing: the concavity of the graph of f, to be absolutely thorough. $f''(x) = \dfrac{8}{(x+1)^3}$, and we make a sign chart for f ", which shows f is concave upward for x in (-∞, -1) and concave upward for x in (-1, ∞). The value x = -1 does not

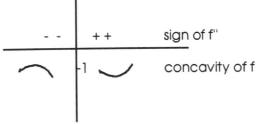

give an inflection point, because f(-1) is not defined there.

◆ The last step is the graph itself:

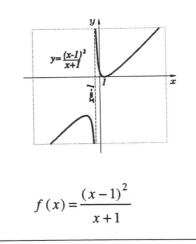

$$f(x) = \frac{(x-1)^2}{x+1}$$

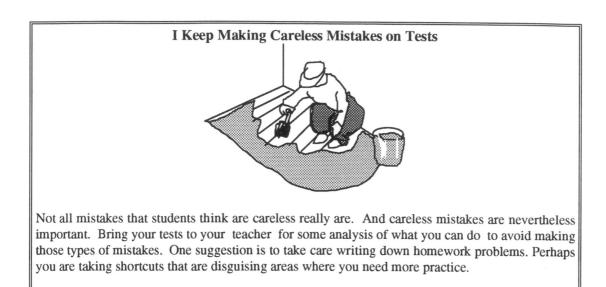

I Keep Making Careless Mistakes on Tests

Not all mistakes that students think are careless really are. And careless mistakes are nevertheless important. Bring your tests to your teacher for some analysis of what you can do to avoid making those types of mistakes. One suggestion is to take care writing down homework problems. Perhaps you are taking shortcuts that are disguising areas where you need more practice.

Self-Check

What skills from this chapter do you feel confident about? Reply in column 2 .
After a test or quiz, come back to this table and fill in the last column with a comment .

TYPE OF PROBLEM	do you feel in control?	after test comments
Exponential Growth and Decay Problems		
Word Problem, using Max/Mins		
Finding Limits as x $\rightarrow \pm\infty$		
Finding Horizontal Asymptotes		
Graphing Problems		
Statements of Theorems (Mean Value Theorem, Max-Min Theorem)		
Finding where Function is Increasing, Decreasing; Using First Derivative Test		
Finding where Function is Concave Upward or Downward; Using Second Derivative Test		
Finding the Function given the Derivative(s), using Initial Conditions		

 # KEEP A LOG OF YOUR STUDY PATTERNS

Determine at what times of day you are likely to put in the most hours, and rate these hours on productivity scale of 1-5 (with 5 being "highly productive"). There is a direct correlation between grades of students and number of *productive* hours spent studying.

CHAPTER 5

"He who pursues fame at the risk of losing himself is not a scholar."...Chuang Tzu

In this chapter we become familiar with the process of integration, which involves the study of the "definite" and "indefinite" integrals. The definite integral makes its appearance when we attempt to find the area under a curve, but it has many more applications. We will be able to define a new function, ln (x) with the definite integral. The indefinite integral is linked with the process of antidifferentiation, where we construct a function whose derivative we know. Amazingly, these two procedures are related by the most important result in calculus, the Fundamental Theorem of Calculus.

Trigonometry is frequently a stumbling block for students in this section of the course. A good ability to differentiate is also important. Review Chapter 1 of this guide, and the chain rule, which appears in Chapter 2.

SECTION 5.1

"The uncomprehended work is beautiful, as on the First Day".....Goethe

Finding Area

LOWER SUM

$L_f(P) = m_1 \Delta x_1 + m_2 \Delta x_2 + \ldots + m_n \Delta x_n$ where m_k is the minimum of f

on the kth subinterval, and Δx_k is the length of the kth subinterval

UPPER SUM

$U_f(P) = M_1 \Delta x_1 + M_2 \Delta x_2 + \ldots + M_n \Delta x_n$ where M_k is the maximum of f

on the kth subinterval, and Δx_k is the length of the kth subinterval

NOTE: The value of the Upper and Lower Sums depends on the partition. The partition is given.

EXAMPLE: Find the Upper and Lower Sums for the function $y = x^2$ on $[-3, 4]$ with the partition P = { -3, -1, 0, 1, 4}.

Answer: Graph $y = x^2$ and the numbers in the partition. Then locate the maximum and minimum values of f on each subinterval. Compute the area of each shaded rectangle, and sum these areas.

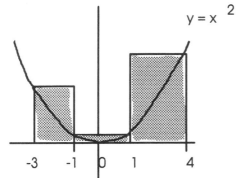

UPPER SUM FOR THE PARTITION

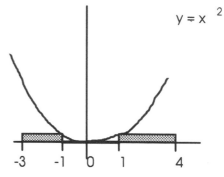

LOWER SUM FOR THE PARTITION

$\Delta x_k = Right - Left =$ $x_k - x_{k-1}$	$\Delta x_1 =$ $-1 - (-3) = 2$	$\Delta x_2 =$ $0 - (-1) = 1$	$\Delta x_3 =$ $1 - 0 = 1$	$\Delta x_4 =$ $4 - 1 = 3$
m_k	$m_1 = f(-1) = (-1)^2 = 1$	$m_2 = f(0) = 0$	$m_3 = f(0) = 0$	$m_4 = f(1) = 1$
M_k	$M_1 = f(-3) = (-3)^2 = 9$	$M_2 = f(-1) = 1$	$M_3 = f(1) = 1$	$M_4 = f(4) = 16$

$$U_f(P) = M_1 \Delta x_1 + M_2 \Delta x_2 + M_3 \Delta x_3 + M_4 \Delta x_4$$

$$L_f(P) = m_1 \Delta x_1 + m_2 \Delta x_2 + m_3 \Delta x_3 + m_4 \Delta x_4$$

$$U_f(P) = (9)(2) + (1)(1) + (1)(1) + (16)(3) = 68$$

$$L_f(P) = (1)(2) + (0)(1) + (0)(1) + (1)(3) = 5$$

Where the maximum and minimum values lie depends on the nature of the function.

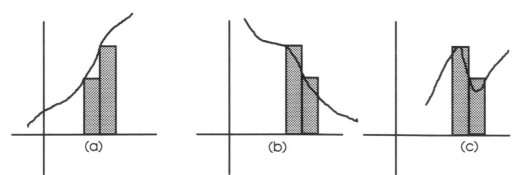

(a): Increasing function: maximum value on an interval occurs at right hand endpoint and minimum value at left hand endpoint. (b): Decreasing function: maximum value occurs at left hand endpoint and minimum value at right hand endpoint (c): Function is not decreasing or increasing: maximum and minimum values can lie anywhere.

EXAMPLE: Find the Upper and Lower Sums for the function $y = |x - 5|$ on $[-1, 8]$ for the partition $P = \{-1, 4, 6, 8\}$.

Answer:

$$L_f(P) = \quad m_1 \quad \Delta x_1 \quad + \quad m_2 \quad \Delta x_2 \quad + \quad m_3 \quad \Delta x_3$$

$$L_f(P) = [-(4-5)](4-(-1)) + [-(5-5)](6-4) \quad + \quad (6-5)(8-6) \quad = \quad 7$$

(1) here $y = -(x-5)$ **(2) here** $y = x - 5$

$$U_f(P) = \quad M_1 \quad \Delta x_1 \quad + \quad M_2 \quad \Delta x_2 \quad + \quad M_3 \quad \Delta x_3$$

$$U_f(P) = [-(-1-5)](4-(-1)) + [-(6-5)](6-4) \quad + \quad (8-5)(8-6) = 38.$$

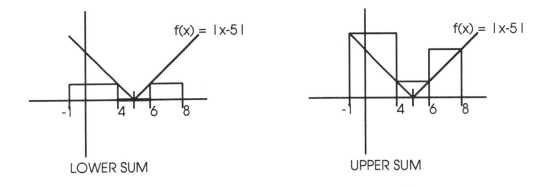

LOWER SUM   UPPER SUM

SECTION 5.2

"The thing of which I have most fear is fear.".....Montaigne

IMPORTANT RESULTS OF THIS SECTION

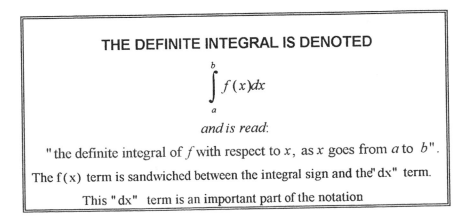

THE DEFINITE INTEGRAL IS DENOTED

$$\int_a^b f(x)\,dx$$

and is read:

"the definite integral of f with respect to x, as x goes from a to b".

The $f(x)$ term is sandwiched between the integral sign and the "dx" term.

This "dx" term is an important part of the notation

THE DEFINITE INTEGRAL AS THE UNIQUE VALUE BETWEEN ALL LOWER AND UPPER SUMS

When f is continuous on a closed interval [a,b], the definite integral, $\int_a^b f(x)dx$

is defined to be the unique value **I** such that $L_f(P) \leq \mathbf{I} \leq U_f(P)$ for all partitions P

of the interval [a,b]. While this definition is intuitively the most obvious, it is difficult to use this definition to compute the definite integral.

THE DEFINITION OF AREA (NOT A DEFINITION OF THE DEFINITE INTEGRAL)

If f is continuous and non-negative on a closed interval [a,b], then the AREA of the region

between the graph of the function f(x) and the x-axis, from a to b is defined to be $\int_a^b f(x)dx$.

(Note that we avoid talking about "negative areas" by insisting that f be non-negative.)

THE DEFINITION OF RIEMANN SUMS

If f is continuous on a closed interval [a,b], then a Riemann sum gives an ***approximation*** to the definite integral. A Riemann sum is a sum of the form:

$$\sum_{k=1}^{n} f(t_k)\Delta x_k.$$

Having formed a partition P, of the interval [a,b],

- the ***product*** $f(t_k)\Delta x_k$ represents the ***"generalized area"*** of *ONE* *rectangle* with height $f(t_k)$ and width Δx_k. Here we use the word "*generalized*" to mean that "area" might be considered to be negative. The shaded region below is the area of one rectangle.

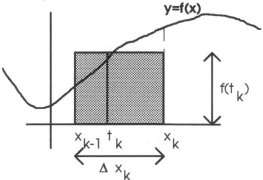

- $\sum_{k=1}^{n} f(t_k)\Delta x_k$ represents the ***sum*** *of the "generalized" areas of the rectangles.*

THE DEFINITE INTEGRAL IS APPROXIMATED BY RIEMANN SUMS

$$\int_a^b f(x)dx \approx \sum_{k=1}^{n} f(t_k)\Delta_{x_k}$$

THE DEFINITE INTEGRAL IS DEFINED AS THE LIMIT OF RIEMANN SUMS . When f is continuous on [a,b], this limit exists.

$$\int_a^b f(x)dx = \lim_{\|P\|\to 0} \sum_{k=1}^{n} f(t_k)\Delta_{x_k}$$

The quantity on the right is the limit of the Riemann sums as the "norm" of the partition goes to zero. The "norm" of the partition is the length of the largest subinterval. This is the definition that is most practical to work with.

CONCEPT CHECK

1. Write out both definitions of the definite integral as they appear in your book. Be prepared to explain all notation. Check back with the book's definition. Are you forgetting something? Saying too much? A definition should be "just that and nothing more".

2. Write up the theory of this section in two pages, in your own words. Exchange your paper with a friend and suggest corrections.

3. Why do the authors speak of refining the partition in a way which forces each new partition to include the points of the previous partition?

4. How would you go about finding the area under the graph of a discontinuous function such as
$$f(x) = \begin{cases} x \text{ for } 0 \le x \le 1 \\ 1 \text{ for } 1 \le x \le 3 \end{cases}$$, and above the x-axis? (You need two definite integrals.)

5. How can you justify that the area under a continuous function on a closed interval may not be zero when the "area" under a point on the graph is zero?

6. In order for the definite integral $\int_a^b f(x)dx$ to exist, what is required of f? of a and b?

7. Under what circumstances will the definite integral be negative?

MEMORY MATRIX: Fill in the grid below

Concept	Definition	Purpose	An example
Upper and Lower Sums			
Riemann Sum			
Definite Integral			
Comparison Property			
Area			

EXAMPLE:

Find an approximation to $\int_{-10}^{4} e^{-x^2} dx$ where $P = \{-10, 5/2, 10/3, 4\}$.

Answer: We use that $\int_{-10}^{4} e^{-x^2} dx = \lim_{\|P\| \to 0} \sum_{k=1}^{n} e^{-t_k^2} \Delta x_k$. We will chose the t_k to be the

left hand endpoint of the kth subinterval in the partition.   This will not give an upper or
lower sum, but one of many possible Riemann sums.

Δx_k	$\Delta x_1 =$ $5/2 + 10 = 25/2$	$\Delta x_2 =$ $10/3 - 5/2 = 5/6$	$\Delta x_3 =$ $4 - 10/3 = 2/3$
f (left hand endpoint)	$f(-10) = e^{-10^2}$	$f(5/2) = e^{-(5/2)^2}$	$f(10/3) = e^{-(10/3)^2}$

Now $\int_{-10}^{4} e^{-x^2} dx \approx e^{-100} (25/2) + e^{-25/4} (5/6) + e^{-100/9} (2/3)$. (Find on calculator.)

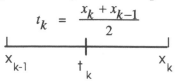

$f(x) = e^{-x^2}$

-10   4

EXAMPLE: Given the table below , find an approximation to $\int_{a}^{b} f(x)dx$, where $a = 3$

and $b = 5$. Use a Riemann sum with left hand endpoints.

x	3	6	7	9
$f(x)$	2	0	4	-1

Answer:
The left hand endpoints of the partition are 3, 6, 7. Thus

$\int_{3}^{9} f(x)dx \approx f(3)(6-3) + f(6)(7-6) + f(7)(9-7) = 2(3) + 0(1) + 4(2) = 14.$

EXAMPLE:

Find an approximation, using Riemann Sums, to the area of the region between the graph of

the function $y = \ln x$ and the x-axis, on the interval [1,2], using the midpoints of the

partition, $P = \{1, 1.4, 1.6, 1.8, 2\}$.

Answer:

The midpoints, t_k, of each subinterval are given

by the "average" of the endpoints:

$$t_k = \frac{x_k + x_{k-1}}{2}$$

x_{k-1} t_k x_k

interval	[1, 1.4]	[1.4, 1.6]	[1.6,1.8]	[1.8,2]
Δx_k	0.4	0.2	0.2	0.2
$f(t_k)$=f(midpt.)	ln1.2	ln1.5	ln1.7	ln1.9

Thus $\int_{1}^{2} \ln x \, dx \approx [\ln 1.2][0.4] + [\ln 1.5][0.2] + [\ln 1.7][0.2] + [\ln 1.9][0.2]$ (Find on

calculator.)

EXAMPLE: Why doesn't the integral $\int_{-1}^{1} \frac{1}{x} dx$ fit the mold of Definition 5.2?

Answer: The function f(x) = 1/x is not continuous on [-1,1].

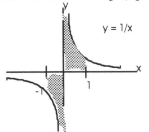

SECTION 5.3

"I'd rather see that done on paper"..(Humpty Dumpty) Lewis Carroll

EXAMPLE: Use the rectangle property to find $\int_{1}^{7} 5 dx$.

Answer: This is the area of the region in Fig. (1) below : $\int_{1}^{7} 5 dx = 5(7-1)=30$.

area of rectangle =(base)(height)
= 5(6-1) = 25
Fig. (1)

area of trapezoid=(base)(sum of heights)/2
= (b-a)(3a+3b)/2 = (3/2)(b²- a²)
= $(b-a)(3a+3b)/2 = (3/2)(b^2 - a^2)$
Fig. (2)

EXAMPLE: Find $\int_a^b 3x\,dx$. See Fig. (2) above.

Answer: $\int_a^b 3x\,dx \;=\; 3\int_a^b x\,dx \;=\; \int_a^b 3x\,dx \;=\; \dfrac{3(b^2 - a^2)}{2}$ by the formula in the text.

(This could be called the Trapezoidal Rule--why?)

NOTE: Constants "float " through the integral sign, as above, where we saw
$\int_a^b 3x\,dx \;=\; 3\int_a^b x\,dx$. **Also note:** $\int_a^b dx = \int_a^b 1\,dx = b - a.$

EXAMPLE: Explain why $\int_0^1 x^6\,dx \;\le\; \int_0^1 x^5\,dx$.

Answer: On the interval $[0,1]$, $x^6 \le x^5$.

EXAMPLE: Find the average value of the function f(x)= 2x on the interval [2, 5]. Find
where the average value occurs. Interpret your answer geometrically.

Answer:

Since the average value, f(c), is given by $f(c) = \dfrac{1}{5-2}\int_2^5 2x\,dx$, we have

$f(c) = \dfrac{(5^2 - 2^2)}{(5-2)} = 7$. To find that value of c, we use f(c) = 2c = 7, so c = 7/2.   The

geometrical interpretation is that the area of the region between the graph of the function

f(x) and the x-axis, on the interval [2,5], is exactly that of a rectangle with height f(c) = 7

and base (5-2) = 3.

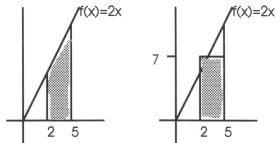

Area under curve equals area of rectangle
when height of rectangle equals average value of f on the interval

---

EXAMPLE: Find the value of b that makes the following equation true:
$\int_1^b f(x)\,dx \;=\; \int_1^7 f(x)\,dx \;-\; \int_5^7 f(x)\,dx$.

Answer: Drawing a line with the numbers 1,5,7:

$$\xleftarrow{\quad}\xleftarrow{\quad}\qquad 7 \text{ to } 5 \quad \xleftarrow{\quad}\xleftarrow{\quad}\xleftarrow{\quad}\xleftarrow{\quad}$$

$$1_____5_____7$$

$$\rightarrow\rightarrow\rightarrow \ 1 \text{ to } 7 \ \rightarrow\rightarrow\rightarrow\rightarrow\rightarrow\rightarrow\rightarrow\rightarrow\rightarrow\rightarrow\rightarrow\rightarrow\rightarrow$$

we see that x in the resulting integral goes from 1 to 5; i.e. b = 5.

EXAMPLE: Find the values of a and b that makes the following equation true:

$$\int_1^2 f(x)dx \ + \ \int_3^1 f(x)dx \ + \ \int_a^b f(x)dx \ .$$

Answer: Drawing a line with the numbers (taken from the limits of integration) 1,2,3:

$$\xleftarrow{}\xleftarrow{}\xleftarrow{}\xleftarrow{}\xleftarrow{}\xleftarrow{} \quad 3 \text{ to } 1 \quad \xleftarrow{}\xleftarrow{}\xleftarrow{}\xleftarrow{}\xleftarrow{}\xleftarrow{}\xleftarrow{}$$

$$1_____2_____3$$

$$\rightarrow\rightarrow\rightarrow \ 1 \text{ to } 2 \ \rightarrow\rightarrow\rightarrow$$

we see that the integral from "2 to 1" will cancel that from "1 to 2" . So the values of x in the resulting integral will go from 3 to 2; i.e., a = 3, b = 2.

EXAMPLE: Estimate $\dfrac{1}{\pi}\displaystyle\int_{-2}^{4}\dfrac{1}{1+x^2}\,dx$ with an upper and lower limit .

Answer: The maximum of the function is taken on when x = 0, where f(0) = 1 . The minimum of the function occurs at x = 4, where f(4) = 1/17. So we have:

$$m(b-a)\le \ \int_a^b f(x)dx \ \le M(b-a), \quad \text{where } m = \frac{1}{17\pi}, \ M = \frac{1}{\pi}, \text{ and } (b\text{-}a) = 6:$$

$$\frac{6}{17\pi} \ \le \ \frac{1}{\pi}\int_{-2}^{4}\frac{1}{1+x^2}\,dx \ \le \ \frac{6}{\pi}.$$

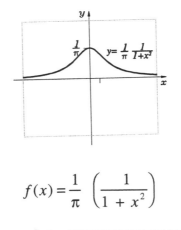

$$f(x) = \frac{1}{\pi}\left(\frac{1}{1+x^2}\right)$$

CHRONOLOGY OF THE SUBJECT: Fill in the blanks.

How did the theory develop so far? Recall, we introduced Upper and Lower Sums, which were ----
We defined the definite integral to be the unique-------. Then we introduced Riemann Sums, which
were --------. They were more general than Upper or Lower Sums in the sense that--------. The
definite integral could then be defined with Riemann Sums as a -------. This definition proved
practical to work with. We found there were a number of rules regarding the definite integral.
(List as many as you can now, and then look back.)   We saw the *Fundamental Theorem of
Calculus* which states that when f(x) is ------on --------, then ------. Another "version" tells how to
differentiate a particular definite integral with x in the upper limit: this result specifically states
that ----. In general, when using this result, we encounter more complicated integrals that require
the chain rule, and an example would be finding the derivative of -------. The average value of a
continuous function on a closed interval is defined as ------, the geometrical meaning being that ----
---. When we cannot evaluate an integral, we can always use-------. The Comparison Property
states -------. The most important components of the theory developed in this section, in my mind,
are ------.

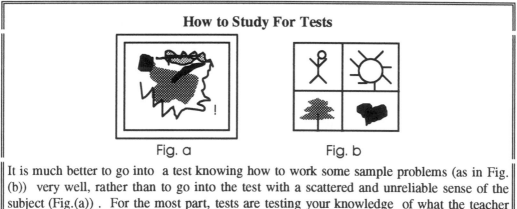

How to Study For Tests

Fig. a   Fig. b

It is much better to go into a test knowing how to work some sample problems (as in Fig.
(b)) very well, rather than to go into the test with a scattered and unreliable sense of the
subject (Fig.(a)) . For the most part, tests are testing your knowledge of what the teacher
thinks is basic and important. Try to find categories of problems as you go along. Work a
few types from each category more than once, perhaps, or until you are sure about exactly
what you are doing. The danger of going into a test with a little knowledge about a lot of
things is that you may not have the focus on the test to succeed with any of the problems.

SECTION 5.4

"The only miracle is that there are no miracles"....Einstein

A REVIEW OF IDEAS SO FAR

♦ DUMMY VARIABLE

The variable in the integrand of a definite integral is called a "dummy variable" because its
name is unimportant; it's more of a place holder. Satisfy yourself that the integrals below
are the same.

$$\int\limits_{a}^{b} t^2 dt = \int\limits_{a}^{b} u^2 du = \int\limits_{a}^{b} x^2 dx = \frac{b^2 - a^2}{2}.$$

♦ WHEN A VARIABLE IS NOT A DUMMY VARIABLE

The "dx " term tells us that the integration takes place with respect to x and any other letter is treated as a constant with respect to x unless we would have other information (as we will in Section 5.6, with the u-substitution.)

$$\int\limits_{a}^{b} \pi \ r^2 dx = \pi \ r^2 \int\limits_{a}^{b} dx = \pi \ r^2 \ x \Big|_{a}^{b} = \pi r^2 (b-a).$$

♦ THE DEFINITE INTEGRAL WITH CONSTANT LIMITS IS A REAL NUMBER

$\int\limits_{a}^{b} f(x)dx$ is a real number. For example, $\int\limits_{2}^{3} x^2 dx = \dfrac{3^3 - 2^3}{3} = \dfrac{19}{3}$.

♦ WHEN A VARIABLE IS A LIMIT OF INTEGRATION

When a variable (such as x) is in a limit of integration we have a definite integral which is a function of x. For example: $\int\limits_{1}^{x} t^2 dt$ expresses the area under the function $f(t) = t^2$ as t goes from 1 to x. The area is now a function of x. Find values of $G(x) = \int\limits_{1}^{x} t^2 dt$ when x = 3, x = 4.   [Answer; 26/3, 63/3].

♦ WHAT IS AN ANTIDERIVATIVE?

An antiderivative for f is a function F whose derivative is f. For example, if f(x) = sin x, an antiderivative is F(x) = - cos x or F(x) = - cos x + 3, or F(x) = - cos x - $\sqrt{2}$. We can write the most general antiderivative as F(x) = - cos x + C , where C stands for any real valued constant.

REVIEW OF TRIGONOMETRIC DERIVATIVES
Complete this table "forward" and "backward" and memorize .

$f(x)$	$f'(x) = ?$	$f'(x)$	$f(x) = ?$
$\cos x$		$-\sin x$	
$\sin x$		$\cos x$	
$\tan x$		$\sec^2 x$	
$\sec x$		$\sec x \tan x$	
$\csc x$		$-\csc x \cot x$	
$\cot x$		$-\csc^2 x$	

THE FUNDAMENTAL THEOREM OF CALCULUS

THEOREM 5.13

If f(x) is continuous on the closed interval [a,b], then f has an antiderivative on [a,b] and

$$\int_a^b f(x)dx = F(b) - F(a), \text{ where } F(x) \text{ is any antiderivative of } f(x); \text{ that is, } F'(x) = f(x).$$

EXAMPLE: Evaluate. $\int_1^4 x^2 + 3dx$.

Answer:

$$\int_1^4 x^2 + 3dx = \left.\frac{x^3}{3} + 3x\right|_1^4 = \left(\frac{4^3}{3} + 3(4)\right) - \left(\frac{1^3}{3} + 3(1)\right) = 30.$$

NOTATION   plug in x = 4   MINUS   plug in x = 1

✋**CAUTION:** Be sure to make abundant use of parentheses when evaluating integrals! Always use all steps when evaluating an integral. It is dangerous to skip any of the steps, shown above. Some students like to evaluate an integral this way, for example,

$$\int_1^4 x^2 + 3dx = \left(\frac{4^3}{3} + 3(4)\right) - \left(\frac{1^3}{3} + 3(1)\right) = 30,$$

skipping the stage where we write

$$\int_1^4 x^2 + 3dx = \boxed{\left.\frac{x^3}{3} + 3x\right|_1^4} = \left(\frac{4^3}{3} + 3(4)\right) - \left(\frac{1^3}{3} + 3(1)\right) = 30.$$

But this is the very step where we can check our answer by differentiating! Do this before evaluating.

EXAMPLE: Find the area of the region between the graph of y = sin 2x and the x-axis, from 0 to π..

Answer: We seek $\int_0^\pi \sin 2xdx$. First we find an antiderivative of sin 2x, namely - (cos2x)/2. Then

$$\int_0^\pi \sin 2xdx = \left. -\frac{\cos 2x}{2}\right|_0^\pi = -\frac{\cos 2\pi}{2} - \left(-\frac{\cos 0}{2}\right) = -1/2 + 1/2 = 0.$$

✋**CAUTION:** We will not always be able to find an antiderivative for every integrand.

For example, what is the antiderivative for the integrand in $\int_2^\pi \sin x^2 dx$? We don't know.

THEOREM 5.12; A RELATED, IMPORTANT RESULT

If f(x) is continuous on a closed interval I, and c is in I,

$$\text{then } \frac{d}{dx} \int_c^x f(t)\, dt = f(x).$$

We are seeing above that $\frac{d}{d\mathbf{X}} \int_c^{\mathbf{X}} f(t)\, dt = f(\mathbf{X}).$

(In other words, **the derivative of this integral is the integrand.**)

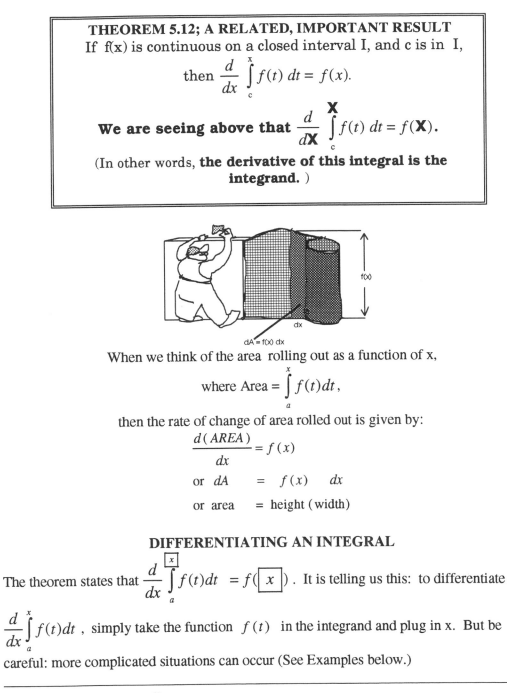

When we think of the area rolling out as a function of x,

$$\text{where Area} = \int_a^x f(t)dt,$$

then the rate of change of area rolled out is given by:

$$\frac{d\,(AREA)}{dx} = f(x)$$

$$\text{or } dA \quad = \quad f(x) \quad dx$$

$$\text{or area} \quad = \quad \text{height}\,(\text{width})$$

DIFFERENTIATING AN INTEGRAL

The theorem states that $\dfrac{d}{dx} \int_a^{\boxed{x}} f(t)dt = f(\boxed{x})$. It is telling us this: to differentiate

$\dfrac{d}{dx} \int_a^x f(t)dt$, simply take the function $f(t)$ in the integrand and plug in x. But be

careful: more complicated situations can occur (See Examples below.)

EXAMPLE: Find $\dfrac{d}{dx} \int_1^x \sin\!\left(\dfrac{1}{t}\right) dt$.

Answer: Do not try to find the antiderivative of this function; we do not know it. However, we use the result of the Fundamental Theorem of Calculus that says the derivative and integral cancel each other (in some sense). Thus

$$\frac{d}{dx} \int_1^{\boxed{\mathbf{x}}} \sin\left(\frac{1}{t}\right) dt = \sin\left(\frac{1}{\boxed{\mathbf{x}}}\right) = \sin\left(\frac{1}{x}\right).$$

EXAMPLE: Find $\dfrac{d}{dx} \displaystyle\int_1^x \cos^2 \sqrt{t+3}\; dt$.

Answer: We cannot integrate the function. But we only intend to differentiate the integral, and these operations will cancel each other. We have:

$$\frac{d}{dx} \int_1^{\mathbf{x}} \cos^2 \sqrt{\mathbf{t}+3}\; dt = \cos^2 \sqrt{\mathbf{x}+3}\;.$$

✋ **CAUTION:** The method above works only for problems of the form $\dfrac{d}{d\mathbf{x}} \displaystyle\int_{\mathbf{a}}^{\mathbf{x}} f(t)\,dt$

where the lower limit is a constant, and the upper limit is merely x. We examine other cases.

EXAMPLE: Find $\dfrac{d}{dx} \displaystyle\int_x^2 \cos t^2 dt$.

Answer: First we must put the variable x in the upper limit. We write

$$\int_x^2 \cos t^2 dt = -\int_2^x \cos t^2 dt. \quad \text{Then} \quad \frac{d}{dx}\int_x^2 \cos t^2 dt = -\left(\frac{d}{dx}\int_2^x \cos \mathbf{t}^2 dt\right) = -(\cos x^2).$$

EXAMPLE: Find . $\dfrac{d}{dx} \displaystyle\int_1^{x^2} \cos^2 \sqrt{t+3}\; dt$.

Answer: This situation is different. Now the x^2 in the top limit must be dealt with by means of the Chain Rule.

(1) Plug in x^2 wherever you saw t in the function in the integrand.

(2) Multiply this result by the derivative of x^2 .

$$\frac{d}{dx} \int_1^{\boxed{x^2}} \cos^2 \sqrt{\mathbf{t}+3}\,dt = \cos^2 \sqrt{\boxed{x^2}+3} \cdot \frac{d}{dx}\boxed{x^2} = \left(\cos^2\sqrt{x^2+3}\right)(2x)$$

$$\underset{\text{Plug } x^2 \text{ in for } t \text{ , times derivative of } x^2}{}$$

EXAMPLE: Find $\dfrac{d}{dx} \displaystyle\int_{-5}^{\sin x} \sqrt{t^2 + 3t}\,dt$

Answer:

$$\frac{d}{dx} \int_{-5}^{\sin x} \sqrt{t^2 + 3t}\,d = \underset{\text{plug in } \sin x \text{ for } t \text{ and multiply by derivative of } \sin x}{\left(\sqrt{(\sin x)^2 + 3\sin x}\right)(\cos x)} = \qquad \text{Simplify!}$$

EXAMPLE: Find $\dfrac{d}{dx} \displaystyle\int_{-x}^{x^2} \sin t^3 \, dt$.

Answer: We use a trick to write

$$\int_{-x}^{x^2} \sin t^3 \, dt = \int_{-x}^{c} \sin t^3 \, dt + \int_{c}^{x^2} \sin t^3 \, dt \text{ where c is any number between - x and } x^2.$$

$$\int_{-x}^{x^2} \sin t^3 \, dt = -\int_{c}^{-x} \sin t^3 \, dt + \int_{c}^{x^2} \sin t^3 \, dt, \text{ so}$$

$$\frac{d}{dx} \int_{-x}^{x^2} \sin t^3 \, dt = \frac{d}{dx} \left(-\int_{c}^{-x} \sin t^3 \, dt \right) + \frac{d}{dx} \left(\int_{c}^{x^2} \sin t^3 \, dt \right)$$

$$\frac{d}{dx} \int_{-x}^{x^2} \sin t^3 \, dt = -(\sin(-x)^3)(-1) + \sin(x^2)^3 (2x) = -\sin x^3 + 2x \sin x^6.$$

$$\uparrow \qquad\qquad \uparrow \qquad\quad \uparrow \qquad\quad \uparrow$$

$$\text{[plug in (-x)]} \quad \text{[deriv of - x]} \quad \text{[plug in } x^2 \text{]} \quad \text{[deriv of } x^2 \text{]}$$

EXAMPLE: Evaluate $\displaystyle\int_{1}^{3} \frac{1}{x} \, dx$.

Answer: $\displaystyle\int_{1}^{3} \frac{1}{x} \, dx = \ln|x| \, \Big|_{1}^{3} = \ln |3| - \ln |1| = \ln |3| = \ln 3$. Why would we expect:

$$\int_{1}^{3} \frac{1}{x} \, dx = -\int_{-3}^{-1} \frac{1}{x} \, dx ? \quad \text{(Evaluate both integrals. Graph the function } y = 1/x.)$$

EXAMPLE: Evaluate $\displaystyle\int_{0}^{1} e^x \, dx$.

Answer: $\displaystyle\int_{0}^{1} e^x \, dx = e^x \, \Big|_{0}^{1} = e^1 - e^0 = e - 1.$

EXAMPLE: Evaluate $\displaystyle\int_{e}^{e^2} \frac{1}{x} \, dx$

Answer: $\displaystyle\int_{e}^{e^2} \frac{1}{x} \, dx = \ln|x| \, \Big|_{e}^{e^2} = \ln e^2 - \ln e = 2 \ln e - \ln e = 2 - 1 = 1.$

EXAMPLE: What's wrong with the following evaluation of an integral?

$$\int_{-1}^{1} \frac{1}{x^2} \, dx = \int_{-1}^{1} x^{-2} \, dx = -x^{-1} \, \big|_{-1}^{1} = -2$$

Answer: This cannot be, because $f(x) = \dfrac{1}{x^2}$ is positive and we should have a positive answer. The Fundamental Theorem cannot be used, because f(x) is not continuous, much less defined on [-1, 1].

DISTANCE, VELOCITY, ACCELERATION

The derivative of distance with respect to time is velocity.
(**think:** meters per second, or miles per hour)

The derivative of velocity with respect to time is acceleration.
(**think :** meters per seconds-squared)

The antiderivative of velocity with respect to time is distance.
The antiderivative of acceleration with respect to time is velocity.

Velocity of Thrown or Dropped Object

The acceleration due to gravity is -9.8 m/sec^2 (or -32 ft/sec^2). From this, the formula for the distance from the ground, s(t), of a thrown or dropped object can be determined as a function of time. It's worthwhile to memorize this formula, although it can be derived. (Try it!) Distance is given by :

$$s(t) = -4.9t^2 + v_0 t + s_0$$

where s_0 is the height from which the object is dropped or thrown, and v_0 represents the "initial velocity"--the velocity with which it was thrown . The distance s(t) is measured **upwards** from the ground. Thus initial velocity , v_0 , is negative if an object is thrown downwards, zero if it is dropped, and positive if thrown upwards.

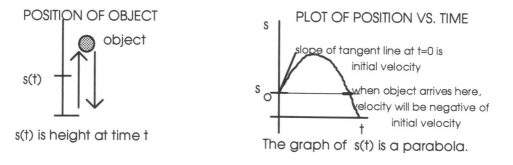

POSITION OF OBJECT

object

s(t)

s(t) is height at time t

PLOT OF POSITION VS. TIME

slope of tangent line at t=0 is initial velocity

when object arrives here, velocity will be negative of initial velocity

The graph of s(t) is a parabola.

EXAMPLE: A stone is dropped from a height of 100 m.
(a) Find its velocity at time t = 1. (b) When does it hit the ground? (c) What is its velocity when it hits the ground? (d) At time t = 1, how high is it off the ground?

Answer: (Note initial velocity v_0 is 0, initial distance s_0 is 100.)
(a) Acceleration is a constant (due to gravity) in cases of free-fall.

$$v(t) - v(0) = \int_0^t a(t)dt, \text{ so } v(t) = \int_0^1 a(t) \ dt = \int_0^1 -9.8 \ dt = \ = -9.8t\Big|_0^1 = -9.8m/\sec.$$

(b) We can find s(t) in the same way: $s(t) - s(0) = \int_0^t v(t)dt = \int_0^t -9.8dt$. This means

$$s(t) - \mathbf{100} = \int_0^t v(t)dt = \int_0^t -9.8dt \text{ or } \mathbf{s(t)} = \int_0^t \mathbf{-9.8dt + 100} \text{ or }$$

$\mathbf{s(t) = -4.9t^2 + 0 + 100}$. Or we can remember from the formula:
$s = -4.9t^2 + v_0 t + s_0$, that $\mathbf{s(t) = -4.9t^2 + 0 + 100}$.

Now we ask: when is s = 0? Solving, we find $4.9t^2 = 100$, or t = $10/\sqrt{4.9}$ sec.

(c) When the stone hits the ground (we have just found t for this event), the velocity is

$$v = \int_0^{10/\sqrt{4.9}} -9.8dt = -9.8(\frac{10}{\sqrt{4.9}} - 0) = -98/\sqrt{4.9} \text{ m/sec.}$$

(d) At t = 1, we use the formula $s = -4.9t^2 + v_0 t + s_0 = -4.9t^2 + 0 + 100$ with t = 1, and conclude that s = 95.1 m.

EXAMPLE: A buggy starting at a distance 50 miles from home moves in a straight line toward home. If the velocity is 10 t mph, where t represents the hours of travel, find the buggy's position 3 hrs. later.
Answer: Let s(t) represent the number of miles from home at time t. Then

$$s(t) - s(0) = s(3) - 50 = \int_0^3 v(t)dt = \int_0^3 -10t \ dt = -10 \cdot \frac{t^2}{2}\Big|_0^3 = -45 \text{ so } s(3) = 5 \text{ miles.}$$

SECTION 5.5

" Cudgel thy brains no more about it."....Shakespeare

THE INDEFINITE INTEGRAL of f (x) with respect to x is
$$\int f(x)dx = F(x) + C$$
where F(x) is an antiderivative of f(x) and C represents any constant.

EXAMPLE: Find $\int \sin 3x\, dx$.

Answer: By some trial and error, we find that $\int \sin 3x\, dx = -(\cos 3x)/3 + C$.

EXAMPLE: Find $\int (4x+2)^2 dx$.

Answer: Expand the integrand.
$$\int (4x+2)^2 dx = \int 16x^2 + 16x + 4\, dx = 16x^3/3 + 8x^2 + 4x + C \ .$$

EXAMPLE: Find $\int \sqrt{x} + \dfrac{2}{\sqrt{x}}\, dx.$

Answer: $\int \sqrt{x} + \dfrac{2}{\sqrt{x}}\, dx = \int x^{1/2} + 2x^{-1/2}\, dx = \dfrac{x^{3/2}}{(3/2)} + \dfrac{2x^{1/2}}{(1/2)} + C = \dfrac{2}{3}x^{3/2} + 4\sqrt{x} + C.$

EXAMPLE: Find $\int_1^3 \dfrac{x-2}{x} dx.$ Hint: rewrite as $\int_1^3 1 - \dfrac{2}{x}\, dx.= (x - 2\ \ln|x|)\Big|_1^3 = ?$

NOTE: The "dx" term determines the variable. Anything that is not "x" is considered to be a constant.

EXAMPLE: Find $\int_0^1 \pi r^2 dx$.

Answer: In this integral, π and r are regarded as constants with respect to x. Thus
$$\int_0^1 \pi r^2 dx = \pi r^2 \int_0^1 dx = \pi r^2 \int_0^1 dx = \pi r^2 x \Big|_0^1 = \pi r^2 (1-0) = \pi r^2 .$$

SECTION 5.6

"Curiouser and curiouser"....Lewis Carroll

NOTE: Always check your answer to an indefinite integral by differentiating your result. The derivative of your answer should be the integrand.

EXAMPLE: To say $\int \sec^2 3x\, dx = \tan 3x + C$ is wrong, because

$\dfrac{d}{dx} \tan 3x = 3\sec^2 3x$, not $\sec^2 3x$. By the method of trial and error, we see the correct

answer is $\int \sec^2 3x\, dx = (\tan 3x)/3 + C$.

> **USEFUL HINT: Memorize these derivatives**
> $$\frac{d}{dx}\, \sqrt{x} = \frac{1}{2\sqrt{x}}$$
> $$\frac{d}{dx}\left(\frac{1}{x}\right) = -\frac{1}{x^2}$$

METHOD OF SUBSTITUTION: This is something that takes getting used to. Just as long division is the reverse of multiplication, integration is the reverse of differentiation. Recognizing quickly which substitutions work is something that takes a lot of practice.

EXAMPLE: $\int x\sqrt{x^2 +1}\, dx$.

Answer:

Let $u = x^2 + 1$. Then $\dfrac{du}{dx} = 2x$, so $du = 2\, x\, dx$ so that $x\, dx = du/2$.

Now $\int x\sqrt{x^2 +1}\, dx = \int \sqrt{x^2 +1}\,(x\, dx) = \int \sqrt{u}\left(\dfrac{du}{2}\right) = \left(\dfrac{1}{2}\right)\int u^{1/2}\, du = \left(\dfrac{u^{3/2}}{3/2}\right) + C$

$= \dfrac{u^{3/2}}{3} + C = \dfrac{(x^2 +1)^{3/2}}{3} + C.$

NOTICE THE PARTS OF THIS PROBLEM

1. Choose a u - substitution : let u = ...

2. Find dx in terms of du, and replace ALL x - terms with u - terms

3. Find the integral, in terms of u. Don't forget the "**+ C**" term!

4. Replace the u's in your answer with x 's .

EXAMPLE: Find $\int x^2 \sqrt{x+5}\, dx$.

Answer: Let u = x + 5. Then du = dx. Also, x = u-5. Replacing all x-terms by u-terms,

$\int x^2 \sqrt{x+5}\, dx = \int (u-5)^2 \sqrt{u}\ du = \int (u^2 - 10u + 25) u^{1/2} du =$

$\int u^{5/2} - 10 u^{3/2} + 25 u^{1/2} du =$

$\frac{2}{7} u^{7/2} - 10 \left(\frac{2}{5}\right) u^{5/2} + 25 \left(\frac{2}{3}\right) u^{3/2} + C = \frac{2}{7}(x+5)^{7/2} - 4(x+5)^{5/2} + \left(\frac{50}{3}\right)(x+5)^{3/2} + C$

> **GOOD SUBSTITUTION IDEAS: Let u = things**
> in denominators
> in parentheses
> under radicals
> raised to powers
> that look complicated
> whose derivative you see as a factor in the numerator

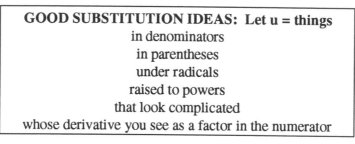

COMMON ERRORS

♦ *MULTIPLICATION ERROR:* The product of functions in the integrand does not give a product of integrals. There are rules for sums and differences of integrals, but not for products.

For example , to say $\int f(x)\, g(x)\, dx = \int f(x)\, dx \int g(x)\, dx$ is wrong.

E.g., $\int (x^{1/2} + 3)\, x^2\, dx = \int (x^{1/2} + 3)\, dx \int x^2\, dx$ is incorrect. Instead, multiply out the integrand: $\int (x^{1/2} + 3)\, x^2\, dx = \int x^{5/2} + 3x^2\, dx$, and solve, where $x^{1/2 + 2} = x^{5/2}$.

♦ *FORGETTING TO CHANGE u BACK TO x:*

$\int x^2 \cos x^3 dx$. Let $u = x^3$. Then $du = 3x^2 dx$. Now we have: $\int \cos u\, du = \frac{1}{3}\sin u + C$.

This is incorrect because it is unfinished: we must now write this as $(\sin x^3)/3 + C$.

♦ *FORGETTING TO REMOVE THE INTEGRAL SIGN AFTER INTEGRATING:*

$\int x^2 \cos x^3\, dx = ?$ Let $u = x^3$, so $du = 3x^2 dx$. Then $\int \frac{\cos u}{3} du = \int \frac{1}{3}\sin u + C$.

This is incorrect: at the integration stage, the integral sign should vanish. The correct answer is found in the previous problem.

♦ FORGETTING THE + C TERM:

$\int x^2 \cos x^3 dx = \dfrac{1}{3} \sin x^3$. The answer should be $\int x^2 \cos x^3 dx = \dfrac{1}{3} \sin x^3 + C$.

♦ MIXING OF x AND u TERMS:

$\int x\sqrt{x+1}\, dx = ?$ Let u = x+1 so $du = dx$. Then we have $\int x\, u\, du =$

$x \int du = xu + C...$

This is incorrect, because once we are committed to a u-substitution, we must express the entire integral in terms of u. The term u is no longer a constant with respect to x.

♦ du-TERM IN THE DENOMINATOR IS INCORRECT:

$\int \dfrac{x^3 + 5}{x^2}\, dx = ?$ Let u = $x^3 + 5$, so du = $3x^2 dx$, and we have $\int \dfrac{3u}{du} = 3u^2 / 2 + C..,$

etc.. No substitution that yields a du-term in the denominator can work. In this case, divide

and obtain, $\int \dfrac{x^3+5}{x^2}\, dx = \int x + \dfrac{5}{x^2}\, dx = \dfrac{x^2}{2} - \dfrac{5}{x} + C.$

♦ MIXED-UP SUBSTITUTION MAKES INTEGRAL WORSE:

Try another substitution.
This happens to all of us!
???????

$\int \dfrac{x}{(x^2+1)^5}\, dx$. Let u = $(x^2+1)^5$, so du = $5(x^2+1)(2x)dx = (10x^3 + 10x)dx...$

and the integral is ????? This integral is going from bad to worse . Try another substitution.

Try u = $x^2 + 1$.

♦ FREQUENTLY TWO SUBSTITUTIONS ARE NECESSARY OR CONVENIENT.

$\int \dfrac{(1+\sqrt{x})^{20}}{2\sqrt{x}}\, dx$. Let $u = \sqrt{x}$, then $du = \dfrac{1}{2\sqrt{x}}\, dx$, and we have $\int (1+u)^{20} du$.

Now: Let $v = (1+u)$ and $dv = du$, and this latter integral becomes $\int v^{20} dv =$

$\dfrac{v^{21}}{21} + C = \dfrac{(1+u)^{21}}{21} + C = \dfrac{(1+\sqrt{x})^{21}}{21} + C.$

While this is all correct, we could have avoided two substitutions by using as an initial substitution, $u = 1 + \sqrt{x}$.

◆ *CHANGING LIMITS OF INTEGRATION*

Many errors occur here. The following example is done incorrectly.

$\int_{1}^{5}(2x+3)^2dx = $? Let u = 2x + 3, so du = 2dx. Then we have $\frac{1}{2}\int_{1}^{5}u^2\,du$.

This is already wrong, because the limits of the integral refer to x 's limits, not to u' s.

Worse, this may lead to the evaluation of this integral as $\left(\frac{1}{2}\right)\left.\frac{u^3}{3}\right|_{1}^{5} = \left(\frac{1}{2}\right)\left(\frac{5^3-1}{3}\right)$.

Instead, when the u substitution is made, we should write:

$$\int_{1}^{5}(2x+3)^2dx = \frac{1}{2}\int_{x=1}^{x=5}u^2du = \left(\frac{1}{2}\right)\left.\frac{u^3}{3}\right|_{x=1}^{x=5} = \left(\frac{1}{2}\right)\left.\frac{(2x+3)^3}{3}\right|_{1}^{5} = (etc.)$$

This system has the advantage of allowing us to check our answer by differentiating . But in the method usually preferred we change the u - limits. We do this by noting that since u = 2x + 3:

when x = 1, u = 2(1) + 3 = 5

when x = 5, u = 2(5) + 3 = 13, so

$$\int_{1}^{5}(2x+3)^2dx = \frac{1}{2}\int_{5}^{13}u^2du = \left(\frac{1}{2}\right)\left.\frac{u^3}{3}\right|_{5}^{13} = (etc.)$$

EXAMPLE: Find $\int_{0}^{\sqrt{\pi}}x\,\sin x^2dx$.

Answer: Let $u = x^2$, so du = 2x dx and the integral becomes $\left(\frac{1}{2}\right)\int_{0}^{\pi}\sin u\,du = $

$\frac{1}{2}(\text{-cos }\pi + \cos 0) = 1$. Note the change of limits at the change of variable.

PAY, IT'S ACADEMIC: Average annual earnings of college graduates with bachelor's degrees in various academic fields, based on monthly earnings in the spring of 1990".
....Source U.S. Census Bureau. (from the Minneapolis Star Tribune, February, 1993)

Biology	$28,908
Business	$29,364
Education	$18,384
Engineering	$35,436
English/Journalism	$19,284

Home Economics	$10,872
Liberal Arts/Humanities	$19,104
Mathematics and Statistics	$30,828
Psychology	$24,252
Social Sciences	$22,092

A NOTE ON MEMORIZATION

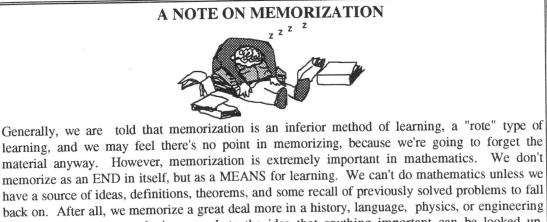

Generally, we are told that memorization is an inferior method of learning, a "rote" type of learning, and we may feel there's no point in memorizing, because we're going to forget the material anyway. However, memorization is extremely important in mathematics. We don't memorize as an END in itself, but as a MEANS for learning. We can't do mathematics unless we have a source of ideas, definitions, theorems, and some recall of previously solved problems to fall back on. After all, we memorize a great deal more in a history, language, physics, or engineering courses. Be sure you don't succumb to the idea that anything important can be looked up. Highlighted areas of the text are not for back-up reference; they should be learned and memorized. If you ROUTINELY memorize important information as you go along, rather than to cram at the last minute, memorization will feel more naturally like part of the learning process, and what you memorize will stay with you.

SECTION 5.7

"I was gratified to be able to answer promptly, and I did. I said I didn't know."...Mark Twain

BE SURE TO REVIEW LOGARITHMIC AND EXPONENTIAL PROPERTIES IN CHAPTER 1 OF THIS GUIDE.

THE NATURAL LOGARITHM FUNCTION

DEFINITION : $\ln x$ is defined as $\displaystyle\int_1^x \frac{1}{t}\, dt$ for $x > 0$.

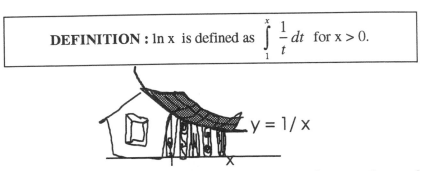

$y = 1 / x$

If f is given by f(t) = 1/t, ln x represents the area of the region between the graph and the x-axis from 1 to x, where x > 0.

IMPORTANT FORMULA

$$\int \frac{1}{x}dx = \ln|x| + C$$

(The reason for |x| is that the derivative of both ln x and ln (-x) is 1/x.)

NOTE: The derivative of ln x is "one over x" . We would like to think: the derivative of ln (anything) is "one over anything", but we must invoke the chain rule.

$$\frac{d}{dx}\ln\left(\;\square\;\right) = \frac{1}{\square}\cdot\frac{d\,\square}{dx}. \text{ This is better said as } \frac{d}{dx}\ln u = \frac{1}{u}\frac{du}{dx}.$$

EXAMPLE: Find the derivative of $y = \ln(\sin x)$.

Answer:

$$\frac{d}{dx}\ln(\sin x) = \frac{1}{\sin x}\cdot(\cos x) = \cot x$$

$$\uparrow \qquad \uparrow$$

riv of ln ( )   deriv of sin ( )

EXAMPLE: Find the derivative of $y = \ln(\ln x)$.

Answer: $y' = [\;(1/\ln x)\;](1/x) = 1/(x\ln x)$.

EXAMPLE: Find the derivative of: $y = \ln\sqrt{\dfrac{x+3}{x-2}}$. Note: this is $y = \ln\left(\dfrac{x+3}{x-2}\right)^{1/2}$.

Answer: For simplicity, we can use a log property to rewrite this equation as

$$y = \frac{1}{2}[\ln(x+3)-\ln(x-2)], \text{ so } \frac{dy}{dx} = \frac{1}{2}\left[\frac{1}{x+3}-\frac{1}{x-2}\right], \text{ which can be simplified as}$$

$$\frac{dy}{dx} = \frac{-5}{2(x-2)(x+3)}.$$

CAUTION! As in the previous problem, we are not differentiating when we write $y = \ln\sqrt{x} = \ln x^{1/2}$ as $y = (1/2)\ln x$. This is merely a property of logarithms.

EXAMPLE: Prove $\ln \dfrac{1}{b} = -\ln b$, where $b > 0$.

Answer:

Let $y = \ln \dfrac{1}{x}$. Now $\dfrac{dy}{dx} = \dfrac{1}{(1/x)} \cdot \left(-\dfrac{1}{x^2}\right) = -\dfrac{1}{x}$. At the same time, if

$z = -\ln x$, $\dfrac{dz}{dx} = -\dfrac{1}{x}$. Thus, $y = z + C$, or $\ln \dfrac{1}{x} = -\ln x + C$. When $x = 1$, $C = 0$,

so $\ln \dfrac{1}{x} = -\ln x$. In particular this is true if $x = b$.

NOTE: Sometimes students wonder why we can't use the usual rules:

$\int \dfrac{1}{x} dx = \int x^{-1} dx$ ...etc. But this gives $\dfrac{x^0}{0} + C$!

EXAMPLE: Find the integral: $\int \dfrac{1}{3x+2} dx$.

Answer:

Let $u = 3x + 2$, $du = 3dx$. So $\int \dfrac{1}{3x+2} dx = \dfrac{1}{3} \int \dfrac{du}{u} = \dfrac{1}{3} \ln|u| + C = \dfrac{1}{3} \ln|3x+2| + C$

Don't forget the absolute value sign that comes in these instances! But it's not needed in the case below.

EXAMPLE: Find the integral: $\int \dfrac{\ln x}{x} dx$.

Answer:

Let $u = \ln x$, then $du = \dfrac{1}{x} dx$. Now $\int \dfrac{\ln x}{x} dx = \int u\, du = \dfrac{u^2}{2} + C = \dfrac{(\ln x)^2}{2} + C$.

No absolute value occurs here, because we never had an integral of the form $\int \dfrac{1}{u} du$.

EXAMPLE: Find the integral : $\displaystyle\int_e^{e^e} \dfrac{\ln(\ln x)}{x \ln x} dx$.

Answer: Let $u = \ln(\ln x)$. Then $du = \dfrac{1}{x \ln x} dx$, by the chain rule.

Now we find the limits of integration : when $x = e^e$, $u = \ln(\ln e^e) = \ln e = 1$

and when $x = e$, $u = \ln(\ln e) = \ln 1 = 0$.

The integral becomes $\displaystyle\int_e^{e^e} \dfrac{\ln(\ln x)}{x \ln x} dx = \int_0^1 u = \dfrac{u^2}{2} \Big|_0^1 = \dfrac{1}{2}$.

CAUTION: Often students think that any integral with a denominator results in

$\ln| \ |$. It is wrong, for example, to say: $\int \frac{1}{\sqrt{x}} dx = \ln|\sqrt{x}|+C$ or $\int \frac{1}{\sin x} dx =$
$\ln|\sin x| + C$. We can only say

$$\int \boxed{\frac{1}{U}} \ \boxed{dU} = \ln| \ U \ | + C.$$

(Here "u" can be f(x), but du = f '(x) dx must be in the numerator!)

EXAMPLE: Integrate $\int \frac{1}{\sqrt{x}(1+\sqrt{x})} \ dx$

Answer:

Let $u = (1+\sqrt{x})$, so $du = \frac{1}{2\sqrt{x}} dx$. The integral becomes $\int \frac{2}{u} du =$

$2\ln | u | + C = 2 \ln (1+\sqrt{x}) + C$, where we dropped the absolute value sign, because
$(1+\sqrt{x}) > 0$.

EXAMPLE: Find the domains:

1. $y = \ln \sqrt{\frac{x+3}{x-2}}$.

Answer: We need $\frac{x+3}{x-2} > 0$. We cannot have $\frac{x+3}{x-2} = 0$. Why? Using
critical numbers $x = 2, -3,$

$$\text{\_ \_ \_ \_ \_ \_ \_ -3 \_ \_ \_ \_ \_ \_ \_ \_ \_ 2 \_ \_ \_ \_ \_ \_ \_ \_ \_}$$

and using as test   - points , $x = -10$, $x = 0$, and $x = 10$, the sign chart for $\frac{x+3}{x-2}$ is

$++++++$ -3 - - - - - - - **2** $+++++++++$,

so we require that $x < -3$ or $x > 2$.

2. $f(x) = \ln (x^2 +5)^7$.

Answer: We simply require that $(x^2 +5)^7 > 0$, or $(x^2 +5) > 0$, which is true for all x.

3. $f(x) = \ln (\cos x)$.

Answer: We require that $\cos x > 0$. This occurs in quadrants I and IV; ie.for
$x \ \varepsilon \ (-\pi/2, \ \pi/2)$. To find all x such that $\cos x > 0$, we add $2n\pi$ to these endpoints, to
obtain, $x \ \varepsilon \ (-\pi/2 + 2n\pi, \ \pi/2 + 2n\pi)$, where $n = 0, \pm1, \pm2,...$

EXAMPLE: Find dy/dx by implicit differentiation of the equation $x + y \ln y = \cos x$.

Answer: It may help to "box" in the y-term, as we did in the section on implicit differentiation.

$x +$ ⬚y⬚ $\ln$ ⬚y⬚ $= \cos x$. Differentiating, $1 + (dy/dx) \ln y + y \ (1/y) \, dy/dx = -\sin x$

And this can be simplified by grouping dy/dx terms on the left, all other terms on the right.

$$\frac{dy}{dx}(\ln y + 1) = -\sin x - 1, \quad \text{and} \quad \frac{dy}{dx} = -\frac{(\sin x + 1)}{(\ln y + 1)}.$$

NOTE: We have $\int_a^b \frac{1}{x}dx = \ln|b| - \ln|a|$. We must never have 0 in the interval [a,b]. Why?

Because of the absolute value signs, we get the result that Area I = - Area II below.

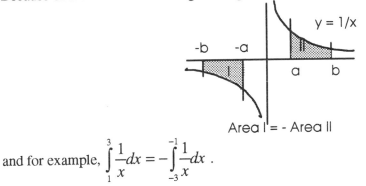

$$y = 1/x$$

Area I = - Area II

and for example, $\int_1^3 \frac{1}{x}dx = -\int_{-3}^{-1} \frac{1}{x}dx$.

EXAMPLE: Logarithmic Differentiation. Assume x is restricted to values that make

$y = \dfrac{(x+2)^3}{\sqrt{x+1}}$ positive. Use logarithmic differentiation to find dy/dx .

Answer:

(1) Take the natural logarithm of both sides: $\ln y = \ln\left(\dfrac{(x+2)^3}{\sqrt{x+1}}\right)$

(2) Simplify, using logarithm properties:
$$\ln y = \ln(x+2)^3 - \ln\sqrt{x+1} = 3\ln(x+2) - (1/2)\ln(x+1).$$

(We have assumed that (x + 2) and (x +1) are positive, so all terms are defined.)

☞ **Note! At this stage we did not differentiate; we merely used ln properties.**

---

(3) Now differentiate implicitly. $\dfrac{1}{y}\dfrac{dy}{dx} = \dfrac{3}{x+2} - \dfrac{1}{2(x+1)}$. We want dy/dx.

(4) Solve: $dy/dx = y\left(\dfrac{3}{x+2} - \dfrac{1}{2(x+1)}\right) = \dfrac{(x+2)^3}{\sqrt{x+1}}\left(\dfrac{3}{x+2} - \dfrac{1}{2(x+1)}\right).$

160 Chapter 5

MEMORIZE THESE INTEGRALS AND THE METHOD FOR THEIR SOLUTION:

$$\int \sec x\, dx = \int \sec x \cdot \left(\frac{\sec x + \tan x}{\sec x + \tan x} \right) dx = \int \frac{\sec^2 x + \sec x \tan x}{\sec x + \tan x} dx = \ln|\sec x + \tan x| + C$$

$$\int \tan x\, dx = \int \frac{\sin x}{\cos x} dx = -\ln|\cos x| + C \quad (\text{using } u = \cos x)$$

EXAMPLE: Find $\int \csc x\, dx$.

Answer: There are some variations that look different but are nevertheless the same. Here is one way:

$$\int \csc x\, dx = \int \csc x \cdot \left(\frac{\csc x + \cot x}{\csc x + \cot x} \right) dx = \int \frac{\csc^2 x + \cot x \csc x}{\csc x + \cot x} dx. \quad \text{Let } u = \csc x + \cot x.$$

Then $du = (-\csc x \cot x - \csc^2 x) dx$ and the integral becomes

$$-\int \frac{du}{u} = -\ln|\csc x + \cot x| + C$$

EXERCISE: Now we know the integrals of all the trigonometric functions. Fill in the chart .

$\int \sin x\, dx$	a
$\int \cos x\, dx$	b
$\int \tan x\, dx$	c
$\int \csc x\, dx$	d
$\int \sec x\, dx$	e
$\int \cot x\, dx$	f

Answers: *(a) -cos x + C (b) sin x + C (c) -ln | cos x | + C (d) -ln | csc x + cot x | + C
(e) ln | sec x + tan x | + C (f) ln | sin x | + C*

NOTE: TWO INTEGRALS WILL SHOW UP A LOT. You may as well memorize their solutions now:

$$\int \cos^2 x\, dx = ?.... \text{Use that } \cos^2 x = \frac{1 + \cos 2x}{2}$$

$$\int \sin^2 x\, dx = ?.... \text{Use that } \sin^2 x = \frac{1 - \cos 2x}{2}$$

EXAMPLE: Evaluate the integral $\int \cos^2 3x\, dx$.

Answer: $\int \cos^2 3x\, dx = \int \frac{1+\cos 6x}{2}\, dx = \int \frac{1}{2}\, dx + \int \frac{\cos 6x}{2}\, dx$. You can do both integrals

in your head (although you might want to use u = 6x in the second integral.)

$\int \frac{1}{2}\, dx + \int \frac{\cos 6x}{2}\, dx = \frac{x}{2} + \left(\frac{1}{2}\right)\left(\frac{\sin 6x}{6}\right) + C = \frac{x}{2} + \frac{\sin 6x}{12} + C$.

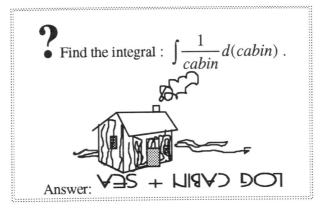

? Find the integral : $\int \frac{1}{cabin}\, d(cabin)$.

Answer: LOG CABIN + SEA

SECTION 5.8

" Reeling and Writhing, of course, to begin with," the Mock Turtle replied, "and the different branches of Arithmetic--Ambition, Distraction, Uglification, and Derision. "....Lewis Carroll

FINDING THE AREA OF THE REGION

> **If f is a function such that f(x) ≥ g(x) on the interval [a,b], then**
> **the area of the region between the graphs of f and g is positive and is given by**
> $$\int_a^b f(x) - g(x)\, dx$$

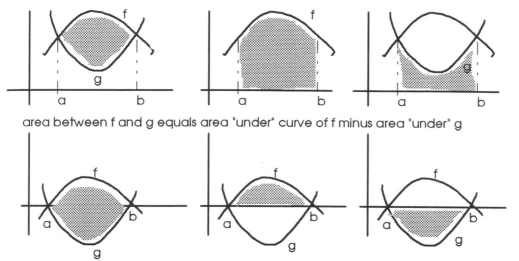

area between f and g equals area "under" curve of f minus area "under" g

area between f and g equals area "under" f minus the negative area "under" g

area between f and g equals negative area "under" f minus negative area "under" g

---

EXAMPLE: Let f(x) = sin x and g (x) = 0 (the x-axis). Find the area of the region between the graphs of f and g on [π, 2π].

Answer: On this interval, the graph of sin x lies below the x-axis.

$$\text{Area} = \int_{\pi}^{2\pi} (0 - \sin x)dx = \cos x \Big|_{\pi}^{2\pi} = 2 \ .$$

Note that this gives us a way of talking about a positive area, even for a negative function.

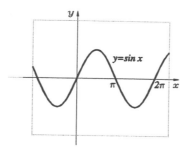

$$f(x) = \sin x \ on \ [\pi \ , \ 2\pi \ ]$$

WORD OF ADVICE: The formula for the area of the region between the graphs of two functions, $\text{AREA} = \int_a^b | f - g | \, dx$, is not very useful, because we need to know more about $| f - g |$. It is INCORRECT to take the absolute value of AREA; i.e. we must not use $\left| \int_a^b (f - g) \, dx \right|$. (Why?) The formula for finding the area between two functions suggests that we must:

(1) Find where f and g intersect and divide the region of integration up into subintervals determined by where these functions intersect.

(2) Using test-points on each subinterval, find whether f or g is the TOP function.

(3) Find the area on each subinterval as follows:
$$\text{AREA} = \int_?^{??} (\text{TOP FUNCTION - BOTTOM FUNCTION}).$$

(4) Sum the areas on each subinterval to give the answer.

AN ALTERNATIVE APPROACH:

On a <u>sub-interval</u> if $\int_a^b f(x) - g(x)dx$ turns out to be negative; we have the wrong function on top, so we can make the result positive by taking absolute value (Case 1). But we cannot simply make the result of $\int_a^b f(x) - g(x)dx$ positive in Case 2 below. Why?

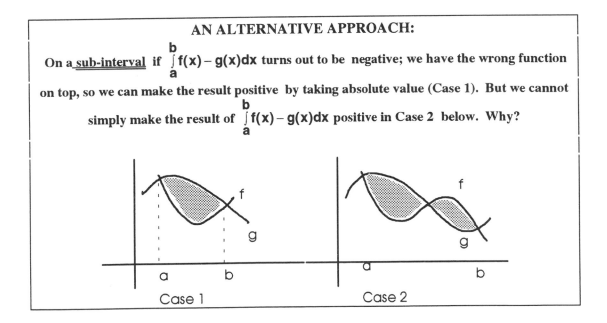

Case 1   Case 2

EXAMPLE: Find the area of the region between the graphs of
$f(x) = 2x^3 + 2x^2$ and $g(x) = 2x^3 - 2x$ on [-5, 3].

Answer:

(1) Solving $f(x) = g(x)$, or $2x^3 + 2x^2 = 2x^3 - 2x$, we have $x^2 + x = 0$, or x = -1, 0.

(2) Draw an interval showing critical points: -5 .... -1......... 0 3

(3) Find test - points in each of these subintervals, summarizing the information below :

interval	(-5,-1)	(-1,0)	(0,3)
test point	-2	-1/2	1
f(test point)	-8	1/4	4
g(test point)	-12	3/4	0
Conclude	f is TOP	g is TOP	f is TOP

(4) Conclude that Area $= \int_{-5}^{-1} \mathbf{f} - \mathbf{g}\ dx + \int_{-1}^{0} \mathbf{g} - \mathbf{f}\ dx + \int_{0}^{3} \mathbf{f} - \mathbf{g}\ dx$. Since now

$f - g = 2x^2 + 2x$ and $g - f = -2x - 2x^2$, we have

Area $= \int_{-5}^{-1} 2x^2 + 2\ dx + \int_{-1}^{0} (-2x - 2x^2)\ dx + \int_{0}^{3} 2x^2 + 2\ dx$ (The rest is left to you.)

Give the Definitive Explanation

☙ Why is it **not** true that $\dfrac{d}{dx} 2^x = x 2^{x-1}$. Don't simply *state* that this is not true.
   Explain WHY.

☙ What is the difference between the functions $y = x^2$ and $y = 2^x$? List as many
   different characteristics of these functions as you can.

☙ Why is it that $\int_{-1}^{1} \dfrac{1}{x} dx = \ln|1| - \ln|-1| = 0$, since this is the answer we might instinctively
   expect anyway? Hint: the answer deals with the fact that the definite integral is the
   limit of a Riemann sum, and this can involve ANY partition. Can you create a way
   of creating partitions of [-1,1] for which the definite integral--the limit of a
   Riemann sum--would not be 0?) **In fact, we should not expect this integral to
   be 0!**

☙ How can it be that the "area under a point" is 0, when the area "under" a function,
   which is comprised of areas under points, is non-zero?

KNOW:

1. The definite integral, its definition as the unique......and the definition using Riemann sums. Be able to approximate a definite integral using some form of sums. Can you give pictures that illustrate upper and lower sums, Riemann sums, midpoint sums, average value of a function, etc.? Know the Comparison Property, the definition of the average value of a function, its geometrical significance, and how to find it.

2. The indefinite integral, the u-substitution. Always check your answer, and watch out for errors. Memorize some of the <u>special</u> <u>integrals</u> that we called attention to. These will occur throughout the course. To review for this, look back at your homework, and rework many problems. Be careful about your form when integrating: small errors in form may look like large errors to your instructor!

3. The ln(x) function, its definition as an integral of a given function, its graph and its algebraic properties (e.g., the logarithm of a product is the sum of the logarithms of the factors, etc.). Know how to find the domain of ln (f(x)) for various f 's , and to differentiate and integrate with the ln(x) function. Be careful not to think that the algebraic properties of ln (x) say things they do not say, but remember that these properties will simplify your work enormously, if you use them properly.

4. The area of the region between the graphs of two functions .

Form a Study Group! Your best learning is usually achieved when you work with a group of others, if you meet on a regular basis.

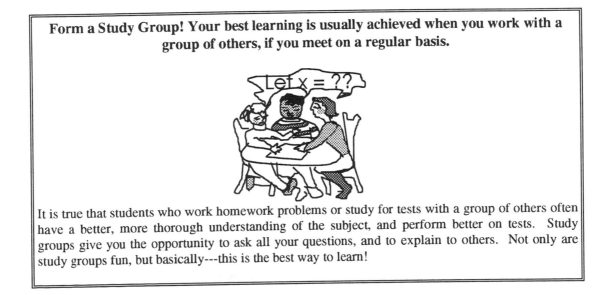

It is true that students who work homework problems or study for tests with a group of others often have a better, more thorough understanding of the subject, and perform better on tests. Study groups give you the opportunity to ask all your questions, and to explain to others. Not only are study groups fun, but basically---this is the best way to learn!

❧ ❧ ❧ ❧ ❧ ❧ ❧ ❧ ❧ ❧ ❧ ❧ ❧ ❧ ❧ ❧

Creating a Review Sheet

Many students do not know about writing review sheets. This is a one or two page summary of important information. Here is part of one student's review sheet for some sections of this chapter. It would have been more helpful to this student if she had written out more of the theorems and definitions and rules. Perhaps she feels she already knows them. Nevertheless, this might have been a good chance to check . What would you add to this sheet?

Know trig identities and derivatives.

Know chain rule.

How to apply u-substitutions.

What to substitute as u

   Ex. things in denominators, things raised to a power, things under square roots, etc.

Remember the +C in indefinite integrals

Definition of the integral and derivative of natural log function.

   Ex. $\ln x = \int_1^x \frac{1}{t} dt$, where x > 0.   d/dx (ln x) = 1/x. Graph:

y-axis, an asymptote

Remember the absolute value in the integral with lns

How to find domain and range with ln functions

Laws of logarithms:

   ln 1 = 0, ln e = 1, ln (1/x) = - ln x,   ln (1/e) = -1, etc.

There is no ln (0)

How to find ln on calculator

Memorize some integrals.

   Ex. $\int \sec x\, dx = \ln|\sec x + \tan x| + C$, $\int \cos^2 x\, dx = \int \frac{1+\cos 2x}{2} dx = ..$, etc.

Definition of definite integral: $\int_a^b f(x)\,dx = \lim_{\|P\|\to 0} \sum_{i=1}^{n} f(t_k)\Delta x_k$

Know Fundamental Theorem of Calculus, how to differentiate an integral.

❧ ❧ ❧ ❧ ❧ ❧ ❧ ❧ ❧ ❧ ❧ ❧ ❧ ❧ ❧ ❧

CHAPTER 6

"I do not know what I may appear to the world; but to myself I seem to have been only like a boy playing on the seashore and diverting myself in now and then, finding a smoother pebble or prettier shell than ordinary, whilst the great ocean of truth lay all undiscovered before me"..Isaac Newton

With the concept of an inverse function, we introduce a fleet of new functions: generalized exponential and logarithmic functions, hyperbolic, inverse hyperbolic and inverse trigonometric functions. It would be helpful to review the exponential and logarithm functions defined in Chapter 1 and developed in Chapter 5. Refresh your trigonometry (See the review preceding Chapter 5) and brush up on finding limits (Chapter 2).

SECTION 6.1

"There never was a war that was not inward"...Marianne Moore

INVERSE FUNCTIONS

NOTE: In order for a function f to have an inverse, f must be one-to-one. We say **f is one-to-one** if $x_1 \neq x_2 \Rightarrow f(x_1) \neq f(x_2)$. This may seem like a strange definition. Essentially this means that the function does not take 2 distinct x-values to a single y-value.

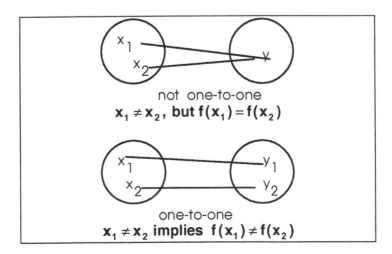

not one-to-one
$x_1 \neq x_2$, but $f(x_1) = f(x_2)$

one-to-one
$x_1 \neq x_2$ implies $f(x_1) \neq f(x_2)$

NOTE: The graph of a one-to-one function does not cross any horizontal line more than once. If f is differentiable, here's a convenient way to see if f has an inverse. If $f' \geq 0$ or $f' \leq 0$ everywhere, then f is one-to-one and has an inverse.

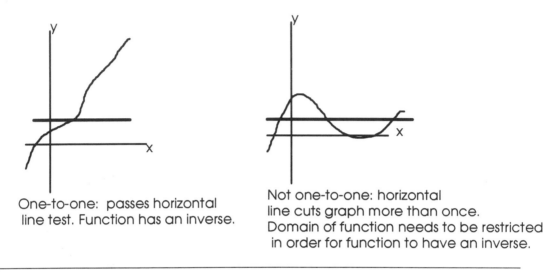

One-to-one: passes horizontal
line test. Function has an inverse.

Not one-to-one: horizontal
line cuts graph more than once.
Domain of function needs to be restricted
in order for function to have an inverse.

EXAMPLE: Prove the function $f(x) = 2x + \sin x$ has an inverse.
Answer: Since $f'(x) = 2 + \cos x > 0$, f is everywhere increasing and has an inverse. (This does not mean we can find the inverse.)

NOTE: We cannot always find the expression giving the inverse of a function; these functions may just not have "names". Soon we will give some names to these inverse functions.

NOTE: We frequently restrict the domain of the function f to a set where f is one-to-one. For example, on the domain $[0, \infty)$, $f(x) = x^2$ has an inverse (f is invertible): the inverse is $f^{-1}(x) = \sqrt{x}$.

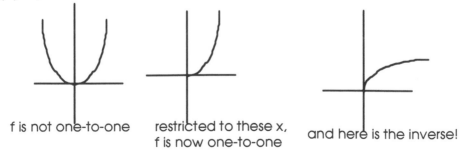

f is not one-to-one   restricted to these x,   and here is the inverse!
                        f is now one-to-one

EXAMPLE: Show $f(x) = \cos x - \sin x$ does not have an inverse.
Answer: This function is not one-to-one. $f(0) = f(2\pi)$. (This is all we need to do--we have found 2 x's that go to the same y.) However, we might have restricted the domain of f to find an inverse.

CHARACTERISTICS OF THE INVERSE FUNCTION

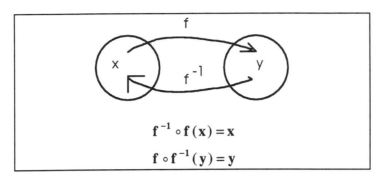

$$f^{-1} \circ f(x) = x$$
$$f \circ f^{-1}(y) = y$$

1. $f^{-1} \circ f$ = identity function on the domain of f AND

   $f \circ f^{-1}$ = identity function on the range of f.

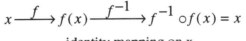

identity mapping on x

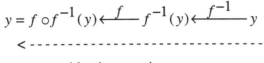

identity mapping on y

EXAMPLE: The cube root and cubing function are inverses.

identity mapping on x

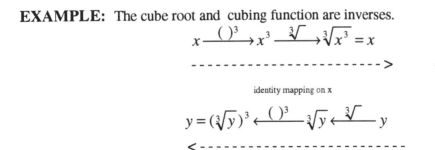

identity mapping on y

2. Graphically, to swap x- and y-coordinates is like exchanging x- and y-axes, or reflecting across the 45°-line.

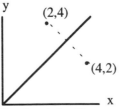

To swap coordinates is to flip across the y = x line

The graph of the inverse function looks like the graph of the original function reflected across the 45°-line.

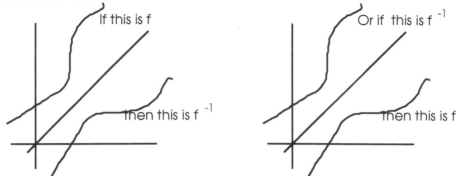

3. Range and domain are swapped. Since the inverse function swaps x and y, it makes sense that the **domain** and **range** of f become, respectively, the **range** and **domain** of f^{-1}. (See Examples below).

4. If f is differentiable and $f' \neq 0$, then (since variables x and y are swapped),

$$(\text{slope of tangent line to } f^{-1} \text{ at } y) = \frac{1}{(\text{slope of tangent line to } f \text{ at } x)} \quad \text{or} \quad \frac{\Delta y}{\Delta x} = \frac{1}{\Delta x / \Delta y}$$

$$\boxed{\left.\frac{df^{-1}}{dx}\right|_{(y,x)} = \frac{1}{\left.df/dx\right|_{(x,y)}}}$$

NOTE: The derivative of the inverse function must be evaluated at **(y,x)** and the derivative of the original function at **(x, y)** in order for this rule to apply.

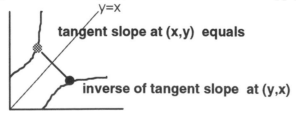

EXAMPLE: Procedure for finding an inverse function. Find the inverse of the function $f(x) = (x^3 + 4)^5$. (We have checked that $f' \geq 0$, so f has an inverse.)

Answer:

1. For simplicity, switch to variables y and x: We have $y = (x^3 + 4)^5$

2. **SWAP variables :** $x = (y^3 + 4)^5$. (This just reflected the function across the 45°-line.)

3. **SOLVE for y :** $(y^3 + 4) = \sqrt[5]{x}$, or $y = \sqrt[3]{\sqrt[5]{x} - 4} = (x^{1/5} - 4)^{1/3}$. (This expresses the inverse function as we are accustomed to seeing it, with y given in terms of x.)

4. We have the inverse function : $f^{-1}(x) = (x^{1/5} - 4)^{1/3}$.

EXAMPLE: Find the inverse of the function $f(x) = \sqrt{x-1}$

Answer:

Rewrite: $y = \sqrt{x-1}$

SWAP: $x = \sqrt{y-1}$

SOLVE $y - 1 = x^2$, or $y = x^2 - 1$, or $f^{-1}(x) = x^2 - 1$

EXAMPLE: **Procedure for finding the domain and range of the inverse function.** If f is given by $f(x) = \sqrt{x-1}$ with $x \geq 1$, find the domain and range of the inverse function.

Answer: Domain and Range are interchanged :

Domain of f = { x such that x ≥ 1 }. *Range of f = { y such that y ≥ 0 }*

Domain of f^{-1} = { x such that x ≥ 0 }. **Range of f⁻¹ = { y such that y ≥ 1 }.**

EXAMPLE: **Procedure for proving a function is the inverse.** (We show $f^{-1} \circ f$ and $f \circ f^{-1}$ are the identity function.) Prove that $f(x) = \sqrt[3]{x-1}$ and $g(x) = x^3 + 1$ are inverse functions.

Answer:

$f \circ g(x) = f(g(x)) = \sqrt[3]{g(x) - 1} = \sqrt[3]{x^3 + 1 - 1} = x$

$g \circ f(x) = g(f(x)) = (f(x))^3 + 1 = (\sqrt[3]{x-1})^3 + 1 = x.$

Variation : If f is its own inverse, this can be shown by showing f(f(x)) = x.

EXAMPLE: Procedure for finding the derivative of an inverse function at x = c.

Find $\dfrac{df^{-1}}{dx}$ (c) where $f(x) = x^3 + 7$ at c = 8.

Answer: We use:

$$\left.\frac{df^{-1}}{dx}\right|_{(8, f^{-1}(8))} = \frac{1}{\left.\dfrac{df / dx}{}\right|_{(f^{-1}(8), 8)}} \quad . \text{ Now } f^{-1}(8) \text{ is the value of x such that}$$

$f(x) = x^3 + 7 = 8$; namely x = 1. Thus

$$\left.\frac{df^{-1}}{dx}\right|_{(8,1)} = \frac{1}{\left. df / dx \right|_{(1,8)}} = \frac{1}{\left. 3x^2 \right|_1} = \frac{1}{3}.$$

To do this type of problem we can also use the inverse function explicitly (if we can find it).

Since $f(x) = x^3 + 7$, we find $f^{-1}(x) = \sqrt[3]{x - 7}$. Then $\left.\dfrac{df^{-1}}{dx}\right|_8 = \dfrac{1}{3}(x - 7)^{-2/3}\Big|_8 = \dfrac{1}{3}$.

===

SECTION 6.2

===

"The basic fact of today is the tremendous pace of change in human life"...Jawaharlal Nehru

THE EXPONENTIAL FUNCTION

NOTE: The derivative of e^x is e^x. With the chain rule, the derivative of

$$\frac{d}{dx}e^{\square} = e^{\square} \cdot \frac{d\square}{dx}, \text{ or } \frac{d}{dx}e^u = e^u \frac{du}{dx}.$$

✋ **CAUTION!** Many students think $\dfrac{de^x}{dx} = xe^{x-1}$, and do not understand why this is

not the case. They are probably remembering the rule : $\dfrac{dx^r}{dx} = nx^{r-1}$ (with r a rational

number). The function f(x) = e^x has the variable in the exponent, and its derivative is different from that of a function which is a power of x. (See this derivative found with the limit definition in your text.)

NOTE: What may contribute to the confusion is that the rule for differentiating e^x gives

us the result that $\dfrac{dx^r}{dx} = nx^{r-1}$ where r is any *real number* (not just rational). This is found

by writing $x^r = e^{\ln x^r}$, and finding the derivative with the chain rule. This is an important result, but it should not to be confused with the derivative of an exponential function.

What is the difference between a rational and real number? A rational number is a number which can be expressed as a quotient of integers (whole numbers); a real number is any number found on the Real Line (no complex numbers allowed).

EXAMPLE: Find the derivative of $f(x) = e^{-x^2}$

Answer:
$$\frac{d}{dx}e^{-x^2} = e^{-x^2} \ (-2x) \ = -2xe^{-x^2}$$

$$\uparrow \qquad \uparrow$$

deriv of $e^{(\ )}$ deriv of ( )

EXAMPLE: Find the derivative of $y = \sin(e^{\sqrt{x}\cos x})$.

Answer:
$$\frac{d}{dx}\sin(e^{\sqrt{x}\cos x}) = (\cos(e^{\sqrt{x}\cos x})) \ (e^{\sqrt{x}\cos x}) \ (\frac{\cos x}{2\sqrt{x}} + \sqrt{x}(-\sin x)) = \ ? \ \text{(Simplify.)}$$

deriv of sin ( )   deriv of $e^{(\ )}$ deriv of $\sqrt{\ }\cos(\ )$ with product rule

EXAMPLE: Find the derivative of $y = x^{\sqrt{3}}$.

Answer: $\dfrac{dy}{dx} = \sqrt{3}\ x^{\sqrt{3}-1}$.

EXAMPLE: Find the integral: $\displaystyle\int (\cos x)e^{\sin x}\, dx$.

Answer:

Let $u = \sin x$, then $du = \cos x\, dx$. Now $\displaystyle\int (\cos x)e^{\sin x}\, dx = \int e^u du = e^u + C = e^{\sin x} + C$

NOTE: What is e? **The number "e" is not defined as 2.718281828...** because we don't really "know" the decimal expansion for "e". In fact, e is the number such that $\ln e = 1$.

SECTION 6.3

" $\dfrac{a + b^n}{n} = x$, thus God exists--reply!"..Euler

GENERAL EXPONENTIAL AND LOGARITHMIC FUNCTIONS:
ANY BASE a > 0

NOTE: Since $e^{(\ )}$ and $\ln(\ )$ are inverses, they cancel each other out when they are in direct proximity:

$$x \xrightarrow{\ e^{(\ )}\ } e^x \xrightarrow{\ \ln(\ )\ } \ln(e^x) = x \ , \text{ back to the start!}$$

$$x \xrightarrow{\ \ln(\ )\ } \ln x \xrightarrow{\ e^{(\ )}\ } e^{\ln x} = x \ , \text{ back to the start!}$$

NOTE: Analogous formulas apply with any a > 0 as a base for an exponential function.

$$\log_a a^x = x$$

$$a^{\log_a x} = x$$

NOTE: These functions can be defined in terms of the familiar function $e^{(\ )}$ and $\ln(\ )$.

$$a^x = e^{\ln(a^x)} = e^{x\ln a}$$

$$\log_a x = \frac{\ln x}{\ln a}$$

✋ **CAUTION:** Notice the above rule does not allow us to say : $e^{3\ln x} = 3\ln x$.

But $e^{3\ln x} = e^{\ln x^3} = x^3$. Similarly $\ln(3e^x) \neq 3e^x$. Rather,
$\ln(3e^x) = \ln 3 + \ln(e^x) = \ln 3 + x$. "Ln" and "e" must "touch" in order to use the rules above.

NOTE: We can use any $a > 0$ as a base. But if we had expressions such as $f(x) = (-3)^x$, (where $a = -3$), $f(x)$ would erratically be positive and negative and there would be an uncountable number of x-values where f would not be defined.

NOTE: To find derivatives and integrals of exponential and logarithmic functions, use the identities that follow:

$$\frac{d}{dx} a^x = \frac{d}{dx} e^{x\ln a} = e^{x\ln a}(\ln a) = a^x \ln a$$

$$\frac{d}{dx} \log_a x = \frac{d}{dx}\frac{\ln x}{\ln a} = \frac{1}{x\ln a}$$

$$\int a^x dx = \int e^{x\ln a} dx = \frac{e^{x\ln a}}{\ln a} = \frac{a^x}{\ln a} + C$$

Integrals involving $\log_a x$ (such as $\int \frac{\log_a x}{x} dx$) use $\log_a x = \frac{\ln x}{\ln a}$

But it is good to learn how to derive these rules in addition to memorizing them.

EXAMPLE: This example offers a technique for deriving that will prove very useful.

$$\boxed{\textbf{Derive the formula :}\ \ \log_a x = \frac{\ln x}{\ln a}.}$$

Answer: We have a logarithm expression we don't like, so we convert to exponential form.

Let $y = \log_a x$. Then $a^y = x$.

Take the natural logarithm of both sides in order to bring the variable y "down to ground level".

$\ln a^y = \ln x$

$y \ln a = \ln x$ (We are not differentiating ; this is a property of the logarithm function.)

$y = \dfrac{\ln x}{\ln a}$

EXAMPLE: Find the derivative: $f(x) = 2^x$.

Answer: There are 3 possible ways to do this.

(a) We can memorize the formula: $f'(x) = 2^x \ln 2$.

(b) We can write: $f(x) = 2^x = e^{\ln 2^x} = e^{x \ln 2}$ so $f'(x) = e^{x \ln 2} \ln 2 = 2^x \ln 2$.

(c) We can let $y = 2^x$, then take the natural logarithm of both sides, $\ln y = x \ln 2$, and differentiating implicitly, $\dfrac{1}{y} \cdot y' = \ln 2$. So $y' = y \ln 2 = 2^x \ln 2$.

EXAMPLE: Find the derivative: $y = x^x$.

Answer: Here are two ways to solve this problem.

(1) Write y as $y = e^{x \ln x}$ and differentiate.

$\dfrac{dy}{dx} = e^{x \ln x} (\ln x + x(1/x)) = x^x (1 + \ln x)$

(2) Take the natural logarithm of both sides.

$\ln y = \ln x^x$. Use the exponential property of the natural logarithm.

$\ln y = x \ln x$. CAUTION! No differentiation has taken place at this step!

$\dfrac{1}{y} \cdot \dfrac{dy}{dx} = (\ln x) + (x)\dfrac{1}{x} = 1 + \ln x$, where we differentiated implicitly .

So $\dfrac{dy}{dx} = y(1 + \ln x) = x^x (1 + \ln x)$.

EXAMPLE: Integrate: $\int 3^x dx$.

Answer: Here are 2 ways to solve this problem.

(1) Remember the formula: $\int 3^x dx = \dfrac{3^x}{\ln 3} + C$.

(2) Rewrite as $\int e^{x \ln 3} dx$. (Note: $\ln 3$ is just a constant.) Letting $u = x \ln 3$, and

$du = \ln 3 \ dx$, we have $\dfrac{1}{\ln 3} \int e^u du = \dfrac{1}{\ln 3} e^u + C = \dfrac{1}{\ln 3} e^{x \ln 3} + C = \dfrac{3^x}{\ln 3} + C$

EXAMPLE: Integrate $\int \dfrac{\log_2 x}{x} dx$.

Answer: Rewrite as $\int \dfrac{\ln x}{x \ln 2} dx = \dfrac{1}{\ln 2} \int \dfrac{\ln x}{x} dx$ Now letting u = ln x, with du = $\dfrac{1}{x} dx$,

$\dfrac{1}{\ln 2} \int u\, du = \dfrac{1}{\ln 2} \left(\dfrac{u^2}{2} \right) + C = \dfrac{1}{2 \ln 2} (\ln x)^2 + C$ which could be rewritten in one way as

$\dfrac{1}{\ln 4} (\ln x)^2 + C.$ (Note: $(\ln x)^2$ is not 2 ln x.)

HOW WOULD YOU...

Define "e"? Define "ln x"? Define x^π? (This surely can't mean x multiplied by
   itself π times!)

Define what it means for a function to be one-to-one?

Find the inverse of a given function?

Prove that ln (ab) = ln a + ln b (where a, b > 0) ?

Prove that $\dfrac{d\,(a^x\,)}{dx} = a^x$ ln a?

Prove that $\log_a x = \dfrac{\ln x}{\ln a}$?

Explain the "absolute value" in the formula $\int \dfrac{1}{x} dx = \ln|x| + C$?

Integrate a function containing a "$\log_a x$" term?

SECTION 6.4

*"I shall be telling this with a sigh, Somewhere ages and ages hence; Two roads diverged in a
wood, and I-- I took the road less traveled by, And that has made all the difference."..Robert Frost*

HYPERBOLIC FUNCTIONS

The functions e^x and e^{-x} occur naturally in these combinations so often, it is worth
recognizing these combinations as new functions.

$$\sinh x = \frac{e^x - e^{-x}}{2}$$

$$\cosh x = \frac{e^x + e^{-x}}{2}$$

These functions have the property:

$$\boxed{\cosh^2 x - \sinh^2 x = 1}$$

EXAMPLE: Find tanh 1:

Answer: $\tanh 1 = = \dfrac{\sinh 1}{\cosh 1} = \dfrac{(e^1 - e^{-1})/2}{(e^1 + e^{-1})/2} = \dfrac{e - \dfrac{1}{e}}{e + \dfrac{1}{e}} = \dfrac{e^2 - 1}{e^2 + 1}$.

EXAMPLE: Find $\dfrac{d}{dx}\coth x$.

Answer: $\dfrac{d}{dx}\coth x = \dfrac{d}{dx}\dfrac{\cosh x}{\sinh x} = \dfrac{(\sinh x)^2 - (\cosh x)^2}{(\sinh x)^2} = \dfrac{-1}{\sinh^2 x} = -\csc h^2 x$.

EXAMPLE: Find $\dfrac{d}{dx}\sinh\sqrt{1 - x^2}$.

Answer: By the chain rule, $\dfrac{d}{dx}\sinh\sqrt{1 - x^2} = (\cosh\sqrt{1 - x^2})\ \dfrac{1}{2\sqrt{1 - x^2}}(-2x)$. Simplify.

EXAMPLE: Find $\dfrac{d}{dx}\sinh^{-1} x$.

Answer: You have seen one derivation of this in your text. Here's another derivation. The expression $y = \sinh^{-1} x$ means sinh y = x. Differentiating implicitly,

$\cosh y\dfrac{dy}{dx} = 1$, or $\dfrac{dy}{dx} = \dfrac{1}{\cosh y}$. However, we do not want dy / dx in terms of y.

Since sinh y = x, and $\cosh^2 y - \sinh^2 y = 1$, we see $\cosh y = \sqrt{1 + \sinh^2 y} = \sqrt{1 + x^2}$. (Note: cosh y > 0, hence the positive square root.) Replacing this in the derivative gives

$\dfrac{dy}{dx} = \dfrac{1}{\sqrt{1 + x^2}}$.

NOTE: When working with cosh x and sinh x, many properties can be proved from the definitions of these functions with e^x and e^{-x} .

EXAMPLE: Derive the formula: sinh 2x = 2 sinh x cosh x.

Answer:

$\sinh 2x = \dfrac{e^{2x} - e^{-2x}}{2} = \dfrac{(e^x)^2 - (e^{-x})^2}{2} = \left(\dfrac{e^x - e^{-x}}{2}\right)\left(\dfrac{2(e^x + e^{-x})}{2}\right) = 2\sinh x \cosh x$

EXAMPLE: Solve for x: cosh x = 6.

Answer: We set cosh x = $\dfrac{e^x + e^{-x}}{2}$ = 6, or $e^x + \dfrac{1}{e^x} = 12$, or $e^{2x} + 1 - 12e^x = 0$. This can

be rewritten with u = e^x, as $u^2 - 12u + 1 = 0$, for which solutions (by the Quadratic

Formula) are: $u = \dfrac{12 \pm \sqrt{144 - (4)(1)}}{2} = \dfrac{12 \pm \sqrt{140}}{2} = \dfrac{12 \pm 2\sqrt{35}}{2} = 6 \pm \sqrt{35}$. Since u = e^x,

which is always positive, we take u = $e^x = 6 + \sqrt{35}$. Solving for x by taking the natural

logarithm, x = ln (6 + $\sqrt{35}$).

The function y = cosh (x) describes a catenary, the curve obtained, for example, when a string (or hammock) is allowed to droop between two points.

SECTION 6.5

"Beware lest you lose the substance by grasping at the shadow"...Aesop

INVERSE TRIGONOMETRIC FUNCTIONS

These functions are written as $\sin^{-1} x$, $\cos^{-1} x$, etc. and sometimes as arcsin x, arccos x, etc. The rationale for the latter notation will help you remember the meaning of these functions. On a circle, arc length is given by s = r θ. If we think of the radius given by r = 1 (think: "one unit"), then s = θ.

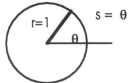

Therefore, if we think of "arc" as representing "angle", the function "arc sin x" reads: *"angle whose sine is x".* Then y = $\sin^{-1} x$ also means **" y is the angle whose sine is x".**

For example, since sin $\pi/2$ = 1, we say $\sin^{-1} 1 = \pi / 2$. (We will see that it will not be possible to say $\sin^{-1} 1 = 5\pi / 2$. If we allowed two values of y for a given x in the expression y = $\sin^{-1} x$, the inverse trigonometric function would not be well-defined.)

NOTE: If sin y = x, we say y = sin^{-1} x , but we must decide what values y can take on. . Each inverse trigonometric function is accompanied by two quadrants that determine where y may lie. We begin by finding a portion of the graph of y = sin x where this function is one-to-one. We may chose x to lie in [-π/2, π/2] (Fig. a) . When we move to the inverse function in Fig. (b), we want y-values in [-π/2, π/2].

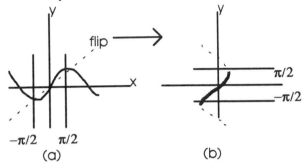

(a)   (b)

These range angles for the inverse trigonometric functions should be memorized.

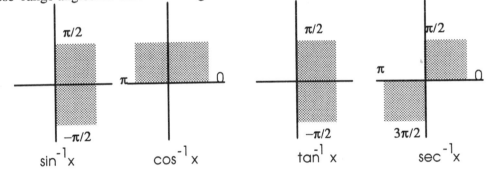

$$\sin^{-1}x \qquad \cos^{-1}x \qquad \tan^{-1} x \qquad \sec^{-1}x$$

Ranges of Inverse Trigonometric functions: these are the angles the inverse trigonometric functions may assume

EXAMPLE:

(a) Find sin^{-1} 3.

   Answer: There is no angle whose sine is 3; this is impossible.

(b) Find $\sin^{-1}(-1/2)$.

   Answer: The reference angle is π/6 and we find our angle in quadrant IV, where sine values are negative. Thus $\sin^{-1}(-1/2) = -\pi/6$.

(c) Find $\cos^{-1}(-1/2)$.

   Answer: The reference angle is π/3, and since cosine is to be negative, we find our angle in quadrant II: $\cos^{-1}(-1/2) = \dfrac{2\pi}{3}$.

EXAMPLE: Find $\tan(\sin^{-1} x)$.

Answer: In Fig. (a) we draw a triangle with angle θ, such that $\theta = \sin^{-1} x$, or

$\sin \theta = \dfrac{\text{opposite side}}{\text{hypotenuse}} = \dfrac{x}{1} = x$.

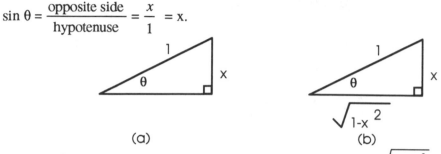

(a)   (b)

Using the Pythagorean Theorem, we fill in the missing side with $\sqrt{1 - x^2}$, as in Fig. (b).

Now $\tan(\sin^{-1} x) = \tan \theta = \dfrac{\text{opposite side}}{\text{adjacent side}} = \dfrac{x}{\sqrt{1 - x^2}}$.

EXAMPLE: Find $\dfrac{d}{dx}(\cos^{-1} x)$.

Answer: First write $y = \cos^{-1} x$. Then $\cos\, y = x$. Differentiating implicitly,
$(- \sin y)\, dy/dx = 1$, or $dy/dx = -1/\sin y = -\csc y$. Now we must express y in terms of x.

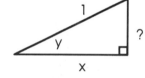

Draw an angle y such that $\cos y = \dfrac{\text{adjacent side}}{\text{hypotenuse}} = \dfrac{x}{1} = x$. Using the Pythagorean Theorem,

the opposite side has length $\sqrt{1 - x^2}$. Thus $\sin y = \dfrac{\sqrt{1 - x^2}}{1} = \sqrt{1 - x^2}$. Replacing this in

the formula for dy/dx above, we have $dy/dx = -\csc y = -\dfrac{1}{\sqrt{1 - x^2}}$.

EXAMPLE: Find $\dfrac{d}{dx}(\tan^{-1} x)$.

Answer: If $y = \tan^{-1} x$ then $\tan y = x$. Differentiating,
$(\sec^2 y)(dy / dx) = 1$, or $dy / dx = \cos^2 y$. We must find y in terms of x. Draw an angle

y such that $\tan y = \dfrac{\text{opposite side}}{\text{adjacent side}} = \dfrac{x}{1} = x$. The hypotenuse of the triangle is $\sqrt{1 + x^2}$.

Thus $dy/dx = \cos^2 y = \left(\dfrac{1}{\sqrt{1+x^2}}\right)^2 = \dfrac{1}{1+x^2}$.

NOTE: Memorize the following integrals :

$$\int \frac{1}{\sqrt{1-x^2}}dx = \sin^{-1} x + C$$

$$\int \frac{1}{1+x^2}dx = \tan^{-1} x + C$$

$$\int \frac{1}{x\sqrt{x^2-1}}dx = \sec^{-1} x + C$$

EXAMPLE: Find $\displaystyle\int \frac{1}{2+x^2}dx$.

Answer: Rewrite this as $\displaystyle\int \frac{1}{2+x^2}dx = \int \frac{1}{2\left(1+\dfrac{x^2}{2}\right)}dx = \frac{1}{2}\int \frac{1}{\left(1+(x/\sqrt{2})^2\right)}dx$. Let $u = x/\sqrt{2}$.

Then $du = dx/\sqrt{2}$ and the integral becomes $\dfrac{1}{\sqrt{2}}\displaystyle\int \frac{1}{\left(1+u^2\right)}du = \frac{1}{\sqrt{2}}\tan^{-1} u + C =$

$\dfrac{1}{\sqrt{2}}\tan^{-1}\dfrac{x}{\sqrt{2}} + C$

EXAMPLE: Find $\displaystyle\int \frac{1}{\sqrt{9-16x^2}}dx$.

Answer: Rewrite this as $\displaystyle\int \frac{1}{\sqrt{9(1-16x^2/9)}}dx = \frac{1}{3}\int \frac{1}{\sqrt{1-(4x/3)^2}}dx$. Now let u = 4x/3,

so du = 4dx/3. Then the integral becomes

$\dfrac{1}{4}\displaystyle\int \frac{1}{\sqrt{1-u^2}}du = \frac{1}{4}\sin^{-1} u + C = \frac{1}{4}\sin^{-1}\left(\frac{4x}{3}\right) + C.$

EXAMPLE: Find $\displaystyle\int_{-2}^{-2/\sqrt{3}} \frac{1}{x\sqrt{x^2-1}}dx$.

Answer: $\displaystyle\int_{-2}^{-2/\sqrt{3}} \frac{dx}{x\sqrt{x^2-1}} =$

$= \sec^{-1} x \Big|_{-2}^{-2/\sqrt{3}} = \sec^{-1}(-2/\sqrt{3}) - \sec^{-1}(-2) = \dfrac{7\pi}{6} - \dfrac{4\pi}{3} = \dfrac{\pi}{6}.$

(This example illuminates why we chose those quadrants for $\sec^{-1} x$.)

KNOW YOUR GRAPHS!

It is important to know some of the basic graphs. Learn the graphs of the exponential and logarithmic functions where $0 < a < 1$ and $a > 1$, for instance. The graphs of the logarithm functions can be obtained by flipping the graphs of the exponential functions across what line? Know growth functions, and decay functions. What effect does the base "a" have on the shape of the graph? Can you graph sinh x, cosh x, and tanh x? How would you go about graphing sech x, if you don't remember that graph? Can you graph the inverse trigonometric functions?

SECTION 6.6

"We are here to add, not get what we can, from life"...William Ostler

L'HOPITAL'S RULE

(See the formal wording for this rule in your text. It is too important to skip!)

Given a limit of the form "$\lim \dfrac{A}{B}$" where A and B are expressed in terms of x, and where

$A \to \pm\infty$ and $B \to \pm\infty$, then l'Hopital's Rule tells us that "$\lim \dfrac{A}{B}$"="$\lim \dfrac{dA/dx}{dB/dx}$", when this

limit exists.   This rule is so pleasant, it is easy to try to overuse it--be careful not to use l'Hopital's Rule where it does not apply!

EXAMPLE: Find $\lim\limits_{x\to\infty} \dfrac{\ln x}{x}$.

Answer: Note that $\ln x \to \infty$ and $x \to \infty$, so the limit is one for which l'Hopital's Rule

applies. Thus taking derivative of numerator and denominator, $\lim\limits_{x\to\infty} \dfrac{\ln x}{x} = \lim\limits_{x\to\infty} \dfrac{1/x}{1} = 0$.

EXAMPLE: Find $\lim\limits_{x\to 0} \dfrac{\cos x}{1-\sin x}$.

Answer: This limit is not of the form for which the rule applies. Had we used l'Hopital's

Rule, we would have $\lim\limits_{x\to 0} \dfrac{\cos x}{1-\sin x} = \lim\limits_{x\to 0} \dfrac{-\sin x}{-\cos x} = 0$, which is wrong. The correct value of

the limit is 1.

EXAMPLE: L'Hopital's Rule can be used if the numerator and denominator

approach either ∞ or $-\infty$. Find $\lim\limits_{x\to -\infty} \dfrac{e^x}{x}$.

Answer: The numerator approaches ∞, the denominator, $-\infty$.   We use:

$\lim\limits_{x\to -\infty} \dfrac{e^x}{x} = \lim\limits_{x\to -\infty} \dfrac{e^x}{1} = 0$.

EXAMPLE: **Sometimes l'Hopital's Rule must be used more than once.** Find $\lim\limits_{x\to\infty}\dfrac{x^2}{e^x}$.

Answer: $\lim\limits_{x\to\infty}\dfrac{x^2}{e^x}=\lim\limits_{x\to\infty}\dfrac{2x}{e^x}=\lim\limits_{x\to\infty}\dfrac{2}{e^x}=0$, where we differentiated above and below twice.

EXAMPLE: **Often l'Hopital's Rule is used with indeterminate "$0\cdot\infty$" forms: i.e., to find "limAB" where $A\to 0$ and $B\to\pm\infty$.** The trick is to express these limits as a quotient. Find $\lim\limits_{x\to 0^+} x\ln(\sin x)$.

Answer: $\lim\limits_{x\to 0^+} x\ln(\sin x)= \lim\limits_{x\to 0^+}\dfrac{\ln(\sin x)}{1/x}$. Note that $\sin x > 0$ when $x\to 0^+$. Now this limit is of the indeterminate form "$\dfrac{-\infty}{\infty}$" and L'Hopital's Rule gives:

$$\lim_{x\to 0^+}\frac{\ln(\sin x)}{1/x}=\lim_{x\to 0^+}\frac{\cos x/\sin x}{-1/x^2}=-\lim_{x\to 0^+}\left(\frac{x^2}{\sin x}\right)\left(\frac{\cos x}{1}\right)=-\lim_{x\to 0^+}x\left(\frac{x}{\sin x}\right)\left(\frac{\cos x}{1}\right)=0.$$

EXAMPLE: **Often l'Hopital's Rule is used with functions having a variable in the exponent.** Find $\lim\limits_{x\to\infty}(1+x)^{1/x}$.

Answer: Since $a=e^{\ln a}$, we have $(1+x)^{1/x}=e^{\ln(1+x)^{1/x}}$. Now by (*) below,

$$\lim_{x\to\infty}(1+x)^{1/x}=\lim_{x\to\infty}e^{\ln(1+x)^{1/x}}=\lim_{x\to\infty}e^{(1/z)\ln(1+x)}=\lim_{x\to\infty}\exp\left(\frac{\ln(1+x)}{x}\right)=$$

$$\lim_{x\to\infty}\exp\left(\frac{-1/(1+x)}{1/x}\right)=\exp(-1)=e^{-1}.$$

(*) We are able to do this by invoking the fact that the exponential function is continuous, and if f is a continuous function of x, then $\lim\limits_{x\to a} f(x)=f(\lim\limits_{x\to a} x)$, or in this case,

$$\lim_{x\to a}e^{g(x)}=e^{\lim\limits_{x\to a}g(x)}.$$

NOTE: A definition of e. L'Hopital's Rule provides us with the definition:

$e=\lim\limits_{x\to\infty}\left(1+\dfrac{1}{x}\right)^x$. In fact, it is useful to know that $e^a=\lim\limits_{x\to\infty}\left(1+\dfrac{a}{x}\right)^x$.

EXAMPLE: Find $\lim\limits_{x\to 3^+}\dfrac{\ln(x-3)}{x-3}$.

Answer: Let $u=x-3$. As $x\to 3^+$, $u\to 0^+$. Now $\lim\limits_{x\to 3^+}\dfrac{\ln(x-3)}{x-3}=\lim\limits_{u\to 0^+}\dfrac{\ln u}{u}$. This does not fit the form for l'Hopital's Rule: the numerator approaches $-\infty$ and the denominator, 0. This limit is $-\infty$. Using l'Hopital's Rule would again give an erroneous limit.

NOTE: Occasionally, l'Hopital's Rule seems to make things no better, or possibly worse. Don't forget other methods for finding limits, such as the Squeezing Theorem!

SECTION 6.7 and 6.8

"Cauliflower is nothing but cabbage with a college education"....Mark Twain

DIFFERENTIAL EQUATIONS

EXAMPLE: Verify that $y = t^2 - 2t$ is the particular solution that satisfies the initial conditions for $\left(\dfrac{dy}{dt}\right)^2 = 4(y+1); \quad y(0) = y(2) = 0.$

Answer: Differentiating $y = t^2 - 2t$ gives dy/dt = $2t - 2$. To verify that $\left(\dfrac{dy}{dt}\right)^2 = 4(y+1)$ we show $(2t-2)^2 = 4((t^2 - 2t)+1)$. Also, we show that y(0) $= 0^2 - 2(0) = 0$ and $y(2) = 2^2 - 2(2) = 0$ to show that the initial conditions are met by this y.

EXAMPLE: Find the general solution for the differential equation: $(1 + y^2)dx = e^x dy$.
Answer: This equation is **separable**, which means we can separate the "x" and "y" terms by means of an equality: $\dfrac{dy}{1+y^2} = \dfrac{dx}{e^x}$. Integrating, $\displaystyle\int \dfrac{1}{1+y^2}dy = \int \dfrac{1}{e^x}dx$, or $\tan^{-1} y = -e^{-x} + C$ (*).

(*) Note: $\tan^{-1} y$, which is an angle, can be defined for various branches which occur at intervals of $n\pi$. The integral $\displaystyle\int \dfrac{1}{1+x^2}dx$ can give results for any of these branches, . We should amend our answer in (*) to $\tan^{-1} y + n\pi = -e^{-x} + C$, , where C' is a new constant. Then, taking the tangent of both sides, $y = \tan(-e^{-x} + C')$.

EXAMPLE: Find a particular solution to the given differential equation with initial condition: $\sqrt{x^3 + 1}\dfrac{dy}{dx} = \dfrac{x^2}{y}; \quad y(2) = 4.$

Answer: Rewrite the equation with y-terms on one side, x-terms on the other:
$y\,dy = \dfrac{x^2}{\sqrt{x^3 + 1}}dx$. Then integrate. $\displaystyle\int y\,dy = \int \dfrac{x^2}{\sqrt{x^3 + 1}}dx = \dfrac{2}{3}(x^3 + 1)^{1/2} + C$, so

$$\frac{y^2}{2} = \frac{2}{3}(x^3+1)^{1/2} + C, \text{ or}$$

$y^2 = \frac{4}{3}(x^3+1)^{1/2} + C, \text{so } y = \pm\sqrt{\frac{4}{3}(x^3+1)^{1/2} + C}$. Now y(2) is positive, so we must have

$y = +\sqrt{\frac{4}{3}(x^3+1)^{1/2} + C}$. Plugging in y(2) = 4,

$4 = +\sqrt{\frac{4}{3}(2^3+1)^{1/2} + C}$, and $4 = \sqrt{4+C}$. Thus C = 12, and $y = \sqrt{\frac{4}{3}(x^3+1)^{1/2} + 12}$.

EXAMPLE: Find the general solution for the first order differential equation

$\frac{dy}{dx} - (\cosh x)y = 0.$

Answer: An antiderivative of $-\cosh x$ is $-\sinh x$. Now multiply both sides of the differential equation by $e^{-\sinh x}$. So

$e^{-\sinh x}\frac{dy}{dx} - e^{-\sinh x}(\cosh x)y = 0 \Leftrightarrow$ (read this as "*if and only if*")

$\frac{d}{dx}\left(e^{-\sinh x}y\right) = 0 \Leftrightarrow$

$e^{-\sinh x}y = C \Leftrightarrow$

$y = \frac{C}{e^{-\sinh x}} = Ce^{\sinh x}.$

Problem... $\lim\limits_{x \to e} e^{5\ln\sqrt[3]{\cosh^{-1}(\ln x)}} = ?$

(Answer: 0)

A FALSE REVIEW

The following statements are all false. Write the corrections that make the statements true.

1. The logarithm function can be defined with any real number, a, as the base.
2. The equation $\log_a x = y$ is equivalent to the equation $a^x = y$.
3. $\frac{\ln A}{\ln B} = \ln A - \ln B$.
4. L'Hopital's Rule can be used to find any limit.
5. The derivative of $\cosh x$ is $-\sinh x$.
6. The derivative of the inverse function at (x,y) is the reciprocal of the derivative of the function at (x,y).
7. $\int_{-1}^{2}\frac{1}{x}dx = \ln 2 - \ln 1 = \ln 2.$

8. The domain of the $\sin^{-1}$ function is $[0,\pi]$.
9. The range of the $\sin^{-1}$ function is $[0,\pi]$.
10. The range of the natural logarithm function is $(0,\infty)$.
11. $\ln 5x^2 = 2 \ln 5x$.
12. $\lim\limits_{x\to 0} x^x$ yields the indeterminate form 0^0, and thus is not defined .
13. The ranges of the inverse cosine and inverse secant function are the same.
14. $\cosh 2x = \cosh^2 x - \sinh^2 x$.
15. The derivative of the inverse sine function is the inverse cosine function.
16. Any function has an inverse function.
17. The functions $y = \ln x$ and $x = e^y$ are inverses.

Formulas! Graphs!

1. Define cosh x and sinh x. What law relates these two functions?
2. What is the derivative of $y = a^x$? Of $y = \log_a x$?
3. What are the inverse properties (there are two) relating the exponential function and the natural logarithm function?
4. Find the derivatives of all the inverse trigonometric formulas.
5. Find the derivatives of all the hyperbolic functions.
6. Find the ranges of all the inverse trigonometric functions.
7. Provide graphs of all the functions mentioned above.
8. Find the following integrals: $\int \dfrac{1}{\sqrt{1-x^2}}dx, \int \dfrac{1}{1+x^2}dx, \int \dfrac{1}{x\sqrt{x^2-1}}dx, \int \dfrac{1}{\sqrt{1+x^2}}dx$.

More Review Could you....

Explain why we have introduced the functions in this section? What use do they serve?
Show how the derivatives of the inverse trigonometric functions are found?
Integrate a variety of integrals combining these integrals with u-substitutions?

$$\int \frac{1}{\sqrt{1-x^2}}dx, \int \frac{1}{1+x^2}dx, \int \frac{1}{x\sqrt{x^2-1}}dx, \int \frac{1}{\sqrt{1+x^2}}dx$$

Solve for x in the equation sinh x = 5? Where does the natural logarithm come into this equation?
Explain why the term "hyperbolic" is used in the definition of cosh x and sinh x?
Prove $e^x = \sinh x + \cosh x$?
Find $\tan(\sin^{-1} x)$, $\sin(\cos^{-1} x)$, etc., using triangles?
Explain what a "separable" differential equation is? Explain what a particular solution and a general solution are? Solve a differential equation by multiplying by an appropriate term?

CHAPTER 7

"Miracle me no miracles"...Cervantes

While it is tempting to rely on tables, calculators, and the computer, we learn integration techniques so that solutions to integrals are at our fingertips. It helps to categorize "types" of integrals and how best to solve each type, rather than to view each integral as an isolated problem. Keep looking backwards and summarizing .

It is useful to review any areas of integration that gave you difficulty in the past, as well as the inverse functions introduced in Chapter 6.

SECTION 7.1

" I know a trick worth two of that"...Shakespeare

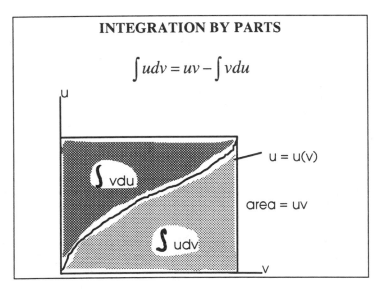

THIS TECHNIQUE IS USEFUL WITH INTEGRALS

1. Having obvious products : $\int e^x x^2 dx$, $\int (\cos x) \, x^2 dx$, $\int (\ln x) \, x^2 dx$, etc.

2. That might use reduction formulas: $\int \sin^5 x dx$.

3. That might come as a surprise: $\int \ln x \, dx$, $\int \tan^{-1} x \, dx$, etc. Memorize these special cases.

NOTE: We usually choose for dv, a factor in the integrand that can be integrated, and for u, a factor that becomes simpler (or no less complicated) when differentiated .

EXAMPLE: Find $\int x \ln x \, dx$.

Answer: Let

u = ln x dx (easy to differentiate) and   **dv = x dx** (easy to integrate). Then

du = 1/x and   **v = x² / 2** . Then, following the formula:

$$\int x \ln x \, dx = uv - \int v \, du = (\ln x)\left(\frac{x^2}{2}\right) - \int \frac{x^2}{2}\left(-\frac{1}{x}\right)dx = (\ln x)\left(\frac{x^2}{2}\right) + \frac{1}{2}\int x \, dx =$$

$(\ln x)\left(\dfrac{x^2}{2}\right) + \dfrac{x^2}{4} + C,$ which can be simplified.

EXAMPLE: The integral may need several applications of this technique : keep using similar substitutions or you will undo your work. (Reduction formulas follow this rule.)
Find $\int x^2 e^x \, dx$.

Answer: Let

$$u = x^2 \qquad dv = e^x dx$$
$$du = 2x \, dx \qquad v = e^x$$

$\int x^2 e^x dx = uv - \int v \, du = x^2 e^x - 2\int e^x x \, dx.$ Keep the substitution in $\int e^x x \, dx$ "similar" (or

we will undo our work). We used $dv = e^x dx$ above, so use this substitution again

$$u = x \qquad dv = e^x dx$$
$$du = dx \qquad v = e^x$$

So $\int x^2 e^x dx = x^2 e^x - 2(xe^x - \int e^x dx) = x^2 e^x - 2xe^x + 2e^x + C$

EXAMPLE: Integral may seem to return to original integral and have a "cyclical" look but can be solved with an equation. (Beware of getting into loops that lead nowhere!)

With our selection of u and dv, the integral we are reduced to solving is at the **same level of complexity** as the original. This is generally a good sign, but can be a problem.

Find $\int e^x \sin x \, dx$.

Answer: Let

$$u = e^x \qquad dv = \sin x \, dx$$
$$du = e^x dx \qquad v = \cos x$$

$$\int e^x \sin x \, dx = e^x(-\cos x) + \int e^x \cos x \, dx.$$

$$u = e^x \qquad dv = \cos x \, dx$$
$$du = e^x dx \qquad v = \sin x$$

$(*) \int e^x \sin x \, dx = e^x(-\cos x) + e^x \sin x - \int e^x \sin x \, dx$. At this point, the new integral we must solve looks on the same level of complexity as the original integral It may seem that we are in a "cycle". However, $(*)$ is like an expression, $X = A - X$. We solve for X to find $2X = A$, and $X = A$. From $(*)$,

$$\int e^x \sin x \, dx = \left(e^x(-\cos x) + e^x \sin x\right) - \int e^x \sin x \, dx$$

$$2\int e^x \sin x \, dx = e^x(-\cos x) + e^x \sin x \, .$$

$$\int e^x \sin x \, dx = \frac{1}{2}\left(e^x(-\cos x) + e^x \sin x\right) + C \, .$$

When we eliminated the integral sign, we added the "+C" term that all indefinite integrals have.

EXAMPLE: Some special integrals, such $\int \ln x \, dx, \int \tan^{-1} x \, dx$, etc. require this method.

Find $\int \ln x \, dx$ using the method of Integration by Parts.

Answer: Let

$$u = \ln x \qquad d v = dx$$
$$du = 1/x \, dx \qquad v = x$$

$$\int \ln x \, dx = x \ln x - \int \frac{x}{x} dx = x \ln x - x + C$$

NOTE: You will see Reduction Formulas derived in this section. For example, reduce expressions like $\int \sin^7 x \, dx$ to expressions involving $\int \sin^5 x \, dx$, which can in turn be reduced to integrals that are easier to evaluate.

SECTION 7.2

"It is easy to despise what is most useful to us."....Aesop

TRIGONOMETRIC INTEGRALS

Integrals with Sine and Cosine

ONE OR MORE ODD POWERS OF SINE OR COSINE

EXAMPLE: **If there is a single cosine or sine term, use it for the "du" term**. Find $\int \sin^5 x \cos x \, dx$.

Hint: Let $u = \sin x$, $du = \cos x \, dx$, etc.

EXAMPLE: **If one power of sine or cosine is odd**, factor out a single power of that term and use it for the "du" term. Find $\int \sin^3 x \cos^5 x \, dx$.

Hint: Factor out a "sin x" term, and rewrite as
$$\int \sin x \sin^2 x \cos^5 x \, dx = \int \sin x (1 - \cos^2 x)(\cos^5 x) dx = \int \sin x \, (\cos^5 x - \cos^7 x) dx$$
Let $u = \cos x$, $du = \sin x \, dx$, etc.

NOTE: Reduction Formulas for integrals such as $\int \sin^5 x \, dx$, $\int \cos^4 x \, dx$ exist, but there may be other techniques for integrating these integrals.

EXAMPLE: Find $\int \sin^5 x \, dx$. (We won't use a Reduction Formula here.)

Hint : $\int \sin^5 x \, dx = \int \sin x \, (\sin^2 x)^2 dx = \int \sin x (1 - \cos^2 x)^2 dx$, which reduces to a sum of integrals of the form: $\int \sin x$ (powers of cos x) dx. Let $u = \cos x$, as above.

BOTH POWERS OF SINE AND COSINE ARE EVEN
These are more trouble. You may need to use the following formulas.

$$\cos^2 x = \frac{1 + \cos 2x}{2}$$
$$\sin^2 x = \frac{1 - \cos 2x}{2}$$
$$\sin 2x = 2 \sin x \cos x$$

EXAMPLE: Find $\int \cos^4 x\,dx$.

Hint : $\int \cos^4 x\,dx = \int (\cos^2 x)^2\,dx = \int \left(\dfrac{1+\cos 2x}{2}\right)^2 dx = \int \dfrac{1}{4}+\cos 2x+\cos^2 2x\ dx$, where

$\int \cos^2 2x\,dx = \int \left(\dfrac{1+\cos 4x}{2}\right)dx$, etc.

EXAMPLE: Find $\int \sin^2 x \cos^2 x\,dx$.

Hint: Rewrite as $\int (\sin x \cos x)^2\,dx = \int \left(\dfrac{\sin 2x}{2}\right)^2 dx = \int \dfrac{\sin^2 2x}{4}\,dx =$

$\dfrac{1}{4}\int \dfrac{1-\cos 4x}{2}\ dx$, which is easier to integrate.

EXAMPLE: Find $\int \sin^4 x \cos^2 x\,dx$.

Hint: Rewrite as $\int (\sin^4 x)(1-\sin^2 x)dx =$ an integral exclusively in powers
of sin x. Then Reduction Formulas developed in a previous section can be applied.

ALTERNATIVELY:

Rewrite $\int \sin^4 x \cos^2 x\,dx$ as $\int (\sin^2 x)^2 \cos^2 x\,dx = \int \left(\dfrac{1-\cos 2x}{2}\right)^2 \left(\dfrac{1+\cos 2x}{2}\right)^2 dx.$

This is an integral involving only powers of cos x, at worst the 4th power. Use

Reduction Formulas, or the method shown in an earlier example to find $\int \cos^4 x\,dx$.

Integrals with Tangent and Secant

NOTE: (Ancient history) Recall that $\int \sec x\,dx = \ln|\sec x + \tan x| + C.$

EXAMPLE: When the integral has an even power of sec x, factor out a $\sec^2 x$ term.
Find $\int \tan x \sec^4 x\,dx$.

Hint: $\int \tan x \sec^4 x\,dx = \int \tan x\,(\sec^2 x)(\sec^2 x\,dx) = \int \tan x\,(1+\tan^2 x)\sec^2 x\ dx$
Express in powers of tan x now with $u = \tan x$, $du = \sec^2 x\,dx$.

EXAMPLE: When the integral has an odd power of tan x, factor out a
"(sec x tan x dx)" term for du. Find $\int \tan^5 x \sec^3 x\,dx$.

Hint: $\int \tan^5 x \sec^3 x dx = \int \tan^5 x \sec x (\sec^2 x)dx = \int \tan^4 x \sec^2 x(\sec x \tan x dx) =$

$\int (\tan^2 x)^2 \sec^2 x(\sec x \tan x \, dx) = \int (\sec^2 x - 1)^2 \sec^2 x \ (\sec x \tan x) \, dx$. Let u = sec x,
du = sec x tan x dx, etc.

EXAMPLE: When the integral has an odd power of sec x, and an even power of tan x. Reduce the integrand to powers of sec x only. Find $\int \tan^2 x \sec^3 x dx$.

Hint: $\int \tan^2 x \sec^3 x dx = \int (\sec^2 x - 1)(\sec^3 x dx)$. Keep reducing. $\int \sec^5 x dx$ is solved using integration by parts.

NOTE: The integral $\int \sec^3 dx$ comes up so often it is worth studying. It is solved using integration by parts. Let u = sec x, dv = $\sec^2 x dx$.

✋ **CAUTION!** An integral that mixes different angles, such as $\int \sin x \cos 2x dx$ requires special attention. Use

sin ax cos bx = (1/2) sin (a-b) x + (1/2) sin (a+b) x.

SECTION 7.3

"Truth lies within a little and certain compass, but error is immense".. Viscount Bolingbroke

TRIGONOMETRIC SUBSTITUTIONS
Hint: Think of the "a" term as a "1" to guide your substitution.

Integrals with $\sqrt{a^2 - x^2}$. Think : **1 - $\sin^2 x = \cos^2 x$.**
Integrals with $\sqrt{x^2 - a^2}$. Think : **$\sec^2 x - 1 = \tan^2 x$.**
Integrals with $\sqrt{a^2 + x^2}$. Think : **1+ $\tan^2 x = \sec^2 x$.**

EXAMPLE: Find $\int \dfrac{1}{\sqrt{25 - x^2}} dx$.

Answer: (We are reminded of $1 - \sin^2 x$ which prompts us to think x will be a "$\sin \theta$" term.)

Write the integral as $\int \dfrac{1}{\sqrt{5^2 - x^2}} dx$. Let $x = 5 \sin \theta$. Then $dx = 5 \cos \theta \, d\theta$. Then

$$\int \frac{1}{\sqrt{5^2 - x^2}} dx = \int \frac{5 \cos \theta}{\sqrt{5^2(1 - \sin^2 \theta)}} d\theta = \int \frac{5 \cos \theta}{5 \cos \theta \, d\theta} d\theta = \theta + C. \text{ Since } x = 5 \sin \theta,$$

$\sin \theta = \dfrac{x}{5}$ and $\theta = \sin^{-1} \dfrac{x}{5}$. Thus $\int \dfrac{1}{\sqrt{5^2 - x^2}} dx = \sin^{-1} \dfrac{x}{5} + C$.

EXAMPLE: This integral involves completing the square. Find $\int \sqrt{x^2 + x} \, dx$.

Partial Answer : $\int \sqrt{x^2 + x} \, dx = \int \sqrt{x^2 + x + (1/4) - (1/4)} \, dx$ (completing the square) =

$\int \sqrt{(x + 1/2)^2 - (1/4)} \, dx = \int \sqrt{u^2 - (1/4)} \, du$ using $u = x + (1/2)$.

The quantity under the radical is of the form that reminds us of $\sec^2 x - 1 = \tan^2 x$.

So let $u = (1/2) \sec \theta$.

Now $\int \sqrt{u^2 - (1/4)} \, du = \int \sqrt{\left(\dfrac{\sec \theta}{2}\right)^2 - \left(\dfrac{1}{2}\right)^2} \cdot (\dfrac{1}{2} \sec \theta \tan \theta \, d\theta) =$

$\int \dfrac{1}{2} \sqrt{\sec^2 \theta - 1} \cdot (\dfrac{1}{2} \sec \theta \tan \theta \, d\theta) = \int \dfrac{1}{2} \cdot \sqrt{\tan^2 \theta} \, (\dfrac{1}{2} \sec \theta \tan \theta \, d\theta) = \dfrac{1}{4} \int \tan^2 \theta \sec \theta \, d\theta$.

This integral reduces to an integral involving $\int \sec^3 \theta \, d\theta$.

When we back-substitute, we need to express θ in terms of u, then u in terms of x!

EXAMPLE: Here are a variety of substitutions that work. (Try to see the rationale for each substitution).

a.) $\int \dfrac{x^2}{\sqrt{16x^2 - 1}} dx$ The term $(4x)^2 - 1$ suggests we let $4x = \sec \theta$.

b.) $\int \dfrac{1}{x\sqrt{x^2 + 9}} dx$ The term $x^2 + 9$ suggests we let $x = 3 \tan \theta$.

c.) $\int \dfrac{x^2}{x\sqrt{x^2 - 9}} dx$ The term $\sqrt{x^2 - 9}$ suggests the substitution $x = 3 \sec \theta$

d.) $\int \dfrac{\sqrt{x^2 - 9}}{x} dx$ The term $\sqrt{x^2 - 9}$, suggests we let $x = 3 \sec \theta$ . .

e.) $\int \dfrac{1}{x^2\sqrt{16-9x^2}}\,dx \;=\; \int \dfrac{1}{x^2\sqrt{16\,(1-\dfrac{9}{16}x^2)}}dx.$

    The denominator suggests $\dfrac{3}{4}x = \sin\theta.$

f.) $\int \dfrac{e^{3x}}{\sqrt{1-3e^{2x}}}dx$. The denominator suggests we let $\sqrt{3}e^x = \sin\theta.$

EXAMPLE: Find the area of a circle of radius r.

Answer: The equation of a circle is $x^2 + y^2 = r^2$.

$\displaystyle\int_{0}^{r}\sqrt{r^2-x^2}\,dx$ gives 1/4 the area of the circle. Let $x = r\sin\theta$. Then $dx = r\cos\theta\,d\theta$. So

$$\int_{0}^{r}\sqrt{r^2-x^2}\,dx = \int_{0}^{\pi/2}\sqrt{r^2-(r\sin\theta)^2}\,r\cos\theta\,d\theta = \int_{0}^{\pi/2} r^2\cos^2\theta\,d\theta = \int_{0}^{\pi/2} r^2\frac{1+\cos 2\theta}{2}\,d\theta$$

$\displaystyle =\frac{r^2}{2}(\theta+\frac{\sin 2\theta}{2})\Big|_{0}^{\pi/2} = (r^2/2)(\pi/2) = \pi r^2/4.$ Multiplying this result by 4, the area of

a circle of radius r is πr^2.

EXAMPLE: Using triangles to solve equations. Find, in the example above, the

indefinite integral $\int\sqrt{r^2-x^2}\,dx$.

Answer: We know: $\int\sqrt{r^2-x^2}\,dx =\dfrac{r^2}{2}(\theta+\dfrac{\sin 2\theta}{2})+C$, but how do we return our answer

in terms of x? Since $x = r\sin\theta, \quad \theta = \sin^{-1}\dfrac{x}{r}.$ We can use that

$\sin 2\theta = 2\sin\theta\cos\theta$, and we have that $\sin\theta = \dfrac{x}{r}$, but what about $\cos\theta$? Using the

triangle technique introduced in Chapter 6, $\cos\theta = \dfrac{\sqrt{r^2-x^2}}{r}.$

Piecing together our answer,

$$\int\sqrt{r^2-x^2}\,dx =\frac{r^2}{2}(\theta+\frac{\sin 2\theta}{2})+C = \frac{r^2}{2}(\sin^{-1}\left(\frac{x}{r}\right)+\left(\frac{x}{r}\right)\cdot\frac{\sqrt{r^2-x^2}}{r})+ C.\;\text{Now}$$

simplify. (This result was known to Archimedes and can found geometrically.)

SECTION 7.4

"The more the marble wastes, the more the statue grows"....Michelangelo

PARTIAL FRACTIONS

NOTE: A "linear" factor of a factor of the form $ax + b$. (The word "line" appears in the word "linear".) We will say a **"quadratic" factor** is a factor of the form $ax^2 + bx + c$.

Some general rules for writing a rational function in terms of partial fractions

Below, (Linear l)(Linear 2) denotes two different linear factors, such as (x+3)(x-2).

$$(1)\frac{1}{(\text{Linear 1})(\text{Linear 2})} = \frac{A}{\text{Linear 1}} + \frac{B}{\text{Linear 2}} \quad (\text{distinct linear factors})$$

$$(2)\frac{1}{(\text{Quadratic})(\text{Linear})} = \frac{Ax + B}{\text{Quadratic}} + \frac{C}{\text{Linear}} \quad (\text{quadratic factor})$$

$$(3)\frac{1}{(\text{Linear 1})^2(\text{Linear 2})} = \frac{A}{\text{Linear 1}} + \frac{B}{\text{Linear 1}} + \frac{C}{\text{Linear 2}} \quad (\text{repeated linear factors})$$

$$(4)\frac{1}{(\text{Quadratic})(\text{Linear})^2} = \frac{Ax + B}{\text{Quadratic}} + \frac{C}{\text{Linear}} + \frac{D}{\text{Linear}}$$

EXAMPLE: Find $\displaystyle\int \frac{5dx}{(x-1)(x+2)}$.

Answer: We follow Rule (1) above.

Write $\dfrac{5}{(x-1)(x+2)} = \dfrac{A}{x-1} + \dfrac{B}{x+2}$ so that $\dfrac{5}{(x-1)(x+2)} = \dfrac{A(x+2) + B(x-1)}{(x-1)(x+2)}$

and $A(x+2) + B(x-1) = 5$. We find A and B by plugging in values of x.

When $x = -2$, we have $A(0) + B(-2-1) = 5$, so $B = -5/3$

When $x = 1$, we have $A(1+2) + B(0) = 5$, so $A = 5/3$.

Thus $\displaystyle\int \frac{5}{(x-1)(x+2)}dx = \int \frac{A}{x-1}dx + \int \frac{B}{x+2}dx$

And this gives:

$$\int \frac{5}{(x-1)(x+2)} dx = \int \frac{5/3}{x-1} dx + \int \frac{-5/3}{x+2} dx = \frac{5}{3}\left(\ln|x-1| - \ln|x+2|\right) + C$$

which by the properties of the logarithm function can be written as $\ln\left|\dfrac{x-1}{x+2}\right|^{\frac{5}{3}} + C$.

When dealing with distinct linear factors, here's a short-cut.

To find A and B in the expression $\dfrac{5}{(x-1)(x+2)} = \dfrac{A}{x-1} + \dfrac{B}{x+2}$

$\dfrac{5}{\cancel{(x-1)}(x+2)}$	Mentally blot out the (x-1) factor and plug in x = 1 to find A = 5/3
$\dfrac{5}{(x-1)\cancel{(x+2)}}$	Mentally blot out the (x+2) factor and plug in x =-2 to find B = -5/3

EXAMPLE: Given a rational function to integrate, **if the degree of the numerator equals or exceeds that of the denominator, divide first.** Find $\displaystyle\int \dfrac{x^2+x+1}{x^2-1}\, dx$.

Answer: (See about the division of polynomials in Chapter 1 of this Guide.)

$$
\begin{array}{r}
1 \quad remainder\ x+2 \\
x^2-1\overline{)x^2+x+1} \\
\underline{x^2\quad -1} \\
-x+2
\end{array}
$$

So $\displaystyle\int \dfrac{x^2+x+1}{x^2-1}dx = \int 1+\dfrac{x+2}{x^2-1}dx = x+\int\dfrac{x+2}{x^2-1}dx$. This integral is left as an example below.

EXAMPLE: Simplifying an answer with the natural logarithm function. Find $\displaystyle\int\dfrac{x+2}{x^2-1}dx$.

Answer: $\displaystyle\int\dfrac{x+2}{x^2-1}dx = \int\dfrac{x+2}{(x+1)(x-1)}dx$. We follow our Rule (1) for partial fractions

decomposition. Write $\dfrac{x+2}{(x+1)(x-1)} = \dfrac{A}{x+1}+\dfrac{B}{x-1}$. Now

$A(x-1)+B(x+1) = x+2$. When $x=1$, $2B = 3$, or $B = 3/2$.
When $x=-1$, $-2A = 1$, or $A = -1/2$.

Now we have $\displaystyle\int\dfrac{-1/2}{x+1}dx+\int\dfrac{3/2}{x-1}dx = (-1/2)\ \ln|x+1| + (3/2)\ln|x-1|+C =$

$\dfrac{1}{2}\ (\ln|x+1|^{-1}+\ln|x-1|^3)+C = \dfrac{1}{2}\ln\left|\dfrac{(x-1)^3}{(x+1)}\right| + C = \ln\sqrt{\dfrac{(x-1)^3}{(x+1)}}+C$

EXAMPLE: Find $\int \dfrac{1}{(x^2+1)(x+2)}dx$.

Answer:
Here we follow our Rule (2) for partial fraction decomposition.

$$\dfrac{1}{(x^2+1)(x+2)} = \dfrac{Ax+B}{x^2+1} + \dfrac{C}{x+2} = \dfrac{(Ax+B)(x+2)+C(x^2+1)}{(x^2+1)(x+2)}, \text{ so that}$$

$(Ax+B)(x+2) + C(x^2+1) = 1.$

With $x = -2$, $C = 1/5$.

With $x = 0$ (to make the arithmetic easy) $2B + C = 1$, and $B = 2/5$.

With $x = 1$ $(A+B)(3) + C(2) = 1$, and $(A+2/5)(3)+2/5 = 1$,

so $A = -3/5$. Thus

$\text{Integral} = \int \dfrac{(-3/5)x+(2/5)}{x^2+1} + \dfrac{(1/5)}{x+2}$ which can be

solved as three separate integrals; $\dfrac{-3}{5}\int \dfrac{x}{x^2+1}dx + \dfrac{2}{5}\int \dfrac{1}{x^2+1}dx + \dfrac{1}{5}\int \dfrac{1}{x+2}dx.$

Solve the first integral with $u = x^2+1$ and the second with a $u = \tan\theta$.

The third integral is $\dfrac{1}{5}$ $\ln|x+2|+C$.

EXAMPLE:

Find the partial fractions decomposition for the integral $\int \dfrac{1}{(x^2+1)(x-1)^3}dx.$

Answer: Use $\dfrac{1}{(x^2+1)(x-1)^3} = \dfrac{Ax+B}{x^2+1} + \dfrac{C}{x-1} + \dfrac{D}{x-1} + \dfrac{E}{x-1}$

REVIEW QUESTIONS TO ASK YOURSELF

Can you list the various integration techniques now available to you, without glancing back? Are you comfortable with completing the square? Remember to factor out the leading coefficient! Can you divide polynomials? Will you remember to divide when the polynomial in the denominator has degree equal or less than that in the numerator? Do you know all the partial fractions forms? Can you solve for A, B, C, etc.? Do you know the integration by parts formula well? (Don't get confused with the minus signs!) Do you know the various approaches you can use with trigonometric integrals? Do you know these special integrals:
$\int \sec x\, dx, \quad \int \sec^3 x dx, \quad \int \sin^2 x dx?$

CREATE AN EXAMPLE

TECHNIQUE FOR INTEGRATION	GIVE FORMULA OR METHOD	GIVE THREE EXAMPLES OF AN INTEGRAL THAT IS SOLVED WITH THIS TECHNIQUE	EXPLAIN HOW YOU WOULD GO ABOUT SOLVING AN INTEGRAL WITH THIS TECHNIQUE
integration by parts			
trigonometric integrals (using sin, cos, sec, tan)			
trigonometric substitutions			
partial fractions			

SECTION 7.6

"What is the answer? [I was silent]. In that case, what is the question?"...Alice B. Toklas

APPROXIMATIONS TO INTEGRALS

EXAMPLE :   The Trapezoidal Approximation

(a) Approximate the integral $\int_0^1 e^x dx$ using the Trapezoidal Rule with 6 sub-intervals.

Answer: $\int_0^1 e^x dx \approx$

$$\frac{1}{12}(\exp(0)+2\exp(1/6)+2\exp(2/6)+2\exp(3/6)+2\exp(4/6)+2\exp(5/6)+\exp(1))$$
$$=1.7222575$$

(b) Find the actual error in this approximation.

Answer: Here the integral can be found; it's value is e -1.   We find | (e-1)-(1.722257493)|, and $E_6^T =$ 0.0039757

(c) Usually we use this rule because we *cannot* evaluate the integral. Find an upper bound for the error if the Trapezoidal Rule is used as above.

Answer: We note that $f''(x) = e^x$ has a maximum on the interval $[0,1]$ when $x = 1$, where $f(1) = e$. So $E_6^T \leq \dfrac{K_T}{12n^2}(b-a)^3 = \dfrac{e}{12(36)}(1-0)^3 = 0.0062923$.

EXAMPLE: Simpson's Rule

(a) Find an approximation to the integral in the exercise above, using Simpson's Rule, with $n = 6$.

Answer: $\displaystyle\int_0^1 e^x dx \approx$

$\dfrac{1}{18}(\exp(0) + 4\exp(1/6) + 2\exp(2/6) + 4\exp(3/6) + 2\exp(4/6) + 4\exp(5/6) + \exp(1))$
$= 1.718289170$.

(b) Find the actual error incurred with Simpson's Rule.
Answer: This is, as above, $E_6^T = |\, e - 1 - 1.718289170\,| = 0.7342 \times 10^{-5}$.

(c) Since we normally use this rule when we cannot find the integral, there is a formula for finding the upper bound for the error.
Answer: $E_6^S \leq \dfrac{K_S}{180n^4}(b-a)^5 = \dfrac{e}{180(6)^4}(1)^5 = 0.000011652$. Since

E_6^S is much smaller than E_6^T, we see that Simpson's Rule provides a better approximation than the Trapezoidal Rule in this case.

Are you working in a Study Group?

SECTION 7.7

"A hundred times every day I remind myself that my inner and outer life depend on the labors of others, living and dead, and that I must exert myself in order to give in the same measure as I have received and am still receiving."....Einstein

IMPROPER INTEGRALS

To find a definite integral, $\int_a^b f(x)dx$, we saw in Chapter 5 that **f must be continuous on the closed interval [a,b].** In this section, we explore situations where some of these conditions break down. *Form in solving these integrals is very important.*

An integral may be *improper* because one or both limits of integration are ∞ or $-\infty$. Or an integral may be *improper* because we are integrating over a "bad" spot--a value of x where the integrand fails to be continuous, or even to be defined. We say **an integral converges** if its result, on taking the appropriate limit(s) of the integral, is a finite number. If the limit **does not exist** or is **infinite**, we say **the integral diverges.**

CAUTION: Remember that to write "$\infty -\infty$ " or "$\infty +\infty$" or even "$3+\infty$ " is incorrect, because ∞ is not a real number with which we can do arithmetic. (However, our methodology will allow us to bypass this problem by using N to represent a large number, then allowing N to approach ∞. The results with this approach are often what we anticipate with our intuitive notions of doing arithmetic with ∞ .)

EXAMPLE: Determine if the integral $\int_0^\infty \frac{1}{(x+3)^2}dx$ converges or diverges.

Answer:
The integrand is continuous on[0, b], for b any positive number (It is discontinuous at x = -3, but this x - value is not in our interval of integration.) The only problem is the "∞" which is a limit of integration We rewrite this integral as

$$\int_0^\infty \frac{1}{(x+3)^2}dx = \lim_{N\to\infty}\int_0^N \frac{1}{(x+3)^2}dx = \lim_{N\to\infty}\int_3^{N+3} \frac{1}{u^2}\,du \text{ (where we use u = x+3)} =$$

$$\lim_{N\to\infty} -\frac{1}{u}\Big|_3^{N+3} = \lim_{N\to\infty}(-\frac{1}{N+3}+\frac{1}{3}) = \frac{1}{3} \text{ . This integral converges to } 1/3.$$

EXAMPLE: Does $\int_0^1 \dfrac{1}{\sqrt{x}}\,dx$ converge or diverge?

Answer:

Rewrite this integral as $\lim\limits_{a\to 0}\int_a^1 \dfrac{1}{\sqrt{x}}\,dx = \lim\limits_{a\to 0}\, 2\sqrt{x}\,\Big|_a^1 = \lim\limits_{a\to 0}(2\sqrt{1} - 2\sqrt{a}) = 2.$ This integral

converges to 2. (We will see a short‑cut in the " p‑test " later.)

WHEN IS $\lim A \pm \lim B$ DEFINED?
$\lim A \pm \lim B$ is is a real number if and only if both limits are finite .
$\lim A + \lim B$ is ∞ if and only if one or both of the limits are ∞.
$\lim A - \lim B$ is ∞ if and only $\lim A = \infty$, and $\lim B$ is finite. BUT...
$\lim A - \lim B$ is not defined if both limits are ∞.

EXAMPLE: Find whether the integral $\int_1^\infty \dfrac{1}{\sqrt{x-1}}\,dx$ converges or diverges.

Answer:
The integrand is discontinuous at x = 1, which is one of our limits of integration. Also we
have an infinite limit of integration. We can take care of both of these problems by

choosing an arbitrary point in $(1,\ \infty)$ where $\dfrac{1}{\sqrt{x-1}}$ is defined (we use x = 2), then writing

the integral as a sum of integrals

$\int_1^\infty \dfrac{1}{\sqrt{x-1}}\,dx = \lim\limits_{a\to 1}\ \int_a^2 \dfrac{1}{\sqrt{x-1}}\,dx + \lim\limits_{N\to\infty}\int_2^N \dfrac{1}{\sqrt{x-1}}\,dx.$ We evaluate each integral separately.

Using u = x -1, with du = dx,

$\lim\limits_{a\to 1}\ \int_a^2 \dfrac{1}{\sqrt{x-1}}\,dx = = \lim\limits_{a\to 1}\ \int_{a-1}^2 \dfrac{1}{\sqrt{u}}\,du = \lim\limits_{a\to 1}\ \dfrac{\sqrt{u}}{2}\,\Big|_{a-1}^2 = \dfrac{1}{2}\lim\limits_{a\to 1}(\sqrt{2} - \sqrt{a-1}) = \dfrac{\sqrt{2}}{2}\ .$

Similarly, using the same substitution on the second integral in our sum,

$\lim\limits_{N\to\infty}\ \int_2^N \dfrac{1}{\sqrt{x-1}}\,dx = = \lim\limits_{N\to\infty}\ \int_2^{N-1} \dfrac{1}{\sqrt{u}}\,du = \lim\limits_{N\to\infty}\ \dfrac{\sqrt{u}}{2}\,\Big|_2^{N-1} = \dfrac{1}{2}\lim\limits_{N\to\infty}(\sqrt{N-1} - \sqrt{2}) = \infty\ .$

Since $\int_1^\infty \dfrac{1}{\sqrt{x-1}}\,dx = \lim\limits_{a\to 1}\ \int_a^2 \dfrac{1}{\sqrt{x-1}}\,dx + \lim\limits_{N\to\infty}\int_2^N \dfrac{1}{\sqrt{x-1}}\,dx$ and the second integral is infinite,

this integral diverges to ∞.

✋ **CAUTION!** Your teacher may or may not approve of this short-cut. Ask, before using it. In this method, we treat both limits at once, making sure to use different variables for our limits.

$$\int_1^\infty \frac{1}{\sqrt{x-1}}\,dx = \lim_{N\to\infty,\,a\to 1}\int_a^N \frac{1}{\sqrt{x-1}}\,dx = \lim_{N\to\infty,\,a\to 1}\int_{a-1}^{N-1}\frac{1}{\sqrt{u}}\,du \quad \text{(where we use } u = x-1).$$

$$\text{Integral} = \lim_{N\to\infty,\,a\to 1}\left.\frac{\sqrt{u}}{2}\right|_{a-1}^{N-1} = \frac{1}{2}\lim_{N\to\infty}(\sqrt{N-1}-\sqrt{1}) = \quad . \text{ This limit is } \infty; \text{ thus the integral}$$

diverges to ∞.

❓ **COMMON QUESTION:** Many students want to know why an integral such as:

$\int_{-\infty}^{\infty}\frac{1}{x}\,dx$ cannot be shown, using a limit of Riemann sums to be 0. However, as the diagram below illustrates:

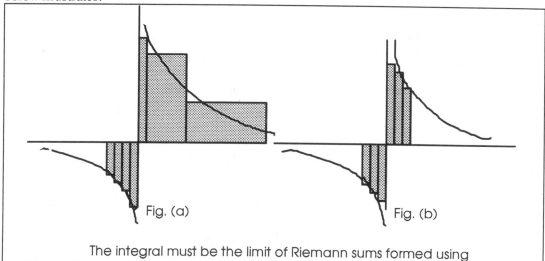

Fig. (a)   Fig. (b)

The integral must be the limit of Riemann sums formed using ANY partition, as in Fig. (a), not just from a symmetrical partitions as in Fig. (b)

EXAMPLE: Evaluate $\int_{-\infty}^{\infty} x^3 e^{x^4}\,dx$.

Answer: Before even attempting to solve this integral, we suspect the limit does not exist. Here's why: the integrand is odd. The integral $\int_{-\infty}^{0} x^3 e^{x^4}\,dx$ will be the negative of

$\int_{0}^{\infty} x^3 e^{x^4}\,dx$, and this latter integral approaches ∞, as the integrand gets extremely large. We will have a "-∞ +∞" situation that will given an undefined answer. Now we prove this.

To underscore that the limits may approach ∞ and $-\infty$ in different ways, we have used two variables for these limits: N and M. However, when we write the integral as a sum of two limits , this is not necessary; we could use N in both integrals.

$$\int_{-\infty}^{\infty} x^3 e^{x^4} dx = \lim_{M\to\infty} \int_{-M}^{1} x^3 e^{x^4} dx + \lim_{N\to\infty} \int_{1}^{N} x^3 e^{x^4} dx = \text{ (letting } u = x^4)$$

$$\lim_{M\to\infty} \frac{e^u}{4} \bigg|_{M^4}^{1} + \lim_{N\to\infty} \frac{e^u}{4} \bigg|_{1}^{N^4}$$

This result is not defined; it looks like "$\infty - \infty$". And as we expected, the integral does not exist: it diverges.

SHORT-CUT?   Again, here is a short-cut in doing the Example above, that you might use only with approval of your teacher. It is important to recognize that we must treat the problem with TWO DISTINCT VARIABLES , if we use ONE INTEGRAL.

$$\int_{-\infty}^{\infty} x^3 e^{x^4} dx = \lim_{N\to\infty, M\to\infty} \int_{-M}^{N} x^3 e^{x^4} dx = \lim_{N\to\infty, M\to\infty} \int_{-M}^{N} x^3 e^{x^4} dx = \lim_{N\to\infty, M\to\infty} \frac{1}{4} \int_{M^4}^{N^4} e^u du =$$

$$\lim_{N\to\infty, M\to\infty} \frac{e^u}{4} \bigg|_{M^4}^{N^4} = \lim_{N\to\infty, M\to\infty} \frac{1}{4} (e^{N^4} - e^{M^4}) \text{ . This limit does not exist (it looks like }$$

"∞"-"∞") and the integral diverges .

EXAMPLE: Does the integral $\int_{-\infty}^{\infty} \frac{1}{1+x^2} dx$ converge or diverge?

Answer:

$\int_{-\infty}^{\infty} \frac{1}{1+x^2} dx$ does not require an evalution with two integrals . Because the integrand is even ,

we can write this integral as $\lim_{N\to\infty} 2\int_{0}^{N} \frac{1}{1+x^2} dx = \lim_{N\to\infty} 2 \arctan x \bigg|_{0}^{N} =$

$\lim_{N\to\infty} 2 \arctan N - 2 \arctan 0 = (2)\frac{\pi}{2} - 0 = \pi$. This integral converges to π.

NOTE: The following result is useful. We will see an analogy to this in Chapter 9. Use it to check your results, or to develop your intuition, but be prepared, on tests, to evaluate integrals in the formal ways shown above.

THE p-TEST

FIRST   RESULT   : $\displaystyle\int_1^\infty \frac{1}{x^p}dx$ $\begin{cases} \text{converges if } p > 1 \\ \text{diverges if } p \le 1 \end{cases}$

Proof of the result:

$$\int_1^\infty \frac{1}{x^p}dx = \lim_{N\to\infty}\int_1^N \frac{1}{x^p}dx = \lim_{N\to\infty}\left.\frac{x^{-p+1}}{-p+1}\right|_1^N = \lim_{N\to\infty}\left(\frac{N^{-p+1}}{-p+1} - \frac{1^{-p+1}}{-p+1}\right).$$

This limit exists if and only if the large N- term lies in the denominator.

This happens if $(-p+1) < 0$, or $p > 1$, as we wished to show.

In fact, with any positive lower limit the result is true. For instance, if $a > 1$, we have

$$\int_1^a \frac{1}{x^p}dx + \int_a^\infty \frac{1}{x^p}dx = \int_1^\infty \frac{1}{x^p}dx.$$ So if $\displaystyle\int_1^\infty \frac{1}{x^p}dx$ is finite, then $\displaystyle\int_a^\infty \frac{1}{x^p}dx$ must

be finite, as well. Similarly, if $\displaystyle\int_1^\infty \frac{1}{x^p}dx$ is infinite, $\displaystyle\int_a^\infty \frac{1}{x^p}dx$ must be also.

S ECOND RESULT   : $\displaystyle\int_0^1 \frac{1}{x^p}dx$ $\begin{cases} \text{diverges if } p > 1 \\ \text{converges if } p \le 1 \end{cases}$

This is proved in the same fashion as the first result.

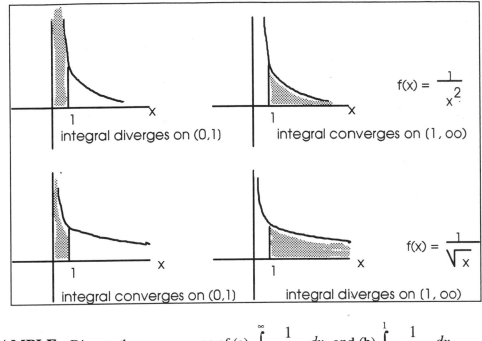

integral diverges on (0,1)

integral converges on (1, oo)

$$f(x) = \frac{1}{x^2}$$

integral converges on (0,1)

integral diverges on (1, oo)

$$f(x) = \frac{1}{\sqrt{x}}$$

EXAMPLE: Discuss the convergence of (a) $\int_0^\infty \frac{1}{\sqrt{x+1}}dx$ and (b) $\int_0^1 \frac{1}{\sqrt{x+1}}dx$.

Answer: (a) Since $\int_0^\infty \frac{1}{\sqrt{x+1}}dx = \int_1^\infty \frac{1}{\sqrt{u}}du$, by the p-test, this integral diverges, since

p = 1/2. In (b), by the same token, the integral converges since p = 1/2 (we should find to what).

Check Yourself on Improper Integrals

Problem: Discuss the convergence or divergence of the integral $\int_{-\infty}^\infty \frac{1}{x-1}dx$ in the formal

terms we have seen presented here. We will discuss the traps that students fall into.

---

1. Did you write $\int_{-\infty}^\infty \frac{1}{x-1}dx = \lim_{N\to\infty} \int_{-N}^N \frac{1}{x-1}dx$? This is wrong on at least two major

counts. You should not use the same "N" in both limits. And the integral has the additional problem that x may not equal 1. If you didn't fall into the trap... Good!✓

2. Did you draw a graph and claim the integral could not exist? This isn't enough. If you realized this....Good! ✓

3. Did you write the integral as $\displaystyle\int_{-\infty}^{\infty}\frac{1}{x-1}dx=\lim_{\substack{N\to\infty\\ \varepsilon\to1}}\left(\int_{-N}^{\varepsilon}\frac{1}{x-1}dx+\int_{\varepsilon}^{N}\frac{1}{x-1}dx\right)$? This

particular format is not O.K. Other formats may be acceptable, if they deal with the "bad

spots" carefully. You may write: $\displaystyle\int_{-\infty}^{\infty}\frac{1}{x-1}dx=\lim_{\substack{N\to\infty\\ \varepsilon\to1}}\int_{-N}^{\varepsilon}\frac{1}{x-1}dx+\lim_{\substack{N\to\infty\\ \varepsilon\to1}}\int_{\varepsilon}^{N}\frac{1}{x-1}dx$, for example.

Although we have used "N" twice, we are using N in two different limits. ✓

4. Did you conclude that the integral diverges ? (It does not diverge to ∞ or to $-\infty$--it
simply diverges!) Good!✓

5. Did you give an argument using the p-test, similar to this?

"Convergence of $\displaystyle\int_{-\infty}^{\infty}\frac{1}{x-1}dx$ requires the convergence of $\displaystyle\int_{1}^{\infty}\frac{1}{x-1}dx$. Letting u = x - 1, we

see that this integral converges if and only if $\displaystyle\int_{0}^{\infty}\frac{1}{u}du$ does. But this integral fails to

converge by the p-test, which states that $\displaystyle\int_{0}^{1}\frac{1}{u^{p}}du$ and $\displaystyle\int_{1}^{\infty}\frac{1}{u^{p}}du$ both diverge to ∞ when

p = 1. Thus the integral $\displaystyle\int_{1}^{\infty}\frac{1}{x-1}dx$ does not converge, and neither does $\displaystyle\int_{-\infty}^{\infty}\frac{1}{x-1}dx$ " .

Well if you did, your instructor will be surprised! (Be sure you **answer the question**
however! Because if the question asked you to *"show all work",* you might want to check
with your instructor that this reply is adequate before you turn in your work. We doubt,
however, that your instructor would begrudge this answer.)

CHAPTER 8

"It is not enough to have a good mind. The main thing is to use it well." ..... Descartes

In this chapter, we see how the definite integral provides us with a means for finding volume, surface area, length of a curve, work performed in moving an object, moments, hydrostatic pressure, etc. We discuss the concept of parametrizing a curve. And we explore the polar coordinate system, which creates something of a break from our classical way of thinking in rectangular, or Cartesian coordinates.

While memorization is important, it is probably a mistake to try to memorize all the formulas in this section. The problems are such, that *variations* on the formulas will be needed. So it is advisable to learn how to create your own formulas.

SECTION 8.1

"Q.E.D....which was necessary to demonstrate."...Euclid

"CRAZY" SOLIDS

To find the volume of some of these solids may seem difficult at first. Many students find these solids more difficult to comprehend than the solids of revolution you will soon see. However, we use these solids to stress the methodology that will be used throughout this chapter.

These solids are generally described by tiny cross-sectional slices that are composed of
(1) a **BASE** (thickness) and (2) an **AREA OF A CROSS-SECTION**.

(1) **Area of a cross section of kth slice:** ΔA_k , usually expressed in terms of x
(2) **Thickness of kth slice:** usually, Δx_k. All other variables in must be expressed in terms of   this variable.

(3) These two quantities produce a **small element of volume of kth slice:** $\Delta V_k = \Delta A_k\, \Delta x_k$.
   (Volume = area · thickness)

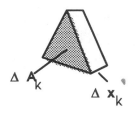

EXAMPLE: Find the volume of the solid whose base is a circle of radius 2, and whose cross-sections perpendicular to this base circle, are equilateral triangles.

Answer: (We will work this problem first slowly, following the theory, then a second time, speeding up the process.)

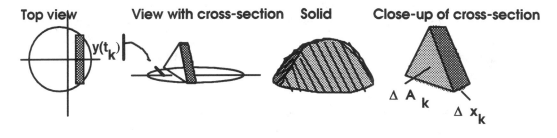

The formula for the area of an equilateral triangle (angles are 60°) whose sides have length s is

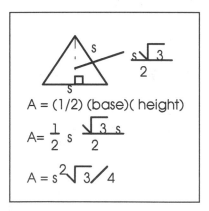

I. Slow and theoretical solution:

From the pictures above, the volume of the kth triangular slice is

$\Delta V_k = (\text{Cross-Sectional Area}) \cdot (\text{Thickness})$ or

$\Delta V_k = A(t_k)\Delta x_k$, where $A(t_k) = \dfrac{\sqrt{3}}{4} y(t_k)^2$ [$y(t_k)$ is seen in the top left picture].

However, the choice of "x" in Δx_k means that we must express "y" in terms of "x".

At this point, we ask: what is the relationship between x and y ?

Since the base is a circle (chosen with center at the origin), the y-values come from the

circle, $x^2 + y^2 = 2^2$. Then $\Delta V_k = \dfrac{\sqrt{3}}{4}(2^2 - t_k^{\;2})(\Delta x_k)$. Think of t_k here as "x".

Recall in our definition of Riemann sums, we used t_k to represent *any* x-value on the

subinterval: $\qquad x_{k-1} \text{_____} t_k \text{\_\_\_\_\_} x_k.$

Now, having found a sample element of volume, we say:

$$\text{Vol} = \lim_{\|\Delta x_k\| \to 0} \sum_{k=1}^{n} \Delta V_k = \lim_{\|\Delta x_k\| \to 0} \sum_{k=1}^{n} \frac{\sqrt{3}}{4}(2^2 - t_k^{\;2})\Delta x_k = \int_{-2}^{2} \frac{\sqrt{3}}{4}(2^2 - x^2)dx$$

(We have noted that on the circle, $x^2 + y^2 = 2^2$, x ranges from -2 to 2.

These values provide the limits of integration.)

II. Speedier solution you will eventually use.
Above we used the "t_k" notation to demonstrate how our integral is obtained from the
Riemann Sums developed in the text. If we drop subscripts and use "x" in place of "t_k",
the solution to the above problem becomes the following.

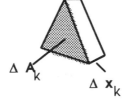

$\Delta V = \Delta A \cdot \Delta x$

$\Delta V = \left(\dfrac{\sqrt{3}}{4} y(x)^2 \right)(\Delta x).$ We write $y(x)$ rather than "y" to emphasize that y is a function

of x. Since y-values come from the circle, $x^2 + y^2 = 2^2$, we have $\Delta V = \dfrac{\sqrt{3}}{4}(2^2 - x^2)(\Delta x).$

We also note that x ranges from -2 to 2. Thus the integral representing the volume is:

$$V = \int_{-2}^{2} \frac{\sqrt{3}}{4}(2^2 - x^2)dx.$$

☞ **NOTE:** From now on, we will drop the use of " t_k " and subscript notation in general. This requires an understanding from the reader that these subscripts are implicit. When we see ΔV, we will interpret this as ΔV_k, the volume of an arbitrary kth slice of the partition . And when we see an expression such as $\Delta V = \dfrac{\sqrt{3}}{4} y(x)^2 (\Delta x)$, it will be understood that Δx is actually Δx_k, the arbitrary thickness of the kth slice , and that "x" represents an x-value from the kth partition, which being arbitrary, was formerly labeled " t_k " . The advantage of dropping the subscripts is that we get a behind-the-scenes (albeit less careful) view of how to set up the appropriate integrals.

NOTE: VARIATIONS in the example above may have the cross-sectional areas of the slices as squares, circles, etc. In this case, only ΔA_k changes.

DISK METHOD FOR EVALUATING VOLUME:
In this text, we are generally rotating an area about the x-axis.
We assume there is no empty space, or hole, in the center of the solid
generated by the rotation.

The volume of a disk (or filled cylinder) is:

$$\text{Volume} = \pi\,(\text{radius})^2\,(\text{thickness})$$

When the volume and thickness are small, this small change will be represented by "Δ". For example, the volume of the small disk below is given as:

$$\Delta V = \pi\,(f(x))^2\,\Delta x$$

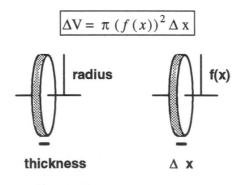

Volume of a solid by the method of disks:
$$V = \int_a^b \pi\,(f(x))^2\,dx$$

NOTE: In the method of disks, the height of the rectangle to be rotated is measured perpendicularly from the axis of rotation.

EXERCISE: Many students have difficulty imagining these solids. As an exercise, draw the solids you would obtain by rotating the shaded areas below, about the x-axis. Draw a sample cylinder that can be used to generate the volume. The first illustration provides a sample drawing.

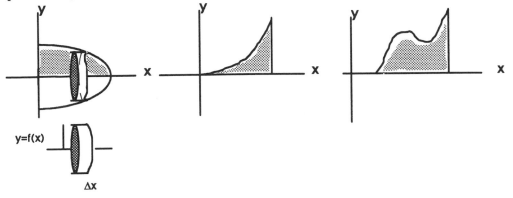

EXAMPLE: Find the integral which expresses the volume of the solid generated by rotating the area between $y = \sqrt{x}$, the x-axis, and the line x = 4, about the x-axis.

Answer: The picture of the area to be rotated, and the resulting solid :

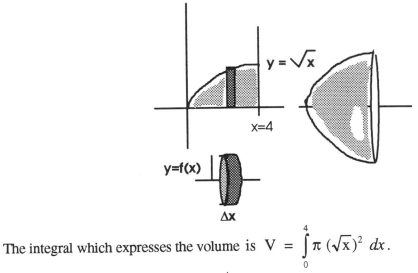

The integral which expresses the volume is $V = \int_{0}^{4} \pi \, (\sqrt{x})^2 \, dx$.

Evaluating, $V = \int_{0}^{4} \pi \, x \, dx = \left. \frac{\pi \, x^2}{2} \right|_{0}^{4} = 8\pi.$

NOTE: The limits of integration refer to the lower and upper bounds on "x" in the **area** we are rotating.

WASHER METHOD FOR EVALUATING VOLUME:
In this method we assume there an empty space, or hole, in the center of the solid.

The "washer method" is equivalent to the "disk method" where we subtract from the volume of the disk, the volume of the hole in the object.

VOLUME OF OBJECT = VOLUME OF LARGE OBJECT – VOLUME OF HOLE

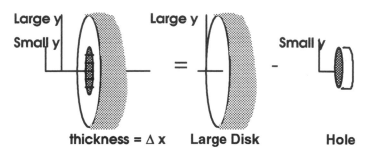

To create a washer: Take volume of large disk and subtract volume of hole.

$$\Delta V = \pi \left((\text{Large } y(x))^2 - (\text{Small } y(x))^2 \right) \Delta x$$

Volume of a solid by the method of washers:

$$V = \int_a^b \pi \left[(\text{Large } y(x))^2 - (\text{Small } y(x))^2 \right] dx \quad \text{or:} \quad \int_a^b \pi \ (f(x))^2 - (g(x))^2 \ dx$$

where f is the function providing the radius of the outer boundary of the object, and g is the function representing the radius of the hole.

EXERCISE: In the drawings below, rotate the shaded area about the x-axis, and draw the resulting solid, with a sample washer that will represent a tiny volume element.

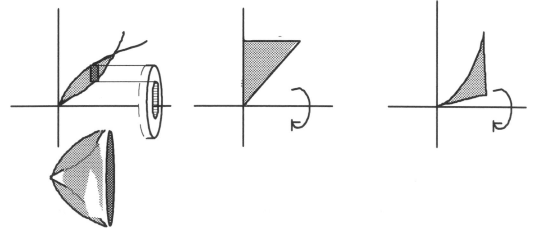

EXAMPLE: Find the integral which expresses the volume of the solid generated by rotating the area between the curves $y = x^2$ and $y = \sqrt{x}$ about the x-axis. Use
(a) The Method of Disks.
(b) The Method of Washers.

These curves intersect at x = 0 and x = 1.

The drawing is given above, leftmost illustration. Here are a few more specifics.

Note: on [0,1], $x^2 \leq \sqrt{x}$.

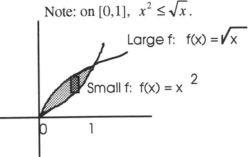

Large f: $f(x) = \sqrt{x}$

Small f: $f(x) = x^2$

Answer (a): Volume by Disks:

Volume = Volume of Larger Object - Volume of Hole

$$V = \int_0^1 \pi \ (\sqrt{x} \ )^2 \, dx \ - \ \int_0^1 \pi \ (x^2 \ )^2 \, dx = \int_0^1 \pi \ x \, dx \ - \ \int_0^1 \pi \ x^4 dx.$$

Answer (b): Volume by Washers:

$$V = \int_0^1 \pi \left[\sqrt{x}^2 - (x^2)^2 \right] dx = \int_0^1 \pi \left[x - x^4 \right] dx.$$

> **The rotation of a region about the x-axis rarely produces the
> same volume as the rotation of the area about the y-axis.**

NOTE: We can find volumes of solids generated by rotating areas about the y-axis easily:
if we exchange all x's and y's in the problem.   For example: finding the volume of the
solid obtained by rotating **y = (3x+2) about the x-axis for x in [0, 4]**   is equivalent to
finding the volume of the solid obtained by rotating **x = (3y+2) about the y-axis for y in
[0,4]** .

SECTION 8.2

*"In some sort of crude sense which no vulgarity, no humor, no overstatement can quite extinguish,
the physicists have known sin, and this is a knowledge which they cannot lose"*.... Oppenheimer

> **THE SHELL METHOD FOR FINDING VOLUMES:**
> **In this text, we are generally rotating about the y-axis. With this method we may or
> may not assume there is a hole in the center of the solid.**

A tiny volume of a shell may be found to be

$$\Delta V = \ 2\pi \ (\text{radius}) \Delta (\text{area of rectangle})$$

You can remember this formula by thinking of taking a little piece of area, ΔArea, for a "spin" about a circle, a trip of length 2 π (radius).

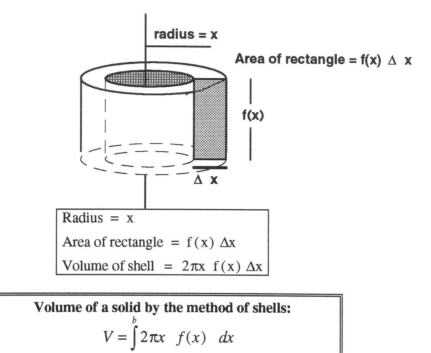

Radius = x

Area of rectangle = f(x) Δx

Volume of shell = 2πx f(x) Δx

Volume of a solid by the method of shells:
$$V = \int_a^b 2\pi x \; f(x) \; dx$$

NOTE: When using the method of shells, the rectangle being rotated must be **parallel** to the axis of rotation .

EXAMPLE: Find the volume of the solid generated by rotating the area between $y = \sqrt{x}$, the x-axis, and the line x = 4 about the y-axis. (The illustration is shown below.)

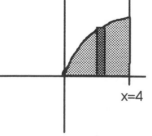

x=4

Hint: The integral which expresses the volume: $\int_0^4 2\pi \, x \, \sqrt{x} \; dx = \left. \frac{\pi \, x^{5/2}}{5/2} \right|_0^4 = \frac{64\pi}{5}.$

EXERCISE: In the drawings on the following page, draw the solid obtained by rotating the shaded area about the indicated axis. Draw a sample shell that represents a tiny volume element.

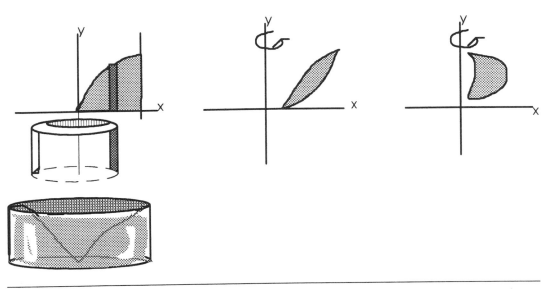

EXAMPLE: rotating about a different axis:

Find the volume of the solid generated by rotating the area of the region between $y = \sqrt{x}$, the x-axis, and the line x = 4, about the line x = 6.

Answer: The axis of rotation is not the y-axis, so we must be careful to insure that our radius, in our formula, reflects the distance from the edge of the shell to the axis of rotation. Draw a picture (see below). In this example, the correct radius is 6-x.

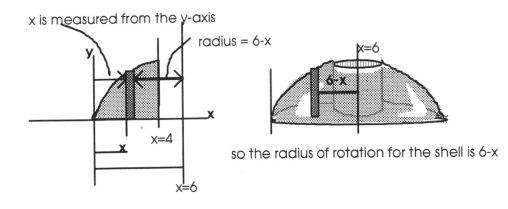

Thus the integral that expresses the volume is

$$\int_0^4 2\pi\,(6-x)\,\sqrt{x}\,dx, \text{ which can be evaluated as } \int_0^4 2\pi\,(6x^{1/2} - x^{3/2})\,dx.$$

MODELING: For perhaps the first time in our lives we are now able to find formulas for spheres, cones, etc. These problems are worked examples in the text. You may be asked to find the volume of a solid of revolution by a particular method, or you may be asked to find the volume of a sphere, a cone, etc. Practice these problems.

PROBLEM	ROTATE WHAT AREA Bounded by what Curve ABOUT WHAT AXIS?
Volume of a Sphere	
Volume of a Cone	
Volume of an Ellipsoid	

Answers: *For volume of a sphere, rotate the top half of a circle,* $y = \sqrt{r^2 - x^2}$ *about the x-axis, x going from -r to r. For volume of a cone of radius r, and height h, rotate the straight line y = (r/h) x about the x - axis, x from 0 to h. For volume of an ellipsoid, see below.*

EXAMPLE: Find the volume of the ellipsoid produced by rotating the top half of the curve $\dfrac{x^2}{a^2} + \dfrac{y^2}{b^2} = 1$ about the x - axis.

Answer: We use the method of disks. Since $[y(x)]^2 = \left[b^2 \left(1 - \dfrac{x^2}{a^2} \right) \right]$,

$V = \pi \displaystyle\int_{-a}^{a} \left[b^2 \left(1 - \dfrac{x^2}{a^2} \right) \right] dx = \pi \ b^2 \displaystyle\int_{-a}^{a} 1 - \dfrac{x^2}{a^2} \, dx$. This integral gives the volume $\dfrac{4}{3} \pi \, a \, b^2$,

which resembles the volume of a sphere $\dfrac{4}{3} \pi \, r^3$, with r^3 replaced by $a \, b^2$.

(Why is b squared ?)

REVIEW OF VOLUMES

	Disks	Washers	Shells
Draw the volume element, with dimensions identified			
Formula for volume element, ΔV			
Integral formula			
Invent a problem			
Your trouble spots with this method			

Rate Your Understanding of Topics on a Scale of 1-5

Using a scale from 1-5 with "1--no understanding", and "5--excellent understanding", rate your understanding on the following:

	Your understanding of concepts involved	Your ability to do problems	Your rating of the understanding of the "class"
Volumes by Disks			
Volumes by Washers			
Volumes by Shells			
Volumes of "Crazy Solids"			
Modeling Problems: finding volume of sphere, cone, etc. (you set up the equations, find the area to rotate, etc.)			

 # PLEASE BE CAREFUL

about forming impressions about what you should know from what others in the class say. Often students will please you by saying they are in complete confusion. Don't go by this that you are not responsible for the material. See your teacher, or a tutor immediately!

SECTION 8.3

" Don't look back. Something might be gaining on you. "...Leroy ("Satchel") Paige

LENGTH OF CURVES (or "ARC LENGTH")

> **Length of curve $y = f(x)$ from a to b:**
> $$\int_a^b \sqrt{1 + [f'(x)]^2}\, dx = \int_a^b \sqrt{1 + [dy/dx]^2}\, dx.$$

When determining the lengths of curves, it is likely that most of your work will be with the algebraic simplifications necessary to reduce the integrand to something integrable.

EXAMPLE: Find the length of the curve $f(x) = \dfrac{2}{3} x^{3/2}$ for x in $[0, 5]$.

Answer: Letting L denote the length of the curve, we have

$$f'(x) = x^{1/2}, \text{ and } L = \int_0^5 \sqrt{1 + (x^{1/2})^2}\, dx = \int_0^5 \sqrt{1 + x}\, dx, \text{ which is easily evaluated,}$$

letting $u = 1 + x$.

EXAMPLE: Find the length of $f(x) = \ln(\cos x)$ on the interval $[0, \pi/4]$.

Answer: Since $f'(x) = \dfrac{1}{\cos x}(\sin x) = \tan x$, we have $L = \int_a^b \sqrt{1 + [f'(x)]^2}\, dx$

$$= \int_0^{\pi/4} \sqrt{1 + \tan^2 x}\, dx = \int_0^{\pi/4} \sec x\, dx = \ln|\sec x + \tan x| \Big|_0^{\pi/4} = \ln(\sqrt{2} + 1).$$

EXAMPLE: Find the length of $f(x) = x^3 + \dfrac{1}{12x}$ on the interval $[1,2]$

Answer: $f'(x) = 3x^2 - \dfrac{1}{12x^2}$. Thus:

$$L = \int_a^b \sqrt{1 + [f'(x)]^2}\, dx = \int_1^2 \sqrt{1 + [3x^2 - \frac{1}{12x^2}]^2}\, dx = \int_1^2 \sqrt{1 + \left(9x^4 - \frac{1}{2} + \frac{1}{144x^4}\right)}\, dx =$$

$$\int_1^2 \sqrt{9x^4 + \frac{1}{2} + \frac{1}{144x^4}}\, dx = \int_1^2 \sqrt{(3x^2)^2 + \frac{1}{2} + \frac{1}{(12x^2)^2}}\, dx = \int_1^2 \sqrt{\left(3x^2 + \frac{1}{12x^2}\right)^2}\, dx =$$

$$\int_1^2 3x^2 + \frac{1}{12x^2}\, dx, \text{ which can now be evaluated.}$$

SECTION 8.4

"There is no royal road to Geometry"....Euclid

SURFACE AREA
Suppose we rotate an arc about the x-axis to generate a surface, for which we want the surface area. Then a tiny element of surface area is:

$$\Delta S = 2\pi \text{ (radius) } \Delta(\text{ length of curve }) =$$
$$\Delta S = 2\pi\, f(x)\, \Delta s$$

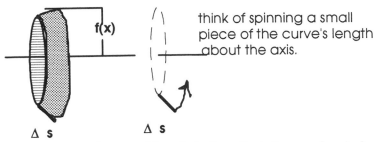

think of spinning a small
piece of the curve's length
about the axis.

Think of taking a little piece of curve of length Δs for a "spin" around a circle of radius f(x).

When rotating about the x-axis, surface area is given by:

$$S = \int_a^b 2\pi\, f(x)\sqrt{1+[f'(x)]^2}\, dx$$

(When rotating about the y-axis, use "x" in place of the "f(x)" term, for the radius.)

EXAMPLE: Find the surface area of the solid generated by rotating the graph of $y = \sqrt{3-x}$ about the x-axis, for x in [0, 1].

Answer:

$$S = \int_a^b 2\pi\, f(x)\sqrt{1+[f'(x)]^2}\, dx = \int_0^1 2\pi\, \sqrt{3-x}\,\sqrt{1+\left[\frac{-1}{2\sqrt{3-x}}\right]^2}\, dx =$$

$$\int_0^1 2\pi\, \sqrt{3-x}\,\sqrt{1+\left[\frac{1}{4(3-x)}\right]}\, dx = \int_0^1 2\pi\, \sqrt{3-x}\,\sqrt{\frac{12-4x+1}{12-4x}}\, dx = \int_0^1 2\pi\, \sqrt{3-x}\,\sqrt{\frac{13-4x}{4(3-x)}}\, dx =$$

$$2\pi \int_0^1 \frac{\sqrt{13-4x}}{2}\, dx = \pi \int_0^1 \sqrt{13-4x}\, dx,\ \text{which can now be evaluated using}\ u = 13-4x.$$

Find the integral that expresses each mathematical construct in the table below.

Surface Area of a Sphere	
Surface Area of a Cone	
Circumference of a Circle	
Circumference of an Ellipse	
Surface Area of an Ellipsoid	

SECTION 8.5

"I don't like work--no man does--but I like what is in work--the chance to find yourself. Your own reality--for yourself, not for others--what no other man can ever know"...Joseph Conrad

WORK

When a constant force F is applied to move an object a distance s, the work done, W , is given by W = F · s. When the force on the object is a function of the distance, which we will now call x, the work done in moving the object from a to b is given by

$$W = \int_a^b F(x)dx$$

EXAMPLE: A 20-lb. weight is carried up 8 ft. Find the work done.

Answer: This represents a constant force applied over a distance. Work = (20)(8) or 160 ft/lbs.

EXAMPLE: A variable force (the wind?) pushes a sand-buggy built in the year 3000 AD. The strange circumstances of this hypothetical problem determine that the force, f, is a function of the distance the sand-buggy has traveled, given as f(x) = 3 x + 1 pounds. How much work is required to send the sand-buggy 500 ft?

Hint: Work $= \int_a^b F(x)dx = \int_0^{500}(3x+1)dx = ?$ ft / lbs.

EXAMPLE: If 7 foot-pounds are required to compress a spring 1 foot from its natural length, find the work, W, necessary to compress the spring one extra foot.

Hint: We have

$\int_0^{-1} kxdx = 7$. But this integral is $\dfrac{kx^2}{2}\Big|_0^{-1} = \dfrac{k}{2}$, so $k = 14$. Now the work needed to

compress the string an *extra foot* (*x* from -1 to -2) is given by $W = \int_{-1}^{-2}14xdx = ?$

SECTION 8.6

" To be able to practice five things everywhere under heaven constitutes perfect virtue...[They are] gravity, generosity of soul, sincerity, earnestness, and kindness"....Confucius

MOMENTS

Try to become familiar with the derivations of the many formulas in this section:

EXAMPLE: Find the center of gravity of the area bounded by f(x) = 3x - 1 and g(x) = x + 2, when x is in the interval [0, 3/2].

Answer:

First, find the bounded area:

$$\text{Area} = \int_0^{3/2} (x+2) - (3x-1)dx. \text{ We note that on this interval, } g > f.$$

Then $\text{Area} = \int_0^{3/2} -2x + 3dx = \left. -x^2 + 3x \right|_0^{3/2} = -\frac{9}{4} + \frac{9}{2} = \frac{9}{4}.$ Now:

$$M_y = \int_0^{3/2} x[(x+2) - (3x-1)]dx = \int_0^{3/2} -2x^2 + 3xdx = \left. \frac{-2x^3}{3} + \frac{3x^2}{2} \right|_0^{3/2} = 9/8 = 1.125$$

$$M_x = \int_0^{3/2} \frac{1}{2}\left\{ (x+2)^2 - (3x-1)^2 \right\}dx = \frac{1}{2}\int_0^{3/2} \left\{ x^2 + 4x + 4 - 9x^2 + 6x - 1 \right\}dx =$$

$$\frac{1}{2}\int_0^{3/2} \left\{ -3x^2 + 10x + 3 \right\}dx = \left. \frac{1}{2}\left(-x^3 + 5x^2 + 3x \right) \right|_0^{3/2} = 27/8 = 3.375$$

Thus $\bar{x} = \frac{M_y}{A} = \frac{9}{8} \cdot \frac{4}{9} = \frac{1}{2}$ and $\bar{y} = \frac{M_x}{A} = \frac{27}{8} \cdot \frac{4}{9} = \frac{3}{2}.$

SECTION 8.7

" Water, water everywhere, and not a drop to drink!"....Coleridge

HYDROSTATIC FORCE

The formula for the hydrostatic force on an object submerged in water is

$$F = \int_a^b (62.5)(c-x)w(x)dx$$

where c is the water level, x is the variable distance from a thin slice of the submerged object (parallel to the water level) to the surface of the water, and w(x) is the width of this thin slice at that point. In these problems "x" is distance measured vertically.

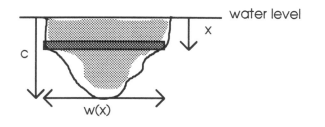

EXAMPLE: A light in the shape of a semi-circle whose diameter is 6' at water level, is submerged in a swimming pool. Find the pressure of the water against this light.

Hint: Note the change of axes in the picture. The water level is given by $c = 0$. And the equation of the semi-circle below the water level is given by: $y = -\sqrt{3^2 - x^2}$, since the radius of the light is 3. Now $w(x) = 2y$, and we have the formula:

$$F = \int_a^b (62.5)(c-x)w(x)dx = \int_0^3 (62.5)(0-x)\left(-2\sqrt{3^2-x^2}\right)dx =$$

$$(62.5)\int_0^3 2x\,\sqrt{9-x^2}\,dx, \text{ which can be evaluated by letting } u = 9 - x^2.$$

If you think you don't get enough partial credit

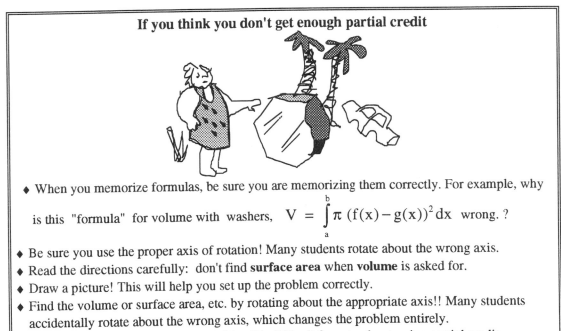

- When you memorize formulas, be sure you are memorizing them correctly. For example, why

  is this "formula" for volume with washers, $V = \int_a^b \pi\,(f(x)-g(x))^2\,dx$ wrong. ?

- Be sure you use the proper axis of rotation! Many students rotate about the wrong axis.
- Read the directions carefully: don't find **surface area** when **volume** is asked for.
- Draw a picture! This will help you set up the problem correctly.
- Find the volume or surface area, etc. by rotating about the appropriate axis!! Many students accidentally rotate about the wrong axis, which changes the problem entirely.
- Explain your work as much as possible. It's difficult for a teacher to give partial credit without understanding how you arrived at your answer.

Concept Review

Explain how all of the concepts we've seen in this chapter are applications of the Definite Integral. Carefully work through each of these concepts, showing how the Definite Integral emerges from a Riemann Sum.

CHAPTER 9

"Why does this magnificent applied science, which saves work and makes life easier, bring us so little happiness? The simple answer runs--because we have not yet learned to make a sensible use of it....Concern for man himself and his fate must always form the chief interest of all technical endeavors.."...Albert Einstein

In this chapter, new ways of thinking about the concept of limit are explored. Remarkable results enable us to approximate certain functions with "series"--which are similar to polynomials, but have an infinite number of terms . The Taylor Series is one of the great results of Calculus. Take care in understanding the Taylor Polynomial in Section 9.1 because the concept resurfaces later.

Concepts used in this chapter that may require review include finding limits, l'Hopital's Rule, the differential, and tangent lines . It helps to have an understanding of the error in an approximation, an idea that came up with Simpson's Rule, for example.

SECTION 9.1

"For let anyone have the clearest demonstration about an equilateral triangle without experience of it, his mind will never lay hold of the problem until he has actually before him [the object]."
...Francis Bacon

TAYLOR POLYNOMIAL APPROXIMATION TO f

We saw in Chapter 3 that a function has a **linear approximation**:

$$f(a+h) \approx f(a) + f'(a)(h) \text{ or with } x = a + h \text{ and } h = x - a, \ f(x) \approx f(a) + f'(a)(x-a)$$

The equation, $y = f(a) + f'(a)(x-a)$, is the equation of the tangent line to f at

x = a. We are saying that f(x) is approximated by y-values on the tangent line. We can get better approximations to f using higher degree polynomials. If f has n derivatives, we have:

---

f is approximated by its nth Taylor PolynomiaL

$f(x) \approx p_n(x)$, or

$$f(x) \approx f(a) + f'(a)(x-a) + \frac{f''(a)(x-a)^2}{2!} + \frac{f'''(a)(x-a)^3}{3!} + \ldots \frac{f^{(n)}(a)(x-a)^n}{n!}$$

A polynomial has a finite number of terms

NOTE: Factorials get very big very fast! For example, 10! =3,268,800. Since factorials are in the denominators in the expression above, we hope the terms in the Taylor Polynomial get smaller and more negligible as n gets larger.

EXAMPLE: Find the 5th Taylor Polynomial approximation, $p_5(x)$, of $f(x) = \sin x$, with a = 0.

Answer: We first compute derivatives, then evaluate each derivative at a = 0.

$f(x) = \sin x$	$f(0) = 0$
$f'(x) = \cos x$	$f'(0) = 1$
$f''(x) = -\sin x$	$f''(0) = 0$
$f'''(x) = -\cos x$	$f'''(0) = -1$
$f^{(4)}(x) = \sin x$ (back to the start!)	$f^{(4)}(0) = 0$
$f^{(5)}(x) = \cos x$. This is as far as we need to go for $p_5(x)$.	$f^{(5)}(0) = 1$

We use the definition of the nth Taylor Polynomial,

$$f(x) \approx f(a) + f'(a)(x-a) + \frac{f''(a)(x-a)^2}{2!} + \frac{f'''(a)(x-a)^3}{3!} + \ldots \frac{f^{(n)}(a)(x-a)^n}{n!}$$

with a = 0 and n = 5, and the derivatives found above. Thus

$$\sin(x) \approx 0 + (x-0) + 0 - \frac{(x-0)^3}{3!} + 0 + \frac{(x-0)^5}{5!} \quad \text{or} \quad \boxed{\sin x \approx x - \frac{x^3}{3!} + \frac{x^5}{5!}}.$$

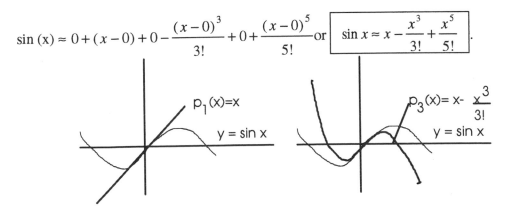

ABOUT ERROR

Suppose we say $\pi \approx 3.14$. (Note: any value we see on calculator for π is just another approximation.) . What is the error of this approximation?

$$\text{Actual Error} = |\text{ Correct Value - Approximate Value }|$$

We can't find the actual error, $|\pi - 3.14|$, because we don't know the exact decimal value of π. However we can find a "bound" on the error. (When $|a(x)| \leq M$, then M is called a "bound" on a(x).) We aren't really finding the "maximum" error, because there would be no limit to how large M might be.

Assuming we don't know the next term in the decimal representation of π, the largest π could be is 3.150 and the smallest it could be is 3.135. Comparing 3.14 to these values, we see the bound on the error is .01. That is,

$$\text{Error} = |\text{ (true value of } \pi) - 3.14| \leq .01$$

TAYLOR'S FORMULA: POLYNOMIAL WITH REMAINDER TERM

We can replace the approximation $f(x) \approx p_n(x)$ by an equality, $f(x) = p_n(x) + r_n(x)$ but there's a hitch! We are not precisely sure what the last term, the remainder term, is! Nevertheless, if we ignore this term, we will know that $|f(x) - p_n(x)| \leq \text{Max } |r_n(x)|$.

Taylor Polynomial (Without the remainder term, we have an approximation):

$$f(x) \boxed{\approx} f(a) + f'(a)(x-a) + \frac{f''(a)(x-a)^2}{2!} + \frac{f'''(a)(x-a)^3}{3!} + ... \frac{f^{(n)}(a)(x-a)^n}{n!}$$

Taylor Formula (With the remainder term, we have equality):

$$f(x) \boxed{=} f(a) + f'(a)(x-a) + \frac{f''(a)(x-a)^2}{2!} + \frac{f'''(a)(x-a)^3}{3!} + ... \frac{f^{(n+1)}(t_x)(x-a)^{n+1}}{(n+1)!}$$

where t_x lies between x and a, and is generally unknown.

NOTES:

◆ **The Remainder Term in the Taylor Formula is** $r_n(x) = \dfrac{f^{(n+1)}(t_x)(x-a)^{n+1}}{(n+1)!}$.

♦ The term $r_n(x) = \dfrac{f^{(n+1)}(t_x)(x-a)^{n+1}}{(n+1)!}$ is called the "remainder after the nth derivative term". It uses the **n+1st** derivative; n does not signal the "n th term" of the series. The value of $r_n(x)$ depends on x and an unknown point t_x which is between a and x. We intend to use the approximation at some value of x. We generally choose "a" to be a point where we have enough information about the function to construct the polynomial. The approximation usually works well for x-values close to a.

♦ **The Maximum Error incurred by using** $f(x) \approx p_n(x)$ without the remainder term is

$$\text{Max } |r_n| = \text{Max } \left| \frac{f^{(n+1)}(t_x)(x-a)^{n+1}}{(n+1)!} \right|$$

EXAMPLE: Find the Taylor Formula for sin x with remainder $r_5(x)$ and a = 0.

Answer: $r_5(x) = \dfrac{f^{(6)}(t_x)(x-a)^6}{6!} = \dfrac{(-\sin t_x)x^6}{6!}$, so $\sin x = x - \dfrac{x^3}{3!} + \dfrac{x^5}{5!} - \dfrac{(\sin t_x)x^6}{6!}$

where t_x is between 0 and x.

EXAMPLE: Finding the error in an approximation for a given x. Find a bound on the error we can incur if we approximate sin $(\pi/7)$ with $p_5(x)$ and a = 0.

Answer: $\sin(\pi/7) = (\pi/7) - \dfrac{(\pi/7)^3}{3!} + \dfrac{(\pi/7)^5}{5!} + \dfrac{(\sin t_x)(\pi/7)^6}{6!}$. Since t_x is between 0 and x = $\pi/7$ and since $|\sin x| \leq 1$ for all x,

$$\text{Max } |r_5| \leq \text{Max } \left| \frac{(\sin t_x)(\pi/7)^6}{6!} \right| \leq \frac{(\pi/7)^6}{6!} = 1.13495462678\,E-5.$$ In other words, using

the approximation, $\sin(\pi/7) \approx (\pi/7) - \dfrac{(\pi/7)^3}{3!} + \dfrac{(\pi/7)^5}{5!} = .433884464752,$ we could be

off by as much as ± 1.13495462678 E-5. So our estimate is:

.433884464752 - 1.13495462678 E-5 $\leq$ sin $(\pi/7)$ $\leq$ .433884464752 + 1.13495462678 E-5

EXAMPLE: How many terms must we use in a Taylor polynomial approximation for sin $(\pi/7)$ to make the remainder (or error) less than .01 ? Use a = 0.

Answer: Taylor's Formula gives

$$\sin x = x - \frac{x^3}{3!} + \frac{x^5}{5!} - \frac{x^7}{7!} + \frac{x^9}{9!} + \ldots \ldots \left(\pm(\sin t_x \text{ or } \cos t_x)\frac{x^{n+1}}{(n+1)!} \right).$$ We need

$$\left| r_n \right| = \left| \pm (\sin t_x \ \text{ or } \ \cos t_x) \frac{(\pi/7)^{n+1}}{(n+1)!} \right| \le .01. \ \text{Since} \ \left| \pm \sin t_x \right| \le 1 \ \text{and} \ |\pm \cos t_x| \le 1 \text{, it}$$

follows that $\left| r_n \right| \le \left| \dfrac{(\pi/7)^{n+1}}{(n+1)!} \right|$. Now we require: $\left| \dfrac{(\pi/7)^{n+1}}{(n+1)!} \right| \le .01$. The smallest such n

can be determined by trial and error with a calculator. We found that when n = 4,

$\dfrac{(\pi/7)^5}{5!} = 1.517316E - 4. \le .01.$ Thus, if we run the approximation up to the 4th

derivative term (actually, this is 0), we obtain $\sin(\pi/7) \approx (\pi/7) - \dfrac{(\pi/7)^3}{3!} + 0.$ With this

approximation, our error is less than .01.

EXAMPLE: Find a Taylor Polynomial for $f(x) = 3x^2 - 2x + 7$.

Answer: We can use a = 0 and compute all derivatives (the third derivative and higher are zero). But the Taylor Polynomial of a polynomial using a = 0 turns out to be the polynomial itself. Now experiment with using a = 1, for example. You will find the Taylor Polynomial for f(x) is expressed in powers (x-1); namely the Taylor Polynomial is 8 + $4(x-1) + 3(x-1)^2$. Multiplying this out, you will see this equals $f(x) = 3x^2 - 2x + 7$, all the same.

NOTE: You can generally tell what "a" has been used in a Taylor Polynomial, because you will see powers of (x - a). For example, if we write $\cos x \approx 1 - x^2/2! + x^4/4!$, the powers of x indicate that a = 0, whereas if we write

$$\ln x \approx (x-1) - (x-1)^2/2 + (x-1)^3/3, \text{ then a = 1.}$$

The highest power of (x-a) indicates at which derivative we stopped the approximation. In the approximation for ln x, we stopped at the 3rd derivative term.

EXAMPLE: Compute $p_5(x)$ for $f(x) = \dfrac{1}{1-x}$.

Answer: Using a = 0, we compute derivatives.

$f(x) = (1-x)^{-1}$	$f(0) = 1$
$f'(x) = -(1-x)^{-2}(-1) = (1-x)^{-2}$	$f'(0) = 1$
$f''(x) = 2(1-x)^{-3}$	$f''(0) = 2$
$f'''(x) = 3!(1-x)^{-4}$	$f'''(0) = 3!$
We see the pattern: $f^{(n)}(x) = n!(1-x)^{-(n+1)}$	$f^{(n)}(0) = n!$

Then $p_5(x) = f(a) + f'(a)(x-a) + f''(a)(x-a)^2/2! + \ldots + f^{(5)}(a)(x-a)^5/5!$ and

replacing terms , $p_5(x) = 1 + x + x^2 + x^3 + x^4 + x^5$. So $f(x) = \dfrac{1}{1-x} \approx p_5(x)$.

EXAMPLE: Approximate $e^{1/3}$ with an error less than .001.

Answer: For $f(x) = e^x$, $r_n(x) = \dfrac{e^{t_x}(x-a)^{n+1}}{(n+1)!}$ (from the text), and since 1/3 is close to

0, it makes sense to use a = 0. Then $r_n(1/3) = \dfrac{e^{t_x}(1/3-0)^{n+1}}{(n+1)!}$ where $0 < t_x < (1/3)$.

We want Max $\left| r_n(1/3) \right| < .001$. But Max $\left| r_n(1/3) \right| = $ Max $\left| \dfrac{e^{t_x}(1/3)^{n+1}}{(n+1)!} \right| = \left| \dfrac{e^{1/3}(1/3)^{n+1}}{(n+1)!} \right|$.

Now we aren't supposed to know $e^{1/3}$. However, we will use a very coarse upper bound and demand that it be small. Since $e^{1/3} < e < 3$, we require:

$$\left| \frac{e^{1/3}(1/3)^{n+1}}{(n+1)!} \right| \le \left| \frac{3(1/3)^{n+1}}{(n+1)!} \right| = \left| \frac{(1/3)^n}{(n+1)!} \right| \le .001.$$

By trial and error, we find n = 4 is the "first" (smallest) value of n to make this quantity smaller than .001. The value of the term with n = 4 is 1.0288 E-4.

So, we say $e^{1/3} \approx 1 + (1/3) + (1/3)^2 / 2! + (1/3)^3 / 3! + (1/3)^4 / 4! = 1.395576....$ Our approximation is off by at most 1.0288...E-4. On calculator (cheating!), we see that $e^{1/3} = 1.395612425....$ and we are satisfied.

EXAMPLE: A Taylor Polynomial using an "a" which is not 0. How many terms are needed in the Taylor Polynomial for f(x) = ln x with a = 1 to give an error less than .01, when we use the Taylor Polynomial with x = 1.8?

Answer: First find the Taylor Polynomial:

f(x) = ln x	f(1) = 0
f '(x) = 1/x	f '(1) = 1
f "(x) = $-1/x^2$	f "(1) = -1
f '''(x) = $2/x^3$	f '''(1) = 2
$f^{(4)}(x) = -3! / x^4$. We see the pattern: $f^{(n)}(x) = \dfrac{(-1)^{n+1}(n-1)!}{x^n}$	$f^{(4)}(1) = -3!$

Using the derivatives computed in the table above, the expression

$$f(x) = f(a) + f'(a)(x-a) + \frac{f''(a)(x-a)^2}{2!} + \frac{f'''(a)(x-a)^3}{3!} + ... \frac{f^{(n+1)}(t_x)(x-a)^{n+1}}{(n+1)!}$$

becomes:

$$\ln x = (x-1) - \frac{(x-1)^2}{2} + \frac{(x-1)^3}{3} - \dots \frac{f^{(n+1)}(t_x)(x-1)^{n+1}}{(n+1)!} \text{ , with}$$

$$|r_n(x)| = \left| \frac{f^{(n+1)}(t_x)(x-1)^{n+1}}{(n+1)!} \right| = \left| \frac{(-1)^{n+2}(n!)(x-1)^{n+1}}{(n+1)!\,t_x^{n+1}} \right| \le \left| \frac{(1.8-1)^{n+1}}{(n+1)\,1^{n+1}} \right| = \frac{(.8)^{n+1}}{n+1}$$

Above, we used that $\dfrac{n!}{(n+1)!} = \dfrac{1}{n+1}$ and that the largest values of $\left(\dfrac{1}{t_x}\right)$ occur when we

have the *smallest* values of t_x on the interval (a, x) or (1, 1.8). Although $t_x = 1$ is not on

the interval itself, we use $t_x = 1$ for our bound, to make $\text{Max}\left(\dfrac{1}{t_x}\right) \le 1$.

Now we must find the n that makes $\dfrac{(.8)^{n+1}}{n+1} \le .01$, as desired. Taking n = 10 with

remainder at the 11th derivative term, we find that the approximation

$$\ln x = (x-1) - \frac{(x-1)^2}{2} + \frac{(x-1)^3}{3} - \frac{(x-1)^4}{4} + \frac{(x-1)^5}{5} - \dots - \frac{(x-1)^{10}}{10}$$

will approximate ln 1.8 with an error less than .01.

SECTION 9.2

" My center is giving way, my right is pushed back, situation excellent, I am attacking"
...Ferdinand Foch, Second Battle of the Marne

SEQUENCES

EXAMPLE:

(1) Write in sequence notation, $\dfrac{1}{2\cdot 3}, \dfrac{1}{3\cdot 4}, \dfrac{1}{4\cdot 5}, \dfrac{1}{5\cdot 6}, \dots$

Answer: Try to identify terms with their counting number, to see a pattern:

$$\frac{1}{2\cdot 3}, \frac{1}{3\cdot 4}, \frac{1}{4\cdot 5}, \frac{1}{5\cdot 6}, \dots$$

$$\uparrow \quad \uparrow \quad \uparrow \quad \uparrow$$

$$n = \quad 1 \quad 2 \quad 3 \quad 4$$

This helps us recognize $\{a_n\}_{n=1}^{\infty} = \left\{ \dfrac{1}{(n+1)(n+2)} \right\}_{n=1}^{\infty}$

(2) Write in sequence notation, $-1, 2^2, -3^2, 4^2, \dots$

Answer: When dealing with alternating signs, $(-1)^n$ or $(-1)^{n+1}$ can be used depending on whether even or odd-numbered terms are positive. Here $\{a_n\}_{n=1}^{\infty} = \{(-1)^n n^2\}_{n=1}^{\infty}$.

CONVERGENCE AND DIVERGENCE OF SEQUENCES

A sequence converges if $\lim\limits_{n\to\infty} a_n$ exists as a finite number.

A sequence diverges if
(1) the terms oscillate, and no limit can be found, or
(2) $\lim\limits_{n\to\infty} a_n = \infty$ or $\lim\limits_{n\to\infty} a_n = -\infty$

NOTE: The term "oscillates" may not mean what you expect-- it does not necessarily mean the sequence bounces back and forth in a rhythmical way. We generally reserve the word to mean the sequence has no limit. Terms of a sequence can bounce around and still converge (see sequence (7) below). When terms go to $\pm\infty$, and also oscillate, we say simply that the sequence oscillates. (see sequence (5) below).

EXAMPLE: Do the following sequences converge or diverge?

(1) $1, \dfrac{1}{2}, \dfrac{1}{4}, \dfrac{1}{8},....$ **Answer:** This sequence converges to 0.

(2) $1, \dfrac{1}{2}, 1, \dfrac{1}{8}, 1, \dfrac{1}{16},...$ **Answer:** This sequence diverges by oscillation.

(3) $1,2,3,4,...$ **Answer:** This sequence diverges to ∞

(4) $-1,-2^2,-3^2,-4^2,...$ **Answer:** This sequence diverges to $-\infty$

(5) $1,0,-1,0,2,0,-2,0,3,0,-3,0,...$ **Answer:** This sequence diverges by oscillation.

(6) $1,0,1,0,0,1,0,0,0,1,0,0,0,0,...$ **Answer:** This sequence diverges by oscillation.

(7) $1,-\dfrac{1}{2},\dfrac{1}{3},-\dfrac{1}{4},\dfrac{1}{5},...$ **Answer:** This sequence converges to 0.

FINDING LIMITS OF SEQUENCES

NOTE: We need to understand what it means for $\lim\limits_{n\to\infty} a_n$ to exist. Before, when we wrote $\lim\limits_{x\to\infty} f(x) = L$, x could "glide" through all real numbers to ∞, whereas now n "hops" by integers to ∞.

The text offers two methods for "verifying" a limit of a sequence: $\lim_{n \to \infty} a_n = L$.

1. Prove this from the definition, using ε's and N's.
2. Compare the sequence to an appropriate continuous function, f(x). Then
   $$\lim_{n \to \infty} a_n = \lim_{x \to \infty} f(x)$$

Using a comparison with a continuous functions we can prove a sequence has a limit more easily than by proving from the definition. But this method relies on the strict theory for functions having limits, a theory that requires ε -δ proofs. So the "continuous function" method should be viewed as intuitively obvious, but not rigorous .

EXAMPLE: Comparing the sequence to a continuous function.

Verify that $\left\{ \sin \dfrac{\pi}{n} \right\}_{n=1}^{\infty}$ converges to 0.

Answer: Since $\lim_{x \to \infty} \sin \dfrac{\pi}{x} = \sin 0 = 0$, then when x is an integer n, $\lim_{n \to \infty} \sin \dfrac{\pi}{n} = 0$.

EXAMPLE: L'Hopital's Rule can be used to evaluate limits.

Find $\lim_{n \to \infty} \dfrac{3n+2}{4n+1}$.

Answer: $\lim_{n \to \infty} \dfrac{3n+2}{4n+1} = \lim_{x \to \infty} \dfrac{3x+2}{4x+1} = \lim_{x \to \infty} \dfrac{d(3x+2)/dx}{d(4x+1)/dx} = \dfrac{3}{4}$.

NOTE : A sequence converges if $\lim_{n \to \infty} a_n = L$, **for some finite number, L. This means :**

given any ε > 0	(This is an error term. We want all terms in the tail end of the sequence to be within this error of the limit L)
there exists an N	(This is a big number--it points to a "cut-off" term on the sequence
such that whenever n ≥ N	(All terms are to the right of that cut-off point...)
$\left\|a_n - L\right\| < \varepsilon$	(... fall within ε of the limit L)

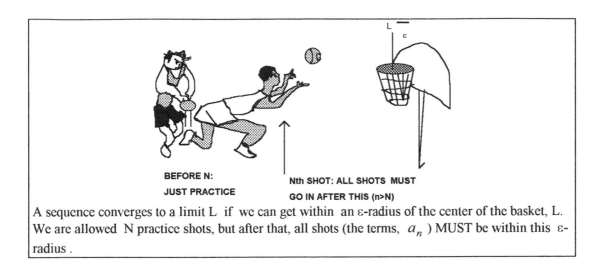

A sequence converges to a limit L if we can get within an ε-radius of the center of the basket, L. We are allowed N practice shots, but after that, all shots (the terms, a_n) MUST be within this ε-radius .

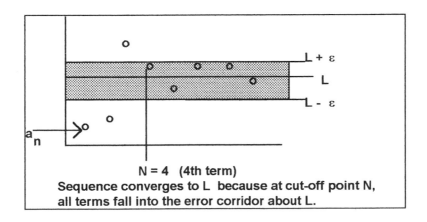

N = 4 (4th term)
Sequence converges to L because at cut-off point N,
all terms fall into the error corridor about L.

EXAMPLE: Using the definition of convergence. Show $\left\{ \dfrac{1}{n^2} \right\}_{n=1}^{\infty}$ converges to 0.

Answer: We must show , given any $\varepsilon > 0$, there exists an N such that

$n \ge N \;\Rightarrow\; \left| a_n - L \right| < \varepsilon$, or (Statement of definition)

$n \ge N \;\Rightarrow\; \left| \dfrac{1}{n^2} - 0 \right| < \varepsilon$, or (Replace formal terms with their values)

$n \ge N \;\Rightarrow\; \dfrac{1}{n^2} < \varepsilon$, or (Clear absolute value)

$n \ge N \;\Rightarrow\; \dfrac{1}{\varepsilon} < n^2$, or (Get expression in the form n >)

$n \ge N \;\Rightarrow\; n > \dfrac{1}{\sqrt{\varepsilon}}$. (Two sides look alike: we can now conclude...)

Take $N \ge \dfrac{1}{\sqrt{\varepsilon}}$. (With this N the previous statements are true.)

NOTE: A sequence diverges to +∞ if

given any M > 0   (Given any arbitrary, large number)

there exists an N   (This is our "cut-off" point)
such that if n ≥ N   (All terms to the right of the cut-off point...)

$a_n > M$ (...are greater than the arbitrary M).

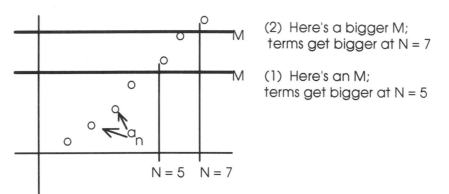

(2) Here's a bigger M;
terms get bigger at N = 7

(1) Here's an M;
terms get bigger at N = 5

Sequence diverges to ∞. Can always find cut-off point N
such that terms to the right get bigger than arbitrary M.

EXAMPLE: Using the definition to show divergence to ∞. Show the sequence
$\left\{n^3\right\}_{n=1}^{\infty}$ diverges to ∞.

Answer: We must show given any M > 0, there exists an N such that

$n \geq N \implies a_n > M$, or   (This is the statement)
$n \geq N \implies n^3 > M$, or   (Replace the formal terms with their values)
$n \geq N \implies n > \sqrt[3]{M}$.   (Get into the form "n > ...")

Take $N \geq \sqrt[3]{M}$.   (Then all previous statements will be true.)

==
SECTION 9.3
==

" Let us be thankful for the fools. But for them the rest of us could not succeed"...Mark Twain

BOUNDED SEQUENCES AND CONVERGENCE

A sequence is bounded above by M if $a_n \leq M$ for all n (for all terms). Since the left hand end of a sequence has a finite number of terms, it is **always bounded** by the largest term. We are only concerned whether the right hand end--the infinite end-- is bounded. If $a_n \geq M$ for all n, the sequence is bounded below.

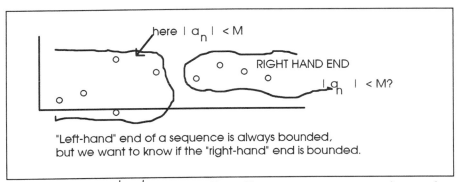

here | a_n | < M

RIGHT HAND END

| a_n | < M?

"Left-hand" end of a sequence is always bounded, but we want to know if the "right-hand" end is bounded.

A sequence is **bounded** if $\left| a_n \right| \leq M$ for all n . If a sequence converges, it must be bounded. But a bounded sequence does not necessarily converge.

We might say a sequence is "eventually" non-decreasing if all terms past some cut-off point are non-decreasing. I.e., there should be some integer N such that for all $n \geq N$, $a_n \leq a_{n+1}$. The sequence may wobble around for a bit, but after some cut-off point, terms satisfy $a_n \leq a_{n+1}$. E.g.: 3, 2, -10, 6,7,8,8, 9,9,10,10...is "eventually" non-decreasing.

A sequence which is **bounded above** and "eventually" **non-decreasing** will converge to the least upper bound of all its terms.

A sequence which is **bounded below** and eventually **non-increasing** (in the sense that $a_n \geq a_{n+1}$) will converge to the greatest lower bound of all its terms.

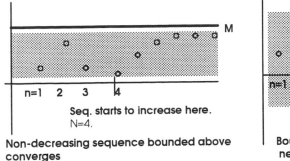

M

n=1   2   3   4

Seq. starts to increase here.
N=4.

Non-decreasing sequence bounded above converges

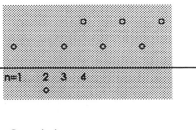

n=1   2   3   4

Bounded sequence may nevertheless diverge.

EXAMPLE: Prove $\left\{\dfrac{4^n}{n!}\right\}_{n=1}^{\infty}$ converges.

Answer: It would be difficult to use the definition, and l'Hopital's Rule does not apply. This sequence is **eventually** decreasing (for $n > 4$) because successive terms are multiplied by $4/n$. And it is bounded below by 0. So it converges (in fact to 0).

"HIERARCHY OF THE n's " (Study Guide term):   There is an order of magnitudes to large numbers that is helpful to know:

When n is very, very large:

$$n^n > n! > 3^n > 2^n > n^3 > n^2 > n > \sqrt{n} > \ln n$$

EXAMPLE: Determine if the following sequences converge.

(1) $\left\{\dfrac{n^n}{n!}\right\}_{n=1}^{\infty}$

Answer: The terms have the form $\dfrac{n}{n} \cdot \dfrac{n}{(n-1)} \cdot \dfrac{n}{(n-2)} \cdots \dfrac{n}{2} \cdot \dfrac{n}{1}$, and the numerator "outweighs" the denominator.   Sequence diverges to ∞.

(2) $\left\{\dfrac{4^n}{n!}\right\}_{n=1}^{\infty}$

Answer: The terms are like $\dfrac{4}{n} \cdot \dfrac{4}{(n-1)} \cdots \dfrac{4}{2} \cdot \dfrac{4}{1}$. As n gets enormous, the denominator "outweighs" the numerator.   Sequence converges to 0.

(3) $\left\{\dfrac{2^n n!}{(2n)!}\right\}_{n=1}^{\infty}$

Answer: The terms have the form $\dfrac{2n}{2n} \cdot \dfrac{1}{(2n-1)} \cdot \dfrac{2(n-1)}{(2n-2)} \cdot \dfrac{1}{(2n-3)} \cdots \dfrac{2(1)}{1}$, and the sequence converges to 0.

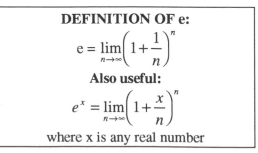

DEFINITION OF e:

$$e = \lim_{n \to \infty}\left(1 + \frac{1}{n}\right)^n$$

Also useful:

$$e^x = \lim_{n \to \infty}\left(1 + \frac{x}{n}\right)^n$$

where x is any real number

SECTION 9.4

"This new development (automation) has unbounded possibilities for good and for evil"...Norbert Wiener

INFINITE SERIES

EXAMPLE: Write in summation notation: $\dfrac{2}{3^2} + \dfrac{3}{4^2} + \dfrac{4}{5^2} + \ldots$

Answer: Index the terms by n = 1, 2, ... and try to find a pattern!

$$\underset{n=1}{\dfrac{2}{3^2}} + \underset{2}{\dfrac{3}{4^2}} + \underset{3}{\dfrac{4}{5^2}} + \ldots$$

There may be many ways to write this series in summation notation. One way is to notice that the terms are $a_n = \dfrac{n+1}{(n+2)^2}$, and thus the series is $\displaystyle\sum_{n=1}^{\infty} \dfrac{n+1}{(n+2)^2}$. (For what n the series begins is important.)

NOTE: A series converges if the infinite sum is a finite number. More formally, the series converges if the sequence of partial sums converges. If a series does not converge, it may diverge to ∞ or $-\infty$ or oscillate. (See examples below).

Series: _____ 1 + 1/10 + 1/100 + 1/ 1000 + 1/ 10000 + ...

Partial Sums: S_n ⌐ 1,   1.1,   1.11,   1.111,   1.1111, ...

The sequence of partial sums converges so the SERIES converges (to 1.1111....= 10/9).

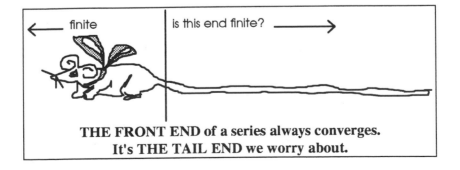

finite ← | is this end finite? ⟶

THE FRONT END of a series always converges.
It's THE TAIL END we worry about.

EXAMPLE: Using partial sums, determine if the series below converge.

(a) $1 + 1/2 + 1/4 + ...$

Answer: The sequence of partial sums is $1, \ 3/2, \ 7/4, \ 15/8,....\dfrac{2^n - 1}{2^{n-1}}$. Since

$\lim\limits_{n\to\infty} \dfrac{2^n - 1}{2^{n-1}} = 2$ (as n gets large, think of the -1 in the numerator as inconsequential), the

series converges to 2.

$$1 + 1/2 + 1/4 + ... = 2$$

(b) $1 + 2 + 3 + ...$ **Answer:** The sequence of partial sums is $1, 3, 6, ...n(n+1)/2$. Series diverges to ∞

(c) $- 1 - 2 - 3 -...$ **Answer:** The sequence of partial sums is the negative of the above. Series diverges to $-\infty$

(d) $1 - 2 + 3 - 4 + 5 ...$ **Answer:** The sequence of partial sums is $1, -1, 1, -1, ...$ Series diverges by oscillation.

(e) $1 - 1 + 1 - 1 +...$ **Answer:** The sequence of partial sums is $1, 0, 1, ...$ Series diverges by oscillation.

CONVERGENCE OF SERIES

Often it is difficult to tell if the sequence of partial sums converges. So we have many, many tests to help us determine if a series converges.

MOST-MISUNDERSTOOD TEST: $\lim\limits_{n\to\infty} a_n = 0$ **TEST** (sometimes called "last term test")

If $\lim\limits_{n\to\infty} a_n \neq 0$, the series $\sum\limits_{n=1}^{\infty} a_n$ diverges. This is a misunderstood theorem because

students often think it says; if $\lim\limits_{n\to\infty} a_n = 0$, the series converges (which is erroneous).

This test says nothing about convergence.

THE HARMONIC SERIES, $1 + 1/2 + 1/3 + 1/4 + ...$, diverges although a_n goes to 0. Read the proof of this in your text!

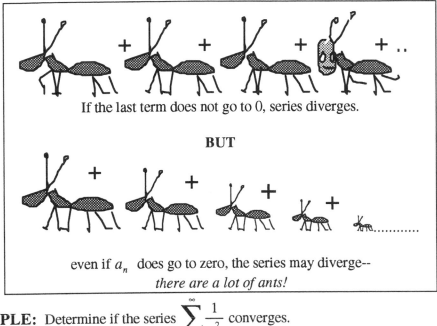

If the last term does not go to 0, series diverges.

BUT

even if a_n does go to zero, the series may diverge--
there are a lot of ants!

EXAMPLE: Determine if the series $\displaystyle\sum_{n=1}^{\infty} \frac{1}{n^2}$ converges.

Answer: a_n goes to 0, but this gives no information. We don't know yet.

EXAMPLE: Determine if $\displaystyle\sum_{n=1}^{\infty} \sin(\frac{\pi}{2}+\frac{1}{n})$ converges.

Answer: Since a_n goes to $\sin\frac{\pi}{2}=1$ and not 0, the series diverges.

EXAMPLE: Determine if (a) the sequence $\left\{\cos(\frac{\pi}{2}+\frac{1}{n})\right\}_{n=1}^{\infty}$ converges,

(b) the series $\displaystyle\sum_{n=1}^{\infty} \cos(\frac{\pi}{2}+\frac{1}{n})$ converges.

Answer: The sequence in (a) converges to 0. In (b), a_n goes to 0, but this says nothing about convergence. We do not know if this series converges.

Quick Review

Many students confuse sequences and series . Where are sequences used in talking about series? Define what it means for a sequence to converge. What are partial sums? What does it mean for a series to converge? State a result about the convergence of sequences involving terms being bounded. Is there a result if the terms of a sequence are not bounded?

State a result about the convergence of a series, referring to the a_n If $\displaystyle\lim_{n\to\infty} a_n$ in $\{a_n\}_{n=1}^{\infty}$, does the sequence converge? (The answer is yes; why?) If $\lim\limits_{n\to\infty} a_n$ exits for the series $\sum\limits_{n=1}^{\infty} a_n$,

do we have a result?

TWO SPECIAL SERIES CONVERGE TO LIMITS THAT CAN BE FOUND

In general, we cannot find what a series converges to. There are some exceptions. Here are two special cases, where we can say WHAT the series converges to, if it converges.

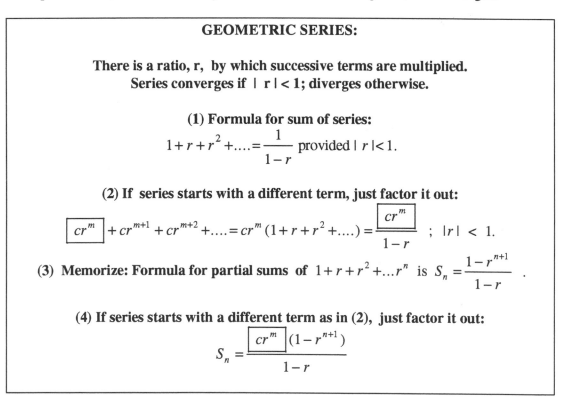

GEOMETRIC SERIES:

There is a ratio, **r,** by which successive terms are multiplied.
Series converges if $|\,r\,| < 1$; diverges otherwise.

(1) Formula for sum of series:
$$1 + r + r^2 + \ldots = \frac{1}{1-r} \text{ provided } |\,r\,| < 1.$$

(2) If series starts with a different term, just factor it out:
$$\boxed{cr^m} + cr^{m+1} + cr^{m+2} + \ldots = cr^m(1 + r + r^2 + \ldots) = \frac{\boxed{cr^m}}{1-r} \;;\; |r| < 1.$$

(3) Memorize: Formula for partial sums of $1 + r + r^2 + \ldots r^n$ **is** $S_n = \dfrac{1 - r^{n+1}}{1-r}$.

(4) If series starts with a different term as in (2), just factor it out:
$$S_n = \frac{\boxed{cr^m}(1 - r^{n+1})}{1-r}$$

EXAMPLE: Find the ratio of the geometric series $\sum\limits_{n=1}^{\infty} \dfrac{2^n}{5^{n+1}}$.

Answer: ALWAYS WRITE OUT A FEW TERMS:

$\sum\limits_{n=1}^{\infty} \dfrac{2^n}{5^{n+1}} = \dfrac{2}{5^2} + \dfrac{2^2}{5^3} + \dfrac{2^3}{5^4} + \ldots$ The ratio, r, is what each term is multiplied by to get the next

term. (You can also find it from $r = \dfrac{a_{n+1}}{a_n}$.) Here we multiply each term by 2/5, so this is

the ratio.

EXAMPLE: Find a formula for the partial sums of the series $\sum_{n=1}^{\infty}\left(\frac{1}{3}\right)^n$ and determine whether the partial sums have a limit. If so, find the limit.

Answer: ALWAYS WRITE OUT A FEW TERMS:

$\sum_{n=1}^{\infty}\left(\frac{1}{3}\right)^n = \frac{1}{3}+\left(\frac{1}{3}\right)^2+\left(\frac{1}{3}\right)^3+...$ We see this is a geometric series with ratio r = 1/3. The

partial sums are given by $S_n = \frac{cr^m(1-r^n)}{1-r^{n+1}} = \frac{(1/3)(1-(1/3)^n)}{1-(1/3)^{n+1}}$. And

$\lim_{n\to\infty} S_n = \lim_{n\to\infty} \frac{(1/3)(1-(1/3)^n)}{1-(1/3)} = 1/2$. So the series converges to 1/2. In fact, we don't

need partial sums to see what this series converges to, since

$$\frac{1}{3}+\frac{1}{3^2}+\frac{1}{3^3}+....=\frac{1}{3}(1+\frac{1}{3}+\frac{1}{3^2}+\frac{1}{3^3}+....)=\frac{1}{3}\left(\frac{1}{1-1/3}\right)=\frac{1}{2}.$$

EXAMPLE: To what does $\sum_{n=1}^{\infty}\frac{(-3)^{n+1}}{4^n}$ converge?

Answer:

$$\underset{A}{\frac{3^2}{4}}-\frac{3^3}{4^2}+\frac{3^4}{4^3}-....=\underset{B}{\frac{3^2}{4}}\underset{C}{\left(1-\frac{3}{4}+\left(-\frac{3}{4}\right)^2+.....\right)}=\underset{D}{\frac{3^2}{4}\left(\frac{1}{1-(-3/4)}\right)}=\frac{9}{7}$$

A: Write out some terms. We recognize this series as geometric.
B: Factor out the first term.
C: We can now see the ratio is -3/4--the ratio is what successive terms are multiplied by.
D: Series converges, since | -3/4 | < 1. Converges to 1/ (1-r), multiplied by 1st term. Thus,

this series converges to $\frac{3^2}{4}\left(\frac{1}{1-(-3/4)}\right)=\frac{9}{7}$.

EXAMPLE: What rational number is represented by the repeating decimal 0.567676767...?

Answer: 0.567676767... =

$$\frac{5}{10}+\frac{67}{10^3}+\frac{67}{10^5}+... = \frac{5}{10}+\frac{67}{10^3}(1+\frac{1}{10^2}+\frac{1}{10^4}+....)=\frac{5}{10}+\frac{67}{10^3}\left(\frac{1}{1-\frac{1}{10^2}}\right)$$

$$=\frac{5}{10}+\frac{67}{10^3}\left(\frac{100}{99}\right)=\frac{281}{495}.$$

EXAMPLE: What rational number is represented by the decimal 2.35678678678678...?
Hint: Write this as

$$2 + 2 + \frac{3}{10} + \frac{5}{100} + \frac{678}{10^5}\left(1 + \frac{1}{10^3} + \left(\frac{1}{10^3}\right)^2 + ...\right) \text{ and proceed as above.}$$

NOTE: Every decimal number that is eventually repeating can be shown to represent a rational number, by the methods above. Every rational number is eventually a repeating decimal. Here's a loose explanation (try to reformulate this argument more precisely!) If we divide the numerator by the denominator to obtain a decimal expression for our rational number, at each division, the remainder term will be some digit from 0 through 9. Eventually the same remainder must pop up, at which point we will repeat our steps. (Try finding the decimal expansion of 1/7.)

EXAMPLE: $\displaystyle\sum_{n=1}^{\infty} e^n = ?$

Answer: Geometric series, but ratio, e is greater than 1. The series diverges.

TELESCOPING SERIES:
In these series, like terms are added and subtracted. When
the last term goes to 0, the middle terms will cancel---and we can find the limit of the series
from what remains after cancellation.

EXAMPLE: Find Partial Sums for $\displaystyle\sum_{n=1}^{\infty} \frac{1}{n(n+1)}$ and determine what the series converges

to, if it converges.

Answer: This series is telescoping. Using **partial fractions:** $\dfrac{1}{n(n+1)} = \dfrac{1}{n} - \dfrac{1}{n+1}$, so

$$\sum_{n=1}^{\infty} \frac{1}{n(n+1)} = \sum_{n=1}^{\infty}\left(\frac{1}{n} - \frac{1}{n+1}\right) = \left(1 - \frac{1}{2}\right) + \left(\frac{1}{2} - \frac{1}{3}\right) + \left(\frac{1}{3} - \frac{1}{4}\right) + ... \text{ (Equation 1)}$$

Thus $S_1 = 1 - 1/2$, $S_2 = 1 - 1/3$, and in general $S_n = 1 - 1/n$. Then

$$\lim_{n\to\infty} S_n = \lim_{n\to\infty}(1 - 1/n) = 1 \text{ and the series converges to 1.}$$

As with geometric series we can usually bypass partial sums, and find the sum from Equation (1).

EXAMPLE: Find what the series $\displaystyle\sum_{n=2}^{\infty} \frac{2}{n(n+2)}$ converges to.

Answer: Write this telescoping series $\displaystyle\sum_{n=2}^{\infty} \frac{2}{n(n+2)}$, using partial fractions, as

$$\sum_{n=2}^{\infty} \frac{1}{n} - \frac{1}{(n+2)} = \left(\frac{1}{2} - \frac{1}{4}\right) + \left(\frac{1}{3} - \frac{1}{5}\right) + \left(\frac{1}{4} - \frac{1}{6}\right) + \left(\frac{1}{5} - \frac{1}{7}\right) + \ldots$$

All terms cancel but 1/2 and 1/3: (1/2)+(1/3)= 5/6. The series converges to 5/6.

SECTION 9.5

"Never put off until tomorrow what should have been done in the early Seventies"...George Ade

SERIES WITH NON-NEGATIVE TERMS

COMPARISON TEST

To use this test, you must sense in advance whether the series converges or diverges in order to make a good comparision. It is useful to know the following series for comparison purposes:

$\bullet \displaystyle\sum_{n=1}^{\infty} \frac{1}{n}$ diverges. This is the harmonic series

$\bullet \displaystyle\sum_{n=1}^{\infty} \frac{1}{n^p} \begin{cases} \text{converges} & \text{if } p > 1 \\ \text{diverges} & \text{if } p \leq 1 \end{cases}$. This is the "p-test", analogous to the "p-test" we saw with

Improper Integrals. We can prove it with the Integral Test, which comes later in this chapter. As an example,

$\bullet \displaystyle\sum_{n=1}^{\infty} \frac{1}{n^2}$ converges. (p = 2. We don't as yet have a way of finding what this series converges to.)

$\bullet \displaystyle\sum_{n=1}^{\infty} \frac{1}{\sqrt{n}}$ diverges. (p = 1/2.)

$\bullet$ Geometric series converge, if $|r| < 1$, and diverge otherwise.

COMPARISON TEST: If the terms of both series are non-negative,

- If $a_n \le b_n$ and $\displaystyle\sum_{n=1}^{\infty} b_n$ converges, then $\displaystyle\sum_{n=1}^{\infty} a_n$ does, too.

  BUT.... if $a_n \le b_n$ and $\displaystyle\sum_{n=1}^{\infty} b_n$ diverges, we can't say anything about $\displaystyle\sum_{n=1}^{\infty} a_n$.

- If $a_n \le b_n$ and $\displaystyle\sum_{n=1}^{\infty} a_n$ diverges to ∞, then $\displaystyle\sum_{n=1}^{\infty} b_n$ does, too.

  BUT.... if $a_n \le b_n$ and $\displaystyle\sum_{n=1}^{\infty} a_n$ converges, we can't say anything about $\displaystyle\sum_{n=1}^{\infty} b_n$.

The Comparison Test tells us when two series diverge or converge together.

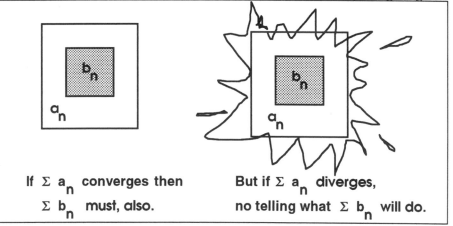

If $\Sigma\ a_n$ converges then $\Sigma\ b_n$ must, also.

But if $\Sigma\ a_n$ diverges, no telling what $\Sigma\ b_n$ will do.

Common Error **This reasoning is false!** Does $\displaystyle\sum_{n=2}^{\infty} \frac{\sqrt{n}}{\ln n}$ converge?

Since $\dfrac{\sqrt{n}}{\ln n} < \dfrac{\sqrt{n}}{n} = \dfrac{1}{n^{1/2}}$ and $\displaystyle\sum_{n=1}^{\infty} \frac{1}{n^{1/2}}$ diverges by the p - test, the series converges .

Showing that a series is term by term less than a divergent series tells us nothing.

EXAMPLE: The comparison in the Comparison Test must take place term by term. But the comparison need not go into effect at the first term, when n = 1 . Determine if the series below converge or diverge.

(1) $\sum_{n=2}^{\infty} \dfrac{2}{\sqrt{n^2 - 1}}$. **Answer:** Now $\dfrac{2}{\sqrt{n^2 - 1}} > \dfrac{2}{n}$ when n $>$ 4, and $\sum_{n=2}^{\infty} \dfrac{2}{n}$ diverges

(harmonic). Thus our original series diverges.

(2) $\sum_{n=1}^{\infty} \dfrac{n}{5^n}$. **Answer:** Since $\dfrac{n}{5^n} < \dfrac{2^n}{5^n} = (2/5)^n$ the series converges by comparison with

the geometric series $\sum_{n=1}^{\infty} (2/5)^n$ which has ratio $2/5$ and converges since $2/5 < 1$.

(3) $\sum_{n=1}^{\infty} \dfrac{1}{n!}$. **Answer:** If $n > 3$, $\dfrac{1}{n!} < \dfrac{1}{n^2}$. Series converges by comparison with $\sum_{n=1}^{\infty} \dfrac{1}{n^2}$.

LIMIT COMPARISON TEST

If terms of our series are similar to those of another series as n gets large , then we may use the

Limit Comparison Test .   Given $\sum_{n=1}^{\infty} a_n$, if terms seem similar to those of

$\sum_{n=1}^{\infty} b_n$ whose convergence we know about , we find $\lim_{n \to \infty} \dfrac{a_n}{b_n}$. If this limit is any

number other than 0 (and not ∞ either), then $\sum_{n=1}^{\infty} a_n$ and $\sum_{n=1}^{\infty} b_n$ converge or diverge

"together". I.e., they both converge or both diverge .If $\lim_{n \to \infty} \dfrac{a_n}{b_n} = 0$ or ∞, the comparison is

not proper . There is no information ; the test fails; try another test .

If two series are term-wise "alike", then they converge or diverge together. But if the series are not "alike", we cannot conclude anything. The Limit Comparison Test determines whether the series behave similarly.

EXAMPLE: Decide if the series below converge or not by the Limit Comparison Test.

(1) $\sum_{n=1}^{\infty} \dfrac{2}{\sqrt{n^2+5}}$. **Answer:** Terms "look like" $\dfrac{2}{n}$, so we suspect divergence. We take:

$\lim\limits_{n\to\infty} \dfrac{2/\sqrt{n^2+5}}{2/n} = 1$. Since this limit is not 0 or ∞, the test holds. Both series

converge or diverge together. Since $\sum_{n=1}^{\infty} \dfrac{2}{n}$ diverges, our series does too.

(2) $\sum_{n=3}^{\infty} \dfrac{\sqrt{n}}{n^2-4}$. **Answer:** Suppose we compare terms to $\dfrac{1}{n^2}$, knowing that $\sum_{n=1}^{\infty} \dfrac{1}{n^2}$

converges. But $\lim\limits_{n\to\infty} \dfrac{1/n^2}{\sqrt{n}/(n^2-4)} = \lim\limits_{n\to\infty} \dfrac{(n^2-4)}{n^2\sqrt{n}} = 0$. A limit of 0 means the

test fails, and we must use a different comparison or test. In fact, our comparison was not "fine" enough. See the solution in (3) below.

(3) $\sum_{n=3}^{\infty} \dfrac{\sqrt{n}}{n^2-4}$. **Answer:** The term $\dfrac{\sqrt{n}}{n^2-4}$ "looks like" $\dfrac{\sqrt{n}}{n^2} = \dfrac{1}{n^{3/2}}$, and we know

$\sum_{n=1}^{\infty} \dfrac{1}{n^{3/2}}$ converges by the p-test ($3/2 > 1$). Now $\lim\limits_{n\to\infty} \dfrac{\sqrt{n}/(n^2-4)}{1/n^{3/2}} = 1$, so the

Limit Comparison Test holds. Since $\sum_{n=1}^{\infty} \dfrac{1}{n^{3/2}}$ converges, so does our series.

NOTE: The Limit Comparison Test is easier to use than the Comparison Test, particularly when dealing with polynomials---as n gets large, polynomials behave like their leading terms. In Exercise (2) above, a comparison test would have been tricky to use.

INTEGRAL TEST

Given $\sum_{n=1}^{\infty} a_n$, if $a_n = f(n)$ for some f, and we can find $\int_{1}^{\infty} f(x)\,dx$, use this test.

Then $\sum_{n=1}^{\infty} a_n$ converges if $\int_{1}^{\infty} f(x)\,dx$ does, and diverges if $\int_{1}^{\infty} f(x)\,dx$ does.

EXAMPLE:

(1) Determine if the series $\sum_{n=1}^{\infty} \dfrac{\ln n}{n}$ converges. **Answer:** We compare $\sum_{n=1}^{\infty} \dfrac{\ln n}{n}$ with the

integral $\int_{1}^{\infty} \dfrac{\ln x}{x} dx$. But this integral diverges , since $\int_{1}^{\infty} \dfrac{\ln x}{x} dx = \lim_{N \to \infty} \dfrac{(\ln x)^2}{2} \Big|_{1}^{N} = \infty$.

Therefore the series diverges. (We could in fact have used the Comparison Test to compare our series with the harmonic series--this would have been easier.)

2) Determine if the series $\sum_{n=1}^{\infty} \dfrac{1}{n(\ln n)^2}$ converges. **Answer:** We compare our series to

$\int_{2}^{\infty} \dfrac{1}{x(\ln x)^2}$. Notice it does not matter that the lower limit is x = 2, not x = 1. Now

$\int_{2}^{\infty} \dfrac{1}{x(\ln x)^2} = -\lim_{N \to \infty} (\ln x)^{-1} \Big|_{2}^{N} = \dfrac{1}{\ln 2}$, which is finite. Since the integral converges, our

series converges.

HINT: Use the Integral Test for series with "continuous" looking terms--series that do not have factorials or powers of n, but series with terms like e^n, $\ln n$, $\sin n$, etc.

NOTE: The series $\sum_{n=1}^{\infty} \dfrac{1}{n^p}$ converges if p > 1, diverges otherwise. We called this the p-test. This result is derived from the Integral Test. See the section on Improper Integrals in this Guide that deals with the p-test for integrals. The results of that test are analogous.

NOTE: Approximating a series with integrals. Sometimes we can approximate the sum of a series by making comparisons with integrals. Consider the series $\sum_{n=1}^{\infty} \dfrac{1}{n^2}$. In Figs. (a) and (c) below, the shaded blocks have base 1, and heights determined by the curve $y = \dfrac{1}{x^2}$; therefore these shaded blocks have areas that represent the terms of our series: $1 + \dfrac{1}{2^2} + \dfrac{1}{3^2} + ...$ In Fig. (a) , where we use a lower sum, we see that

$\sum_{n=1}^{\infty} \dfrac{1}{n^2} \leq$ (area of lst block plus area under curve from 1 to ∞) $= 1 + \int_{1}^{\infty} \dfrac{1}{x^2} dx = 1 + 1 = 2$. And

in Fig. (c), where we use an upper sum, we have that $\sum_{n=1}^{\infty} \dfrac{1}{n^2} \geq \int_{1}^{\infty} \dfrac{1}{x^2} dx = 1$. We can

sharpen this estimate by summing the first terms of the series. Refer to Fig. (c):

$$\sum_{n=1}^{\infty} \dfrac{1}{n^2} \geq 1 + \dfrac{1}{2^2} + \dfrac{1}{3^2} + \int_{4}^{\infty} \dfrac{1}{x^2} dx = 1 + \dfrac{1}{2^2} + \dfrac{1}{3^2} + \dfrac{1}{4}.$$

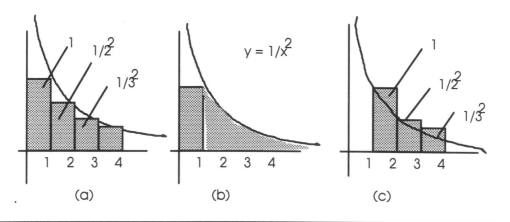

REVIEW

1. Name two types of series for which we can find a sum, when the series is convergent. (What is the sum of each of these types?) Can you *derive* the formula for the partial sum of a geometric series?

2. State the test that uses $\lim_{n \to \infty} a_n$. How is it often misused?

3. State the Comparison Test. State the Limit Comparison Test. How are these two tests related? What number, when we take a limit in the Limit Comparison Test, tells us the test will not work? What are some good series to remember, for use with the Comparison Test?

4. What is the Integral Test? For what types of series might we use this test?

5. Why are these tests only useful for series with non-negative terms? Can any of these tests be "broadened" to apply to series with non-positive terms?

Partial Answers: *(1) The geometric and telescoping series. The geometric series converges to a/(1-r) where a is the first term and r is the ratio. Write out some terms of the telescoping series to see what cancels, and what it converges to. To derive the formula for the partial sums, see your text. This is a popular question. (2) The "last term" test tells us that a series diverges if the "last term" fails to go to 0. It is often misinterpreted to say that the series converges if the "last term" goes to 0. (3) See the above. The Comparison Test makes a comparison of two series term by term. The Limit Comparison Test compares the limit of the ratios of terms of two series. The Limit Comparison Test fails if we see a ratio of 0 (or $\propto$). (4) See the above.*

Review ! Review !

Teachers like questions that require you to review. Does the last term of the series $\displaystyle\sum_{n=1}^{\infty} ne^{-n}$

go to zero? (Use l'Hopital's Rule) Does the series converge? (You can't go by the last term.) If you apply the Integral Test, you will need to use integration by parts .

SECTION 9.6

"Beyond the bright searchlights of science, Out of sight of the windows of sense, Old riddles still bid us defiance, Old questions of Why and of Whence"...William Cecil Whetham

SERIES WITH NON-NEGATIVE TERMS, Continued:

RATIO TEST

$$\lim_{n \to \infty} \frac{a_{n+1}}{a_n} \begin{cases} < 1 & \text{series converges} \\ > 1 & \text{series diverges} \\ = 1 & \text{no information; test fails; try another test} \end{cases}$$

(This test works well with powers and factorials.)

EXAMPLE: Determine if the series below converge.

(1) $\sum_{n=1}^{\infty} \dfrac{n!}{3^n}$. **Answer:** (If you remember our "hierarchy of n's", we see that the last term

does not go to 0 and the series diverges.) We can also use the Ratio Test:

$\lim_{n \to \infty} \dfrac{a_{n+1}}{a_n} = \lim_{n \to \infty} \dfrac{(n+1)!/3^{n+1}}{n!/3^n} = \lim_{n \to \infty} \dfrac{(n+1)n!}{n!} \dfrac{3^n}{3^{n+1}} = \lim_{n \to \infty} \dfrac{(n+1)}{3} = \infty$. Since this limit is

greater than 1, the series diverges.

(2) $\sum_{n=1}^{\infty} \dfrac{n!}{(2n)!}$. **Answer:** $\lim_{n \to \infty} \dfrac{a_{n+1}}{a_n} = \lim_{n \to \infty} \dfrac{(n+1)!/(2(n+1))!}{n!/(2n)!} =$

$\lim_{n \to \infty} \dfrac{(n+1)!}{n!} \dfrac{(2n)!}{(2n+2)(2n+1)(2n)!} = \lim_{n \to \infty} \dfrac{(n+1)}{1} \dfrac{1}{(2n+2)(2n+1)} = 0 < 1$. The

series converges.

(3) $\sum_{n=1}^{n=\infty} \dfrac{1}{n+2}$. **Answer:** $\lim_{n \to \infty} \dfrac{a_{n+1}}{a_n} = \lim_{n \to \infty} \dfrac{1/(n+1+2)}{1/n+2} = \boxed{1}$. The test fails! Use

another test. The Limit Comparison Test with the harmonic series will show divergence.

CAUTION! In the Ratio Test, a limit of 1 means the test fails. A limit of 0 means convergence. In the Limit Comparison Test, however, a limit of 1 means a good comparison. In the Limit Comparison Test, a limit of 0 means a bad comparison.

ROOT TEST

Given $\displaystyle\sum_{n=1}^{\infty} a_n$, where the a_n are non-negative

if $\displaystyle\lim_{n\to\infty} \sqrt[n]{a_n} \begin{cases} < 1 & \text{, series converges} \\ > 1 & \text{, series diverges} \\ = 1 & \text{, no information; test fails; try another test} \end{cases}$

Generally we use this test when a_n is an nth power of something.

HINT: Don't forget $\displaystyle\lim_{n\to\infty} \sqrt[n]{n} = 1$ and $\displaystyle\lim_{n\to\infty} \sqrt[n]{c} = 1$ if $c > 0$.

EXAMPLE:

(1) Determine if $\displaystyle\sum_{n=1}^{\infty} \left(\frac{n}{2n+5}\right)^n$ converges.

Answer: Using the Root Test, $\displaystyle\lim_{n\to\infty} \sqrt[n]{a_n} = \lim_{n\to\infty} n/(2n+5) = 1/2 < 1$. Series converges.

(2) Determine if $\displaystyle\sum_{n=1}^{\infty} \left(1+\frac{1}{n}\right)^n$ converges. **Answer:** By the Root Test

$\displaystyle\lim_{n\to\infty} \sqrt[n]{a_n} = \lim_{n\to\infty}\left(1+\frac{1}{n}\right) = 1$. No information. However $\displaystyle\lim_{n\to\infty} a_n = e$ and not 0, so the series diverges.

SECTION 9.7

" Mathematics takes us still further from what is human, into the region of absolute necessity, to which not only the actual world, but every possible world, must conform"...Bertrand Russell

ALTERNATING SERIES TEST

If a series is strictly alternating and terms decrease to 0, the series converges.

The error incurred by approximating an alternating series by its nth partial sum is less than the first neglected term, in absolute value.

EXAMPLE: Does $1 - \dfrac{1}{2} + \dfrac{1}{3} - \dfrac{1}{4} + \dfrac{1}{5} - \dfrac{1}{6} + \dots$ converge?

Answer: This series converges because it is strictly alternating and terms decrease to 0. Approximating the series by the first 3 terms, the error in this approximation is what we have neglected: $-\dfrac{1}{4}+\dfrac{1}{5}-\dfrac{1}{6}+.....$ But in absolute value, this quantity is less than 1/4 (first neglected term, in absolute value). Thus the error of this approximation is less than 1/4.

EXAMPLE: Determine if the following series converge.

(1) $\displaystyle\sum_{n=1}^{\infty}\dfrac{(-1)^{n}\ln n}{n}$.

Answer: Strictly alternating, terms decrease to 0, converges by Alternating Series Test.

(2) $\displaystyle\sum_{n=1}^{\infty}\dfrac{(-1)^{n}n!}{4^{n}}$.

Answer: Recall our "hierarchy of n's", in Section 9.3. Since a_{n} does not go to 0, the series diverges.

EXAMPLE: Approximating alternating series.

(1) Find an upper bound for the error in approximating $\displaystyle\sum_{n=1}^{\infty}\dfrac{(-1)^{n+1}}{\sqrt{n}}$ by the first 4 terms .

Answer: The error is less than the absolute value of the 5th term, namely $\dfrac{1}{\sqrt{5}}$.

(2) How many terms are needed to approximate $\displaystyle\sum_{n=1}^{\infty}\dfrac{(-1)^{n}}{n^{2}}$ with an error less than. 01?

Answer: The 10th term is 1/100, and the error is less than the absolute value of the first neglected term. If we use n = 10 terms, the error is less than $1/11^{2}$.

ABSOLUTE AND CONDITIONAL CONVERGENCE

There are more tests for the convergence of non-negative series than for series whose signs may not strictly alternate, but can wobble around.

> If a series $\displaystyle\sum_{n=1}^{\infty}a_{n}$ converges absolutely, i.e., if $\displaystyle\sum_{n=1}^{\infty}|a_{n}|$ converges, then the series itself will converge. Absolute convergence implies convergence. If a series converges and does not converge absolutely, but converges only because of plus and minus signs, the series is said to converge conditionally.

EXAMPLE: Discuss the convergence of the following series.

1.) $\displaystyle\sum_{n=1}^{\infty} \frac{(-1)^n}{n^2}$ **Answer:** The series converges absolutely , since $\displaystyle\sum_{n=1}^{\infty} \frac{1}{n^2}$ converges .

Thus the series converges, as well.

2.) $\displaystyle\sum_{n=1}^{\infty} \frac{(-1)^n}{\ln n}$ **Answer :** This series does not converge absolutely because $\displaystyle\sum_{n=1}^{\infty} \frac{1}{\ln n}$ does not

converge. However, because of its alternating signs it converges by the Alternating Series

Test . So our series converges conditionally

NOTE: A series that diverges absolutely does not necessarily diverge.

EXAMPLE: Discuss the convergence of $\displaystyle\sum_{n=1}^{\infty} \frac{\sin n}{n^2}$.

Answer: This series does not match the conditions needed to apply the Alternating Series

test (the signs wobble around; they do not strictly alternate.) But $\dfrac{\sin n}{n^2} \leq \left|\dfrac{\sin n}{n^2}\right| \leq \dfrac{1}{n^2}$, and

$\displaystyle\sum_{n=1}^{\infty} \frac{1}{n^2}$ converges, so our series converges absolutely, by the Comparison Test. Therefore

it converges.

NOTE: When we test a series for absolute convergence, we can use the tests that required
terms of a series be non-negative. Now we have an enormous battery of tests for
convergence of series!

Be careful not to confuse convergence of series with convergence of sequences on a test!!!
Many a student has lost a test because of this easy confusion.

True or False

1. A series converges if and only if the sequence of partial sums converges.
2. If a_n in a series goes to 0, the series converges.
3. If a_n in a sequence goes to 0, the sequence converges.
4. All divergent series diverge to ∞.
5. A series converges if $\displaystyle\lim_{n\to\infty} a_n$ exists.

6. If every term of a series can be shown to be less than the corresponding term of a series known to diverge, then the original series diverges.
7. If a series converges, all terms are bounded by a fixed number, M.
8. If all terms of a series are bounded by a fixed number, M, the series converges.
9. An alternating series converges if the terms decrease.
10. A telescoping series converges.
11. A geometric series converges if $r < 1$.
12. If a series $\displaystyle\sum_{n=1}^{\infty} a_n$ converges, then $\displaystyle\sum_{n=1}^{\infty} |a_n|$ converges.
13. If the Ratio Test yields a ratio of 1, the test fails, and we must try another test.
14. If the Limit Comparison Test yields a ratio of 1, the test fails, and we must try another test.
15. If the Root Test yields a root of 1, the test fails, and we must try another test.
16. The Root Test can be used for any series whose terms are powers of n.
17. The Integral Test can be used for any series whose terms resemble a function we can integrate.
18. If $a_n < b_n$ and $\displaystyle\sum_{n=1}^{\infty} a_n$ converges, then $\displaystyle\sum_{n=1}^{\infty} b_n$ converges.
19. Any series whose terms are bounded below and decreasing will converge.
20. A series converges if the partial sums are always finite.

Answers: (1) T (2) F --the harmonic series! (3) T--we're speaking of sequences here (4) F--some may oscillate or diverge to -∞   (5) F --but true for sequences (6) F--change " less than" to "greater than" and the statement is true. (7) T (8) F (9) F--terms must go to 0 (10) F--last term must go to 0, (11) F--need | r | < 1 (12) F (13) T (14) F (15) T (16) F--terms must be non-negative (17) F--same reason (18) F (19) F (20) F--partial sums are always finite; they must approach a limit.

Why Doesn't the Teacher Scale the Grades?

Grading is not completely in the hands of the teacher. Teachers must certify your mastery of a subject, so you and your teacher can feel confident about your readiness for the next class, which in mathematics, is likely to be linked to this course .   Sometimes entire teams fail, as when they lose a game. And sometimes an entire class performs poorly, in comparison to classes that teacher may have had before. If you do poorly on a test--relearn the material! It is likely to resurface.

Fill in the blanks.

Define a series-----A series converges if --------(watch out! Your answer should use Partial Sums). Some tests that can be used for the convergence of series with non-negative terms are ------- An example of a series that diverges by oscillation is -------Two series whose actual value we can find, when they converge are the ---------- The rule for the convergence of these series is -------. We can approximate the error of an alternating series as follows --- Absolute and Conditional convergence mean ------------

MEMORY TESTER

Make your own grid similar to this, and fill in the spaces.

TEST	USED WHEN	EXAMPLE	CAUTIONS, REMARKS
Comparison Test	non-negative terms; when we see a series similar to one whose convergence we know	$\sum_{n=1}^{\infty} \dfrac{1}{n!}$ converges comparing to $\sum_{n=1}^{\infty} \dfrac{1}{n^2}$	Series must be compared termwise with another series. If terms of series A are smaller than those of B and B converges, then A does....etc......
Limit Comparison Test			
Ratio Test			
Root Test			
Integral Test			
Alternating Series Test			
Absolute Convergence and Conditional			
$\lim_{n \to \infty} a_n$ Test			
Special Series-- telescoping			
Special Series-- geometric			

SECTION 9.8

" 'I can't explain myself, I'm afraid, sir', said Alice, ' because I'm not myself, you see.' 'I don't see,' said the Caterpillar."...Lewis Carroll

POWER SERIES

A power series differs from an ordinary series in that it has powers of a variable, x, which can take on values. A power series defines a **function** on values of x for which the series converges.

ORDINARY SERIES

$$\sum_{n=0}^{\infty} \frac{(-1)^n 5^n}{(n+2)!}$$

Either converges or doesn' t

POWER SERIES

$$\sum_{n=0}^{\infty} \frac{(-1)^n 5^n (\mathbf{x-4})^n}{(n+2)!} \leftarrow \text{power of x}$$

Convergence depends on value of x

A power series $\displaystyle\sum_{n=0}^{\infty} c_n (x-a)^n$ **does one of three things:**

1. **Converges for all x** . The interval of convergence is all Reals, and the radius of convergence is infinite.

2. **Converges in an interval about x = a , say (a-h, a+h).** The radius of convergence is h. **The series may or may not converge at one or both endpoints.**

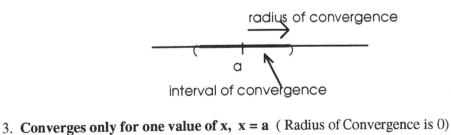

3. **Converges only for one value of x, x = a** (Radius of Convergence is 0)

NOTE: These series often begin with n = 0. Recall: 0! = 1.

EXAMPLE: Write out some terms of the power series $\sum_{n=0}^{\infty} \frac{(-1)^{n+1}}{n!}(x-3)^n$.

Answer: $\sum_{n=0}^{\infty} \frac{(-1)^{n+1}}{n!}(x-3)^n = \frac{-1}{0!}(x-3)^0 + \frac{(-1)^2}{1!}(x-3)^1 + \frac{(-1)^3}{2!}(x-3)^2 + ... =$

$-1+(x-3)-\frac{1}{2}(x-3)^2 + ...$ Note that we keep the series in terms of powers of (x-3).

Note how the series has powers of (x-3) accompanied by coefficients.

EXAMPLE: Does the series above converge if (a) x = 3? (b) x = 5? (c) x = 1?

Answer:

(a) When x = 3, we are summing "0" an infinite number of times, and the series converges to 0.

(b) When x = 5, the terms of the series are $\frac{(-1)^{n+1}(5-3)^n}{n!} = \frac{(-1)^{n+1}2^n}{n!}$, and by the

Alternating Series Test, the series converges.

(c) When x = 1, the terms of the series are

$\frac{(-1)^{n+1}(1-3)^n}{n!} = \frac{(-1)^{n+1}(-2)^n}{n!} = \frac{(-1)^{n+1}(-1)^n(2^n)}{n!} = \frac{(-1)^{2n+1}2^n}{n!} = -\frac{2^n}{n!}$. The terms of

this series are negative, but the series converges absolutely by the Ratio Test, and therefore converges. (We don't know to what, however.)

NOTE: We can't use the method above to find all x for which a power series converges; we want more general methods.

1. The power series $\sum_{n=0}^{\infty} c_n (x-a)^n$ is said to be "expanded about a". The power series

$\sum_{n=0}^{\infty} c_n x^n$ is expanded about x = 0.

2. The power series $\sum_{n=0}^{\infty} c_n (x-a)^n$ always converges if x = a. If there is an *interval* of

convergence, the midpoint of the interval will be x = a. We must test convergence at the endpoints of this interval. (These are x-values for which the Ratio Test is inconclusive.)

3. A power series $\displaystyle\sum_{n=0}^{\infty} c_n(x-a)^n$ converges for values of x in an open interval, (a-h, a+h)

if and only if the series $\displaystyle\sum_{n=0}^{\infty} \left| c_n(x-a)^n \right|$ converges (note absolute value). This makes it easy

to determine convergence of the original series, because we can look at the many tests for series with non-negative terms. We usually use the (Generalized) Ratio Test. We find the x

for which $\displaystyle\lim_{n\to\infty} \left| \dfrac{a_{n+1}}{a_n} \right| < 1$ where the a_n are the terms of the power series, i.e.

$a_n = c_n(x-a)^n$.

EXAMPLE: Convergence in an interval. Testing endpoints. Find the interval of

convergence of $\displaystyle\sum_{n=1}^{\infty} \dfrac{x^{3n}}{n}$. (Notice: this series begins with n = 1.)

Answer:

(1) The series converges at x = 0. Now assume x ≠ 0. We want $\displaystyle\lim_{n\to\infty}\left| \dfrac{a_{n+1}}{a_n} \right| < 1$, or :

$$\lim_{n\to\infty}\left| \frac{x^{3(n+1)}/(n+1)}{x^{3n}/n} \right| = \lim_{n\to\infty}\left| \frac{x^{3n}x^3/(n+1)}{x^{3n}/n} \right| = \lim_{n\to\infty}\left| \frac{x^3 n}{n+1} \right| = |x^3|\lim_{n\to\infty}\left| \frac{n}{n+1} \right| = |x^3| < 1 .$$

This is the case if | x | < 1. This gives that the series converges for x in (-1, 1).

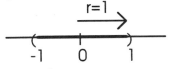

(2) Now we test the endpoints, x = -1 and x = 1 (points where the Ratio Test is inconclusive.)

♦ Plugging in x = -1, our series becomes $\displaystyle\sum_{n=1}^{\infty} \dfrac{(-1)^{3n}}{n}$, which converges by the Alternating

Series Test. So the interval of convergence includes x = -1.

♦ Plugging in x = 1 , our series becomes $\displaystyle\sum_{n=1}^{\infty} \dfrac{(1)^{3n}}{n} = \sum_{n=1}^{\infty} \dfrac{1}{n}$. This is the harmonic series,

which diverges. So the interval of convergence will not include x = 1.

Conclusion: The interval of convergence is [-1, 1), and the radius of convergence is 1.

EXAMPLE: Convergence everywhere. Find the interval of convergence of $\displaystyle\sum_{n=0}^{\infty} \frac{n!x^n}{(2n)!}$.

Answer: The series converges at x = 0. This series converges inside this interval (not necessarily at endpoints) if and only if:

$$\lim_{n\to\infty} \left| \frac{a_{n+1}}{a_n} \right| < 1 \quad \text{(assuming x} \neq 0 \text{)} \quad \text{or}$$

$$\lim_{n\to\infty} \left| \frac{(n+1)!x^{n+1}/(2n+2)!}{n!x^n/(2n)!} \right| < 1 \text{ or}$$

$$\lim_{n\to\infty} \left| \frac{(n+1)x}{(2n+2)(2n+1)} \right| < 1.$$

But $\displaystyle\lim_{n\to\infty} \left| \frac{(n+1)x}{(2n+2)(2n+1)} \right| = 0 < 1$ No "x" left over! Since the limit is always less than

1, and does not depend on the value of x, the series converges everywhere.

Conclusion: The interval of convergence is (-∞,∞) and the radius of convergence is infinite.

EXAMPLE: Convergence only at a point. Find the interval of convergence of

$$\sum_{n=0}^{\infty} nx^n.$$

Answer: The series converges if x = 0. If x ≠ 0, we form $\displaystyle\lim_{n\to\infty} \left| \frac{(n+1)!x^{n+1}}{n!x^n} \right| =$

$\displaystyle\lim_{n\to\infty} \left| (n+1)x \right| = \infty$. We conclude: there are *no x other than x = 0* for which this limit is less than 1, and the series converges only when x = 0.

EXAMPLE: Find the interval of convergence and radius of convergence for $\displaystyle\sum_{n=0}^{\infty} \frac{nx^n}{5^n}$.

Answer: The series converges for x = 0. If x ≠ 0, we require

$$\lim_{n\to\infty} \left| \frac{a_{n+1}}{a_n} \right| = \lim_{n\to\infty} \left| \frac{(n+1)x^{n+1}/4^{n+1}}{nx^n/4^n} \right| = \lim_{n\to\infty} \left| \frac{x}{4} \right| \cdot \left| \frac{n+1}{n} \right| = \left| \frac{x}{4} \right| < 1, \text{ or } |x| < 4. \text{ So}$$

convergence is assured for x in (-4, 4). Now test the endpoints. When x = 4, $\displaystyle\sum_{n=0}^{\infty} \frac{n4^n}{5^n}$

$= \displaystyle\sum_{n=0}^{\infty} n(4/5)^n$ which converges (ratio , root, or comparison tests). When x = -4,

the series will also converge. We can cite the Alternating Series Test, but in fact,

$\displaystyle\sum_{n=0}^{\infty} \frac{n(-4)^n}{5^n}$ will converge because, the series in absolute value, $\displaystyle\sum_{n=1}^{\infty} \frac{n4^n}{5^n}$,

converges.

Conclusion: the interval of convergence is [-4,4] and the radius of convergence is 4.

EXAMPLE: Find the interval of convergence of the power series $\displaystyle\sum_{n=0}^{\infty} \frac{(-1)^n (x-4)^n}{n+1}$.

Answer: Although the interval of convergence will be centered at x = 4, we proceed as above. We know convergence occurs at x = 4. Now assume x ≠ 4. The series will converge for x such that

$$\lim_{n\to\infty}\left|\frac{a_{n+1}}{a_n}\right| = \lim_{n\to\infty}\left|\frac{(-1)^{n+1}(x-4)^{n+1}/(n+2)}{(-1)^n(x-4)^n/(n+1)}\right| = \lim_{n\to\infty}|x-4|\cdot\left|\frac{n+1}{n+2}\right| = |x-4| < 1.$$ I.e. for x in

(-3,5). Now we check endpoints. When x = -3, we have

$$\sum_{n=0}^{\infty} \frac{(-1)^n(3-4)^n}{n+1} = \sum_{n=0}^{\infty} \frac{(-1)^n(-1)^n}{n+1} = \sum_{n=0}^{\infty} \frac{(-1)^{2n}}{n+1} = \sum_{n=0}^{\infty} \frac{1}{n+1},$$ which is the harmonic

series. We have divergence when x = -3. When x = 5, we have

$$\sum_{n=0}^{\infty} \frac{(-1)^n(5-4)^n}{n+1} = \sum_{n=0}^{\infty} \frac{(-1)^n}{n+1},$$ which converges by the Alternating Series Test.

Conclusion: The series converges for x in (-3, 5].

NOTE: *Within* its radius of convergence, a power series can be integrated or differentiated term by term. But the interval of convergence of the differentiated or integrated series may not include the endpoints of the interval of convergence of the original series. You will have to test endpoints all over again.

EXAMPLE: Let $f(x) = \displaystyle\sum_{n=0}^{\infty} \frac{(-1)^n x^{2n}}{n+1}$, which converges for x in [-1, 1]. Find f'(x) and

$\displaystyle\int_0^x f(t)dt$ and their intervals of convergence.

Answer: $f'(x) = \displaystyle\sum_{n=0}^{\infty} \frac{(-1)^n (2n\, x^{2n-1})}{n+1}$, where we differentiated inside the summation

symbol (this is called differentiating term-by-term). And $\displaystyle\int_0^x f(t)dt =$

$$= \int_0^x \sum_{n=0}^{\infty} \frac{(-1)^n t^{2n}}{n+1} dt = \sum_{n=0}^{\infty} \int_0^x \frac{(-1)^n t^{2n}}{n+1} dt = \sum_{n=0}^{\infty} \frac{(-1)^n x^{2n+1}}{(n+1)(2n+1)} , \text{ where we integrated inside}$$

the summation symbol. These series will converge for values of x in (-1,1). Testing

endpoints, we see for f '(x), the interval of convergence is (-1,1), and for $\int_0^x f(t)dt$, the

interval of convergence is [-1,1].

NOTE: A power series is not useful to us where it does not converge.

EXAMPLE: Approximating an integral with a series. Find a value for
$\int_0^{0.2} \sin x^2 dx$ with error < .001. (Note, we have no way to integrate this integral.)

Answer:

$\sin x = \sum_{n=0}^{\infty} \frac{(-1)^n x^{2n+1}}{(2n+1)!}$. We may plug in x^2 in place of x to obtain

$\sin x^2 = \sum_{n=0}^{\infty} \frac{(-1)^n x^{4n+2}}{(2n+1)!}$. The ratio test will show that this power series converges for all x,

in particular for those x in the interval [0, 0.2]. Now we may exchange integral and sum
where the series converges, so

$$\int_0^{0.2} \sin x^2 dx = \int_0^{0.2} \sum_{n=0}^{\infty} \frac{(-1)^n x^{4n+2}}{(2n+1)!} dx = \sum_{n=0}^{\infty} \int_0^{0.2} \frac{(-1)^n x^{4n+2}}{(2n+1)!} dx = \sum_{n=0}^{\infty} \frac{(-1)^n x^{4n+3}}{(2n+1)!(4n+3)} \Bigg|_0^{0.2}$$

$$= \sum_{n=0}^{\infty} \frac{(-1)^n (0.2)^{4n+3}}{(2n+1)!(4n+3)} .$$

This is an alternating series, and we need n large enough to make the first neglected term,
in absolute value, less than .001. I.e. we want the smallest n such that
$\left| \frac{(-1)^n (0.2)^{4n+3}}{(4n+3)(2n+1)!} \right| < .001$. This n is n =1, since $\frac{(0.2)^7}{7 \cdot 6} < .000000305$. So we

approximate our integral with the *0 th* term in the series $\sum_{n=0}^{\infty} \frac{(-1)^n (0.2)^{4n+3}}{(2n+1)!(4n+3)}$, and

obtain $\int_0^{0.2} \sin x^2 dx \approx \frac{(0.2)^3}{3} = .002\overline{6}$.

SECTION 9.9

"The past is but the beginning of the beginning, and all that is, and has been is but the twilight of the dawn"...Herbert George Wells

TAYLOR SERIES

This section relies heavily on the theory that was developed in Section 9.1, for Taylor Polynomials.

IF: f has derivatives of all orders (an infinite number) and the Remainder Term in the Taylor Polynomial goes to 0 as n →∞,

THEN...f is represented by a Taylor Series in its interval of convergence.

$$f(x) = f(a) + f'(a)(x-a) + \frac{f''(a)(x-a)^2}{2!} + \frac{f'''(a)(x-a)^3}{3!} + ... \frac{f^{(n)}(a)(x-a)^n}{n!} + ...$$

We find the Taylor Series in the same way we found Taylor Polynomials. But the Taylor Series has an infinite number of terms, whereas polynomials have a finite number of terms.

NOTE: This is an important result. Once we have a Taylor Series for a function, we can use it in a variety of ways (shown at the end of this section.) The function f must have an infinite number of derivatives to have a Taylor series, but more is required. The Taylor series must converge, and must converge when x = a, to the value of the function, f(a), so that the series actually "represents" the function. Functions for which these conditions are met are said to be "analytic".

EXAMPLE: Review Taylor Polynomials! Find the 3rd Taylor Polynomial for the function f(x) = $\sqrt{x}$ about x = 9. (I.e., a = 9).
Answer: This is the same as finding $p_3(x)$ as we did in Section 9.1.

NOTE: In finding a Taylor Series, choice of x = a is important. If we find the Taylor Series for f(x) = $\sqrt{x}$, we cannot use a = 0. Why? Suppose we use a = 9 to find the Taylor Series for f(x) = $\sqrt{x}$.

Since the interval of convergence is centered at x = 9, and x = 0 forms the left-hand boundary the series will *not* converge for values of x > 18. (As a power series, the series converges in an interval). If we want to use the Taylor Series with an x =18, a good choice for a is a = 16. The Taylor Series gives its best approximations to f(x) for x-values close to x = a, the center of the interval of convergence.

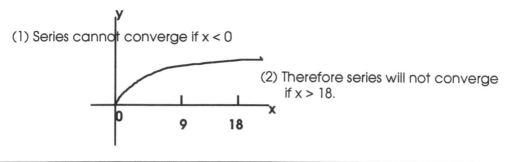

(1) Series cannot converge if x < 0

(2) Therefore series will not converge if x > 18.

EXAMPLE: (1) Find the Taylor Series for f(x) = cos x at x = π/3 (this is "a".)
(2) Show the remainder term goes to 0 ($r_n \rightarrow 0$), so the function converges to its Taylor
Series.

Answer:

f(x) = cos x	f(π/3) = 1/2
f '(x) = -sin x	f '(π/3)= - √3/2
f "(x) = -cos x	f "(π/3)= - 1/2
f '''(x) = sin x	f '''(π/3)= √3/2
$f^{(4)}(x) = \cos x$ (Notice that the derivatives begin to repeat now)	$f^{(4)}(\pi/3) = 1/2$

Using

$$f(x) = f(a) + f'(a)(x-a) + \frac{f''(a)(x-a)^2}{2!} + \frac{f'''(a)(x-a)^3}{3!} + ... \frac{f^{(n)}(a)(x-a)^n}{n!} + ...$$

we have: $\cos x = \frac{1}{2} - \frac{\sqrt{3}}{2}(x - \pi/3) - \frac{1}{2} \cdot \frac{(x-\pi/3)^2}{2!} + \frac{\sqrt{3}}{2} \cdot \frac{(x-\pi/3)^3}{3!} + ...$

It is difficult to write the general term--it can be done with the greatest-integer-less-than
function. We will avoid that approach and look at the possible values for $r_n(x)$.

$$\text{nth term} = \begin{cases} \pm\frac{1}{2} \cdot \frac{(x-\pi/3)^n}{n!}, or \\ \pm\frac{\sqrt{3}}{2} \cdot \frac{(x-\pi/3)^n}{n!} \end{cases} \qquad \text{so } r_n(x) = \begin{cases} \pm\frac{1}{2} \cdot \frac{(t_x-\pi/3)^{n+1}}{(n+1)!}, or \\ \pm\frac{\sqrt{3}}{2} \cdot \frac{(t_x-\pi/3)^{n+1}}{(n+1)!} \end{cases}$$

where t_x is any real number between $\pi/3$ and x.

Using the different definitions of $r_n(x)$, we have:

$$\lim_{n\to\infty} r_n(x) = \lim_{n\to\infty} \pm\frac{1}{2} \cdot \left(\frac{(t_x-\pi/3)^{n+1}}{(n+1)!} \right) = 0 \text{ or } \lim_{n\to\infty} r_n = \lim_{n\to\infty} \pm\frac{\sqrt{3}}{2} \cdot \left(\frac{(t_x-\pi/3)^{n+1}}{(n+1)!} \right) = 0.$$

Since the remainder term goes to 0, the Taylor series represents the function.

NOTE: MEMORIZE SOME TAYLOR SERIES! It is advisable to memorize the Taylor Series for some functions such as

$$e^x, \sin x, \cos x, \ln x, \ln(x-1), \frac{1}{1+x}, \frac{1}{1-x} .$$

Taylor Series contain a world of theory about topics we have seen in calculus.

Function	Taylor Series and Radius of Convergence
e^x	$1 + x + \dfrac{x^2}{2!} + \dfrac{x^3}{3!} + \ldots = \displaystyle\sum_{n=0}^{\infty} \dfrac{x^n}{n!}$. Converges for all reals.
$\sin x$	?
$\cos x$	?
$\dfrac{1}{1-x}$	?
$\dfrac{1}{1+x}$	?
$\ln x$ (expand about $a = 1$)	?
$\ln(x+1)$ (expand about $x = 0$)	?

NOTE: The Taylor Series expansion for a function, given "a", when it exists is unique. We will talk of "the" Taylor Series for f, given "**a**".

EXAMPLE: Doing "tricks" with a Taylor Series.

(1) Find the Taylor Series for $f(x) = e^{x-3}$ expanded about (a) $a = 3$, (b) $a = 0$.
Answer:

(a) We know $e^x = \displaystyle\sum_{n=0}^{\infty} \frac{x^n}{n!}$. Using $a = 3$, we have $e^{x-3} = \displaystyle\sum_{n=0}^{\infty} \frac{(x-3)^n}{n!}$

(b) We write: $e^{x-3} = \dfrac{e^x}{e^3} = \dfrac{1}{e^3} \displaystyle\sum_{n=0}^{\infty} \frac{x^n}{n!}$.

(2) Find the Taylor Series expansion for $f(x) = \sin x^2$ using $a = 0$.

Answer: Since $\sin x = x - \dfrac{x^3}{3!} + \dfrac{x^5}{5!} - \dfrac{x^7}{7!} + \ldots$,

$\sin x^2 = x^2 - \dfrac{x^6}{3!} + \dfrac{x^{10}}{5!} - \dfrac{x^{14}}{7!} + \ldots$ (We have used this result already.)

(3) Use the expansion of sin x above to find $\displaystyle\lim_{x \to 0} \frac{\sin x}{x}$.

Answer: Letting x → 0 above, $\lim\limits_{x\to 0} \dfrac{\sin x}{x} = \lim\limits_{x\to 0} \dfrac{x - \dfrac{x^3}{3!} + \dfrac{x^5}{5!} - \dfrac{x^7}{7!} + \ldots}{x} = 1.$

(4) Find the Taylor Series for cos x expanded about a = 0.

Answer: Differentiating the Taylor Series, $\sin x = \sum\limits_{n=0}^{\infty} \dfrac{(-1)^n x^{2n+1}}{(2n+1)!}$, term by term gives the

Taylor Series for cos x.

(5) Find the tangent line to the function f(x) = sin x at x = 0.
Answer: The tangent line is given by the linear part ("a + bx" part) of the Taylor Series.

Since $\sin x = x - \dfrac{x^3}{3!} + \dfrac{x^5}{5!} - \dfrac{x^7}{7!} + \ldots$, the tangent line is y = x.

(6) Find the Taylor Series for f(x) = $\sin^2 x$ expanded about a = 0.

Answer: We use $\left(x - \dfrac{x^3}{3!} + \dfrac{x^5}{5!} - \dfrac{x^7}{7!} + \ldots \right)^2$ where this unusual squaring operation requires

collecting terms of lowest degree first. I.e.

$$\sin^2 x = \left(x - \dfrac{x^3}{3!} + \dfrac{x^5}{5!} - \dfrac{x^7}{7!} + \ldots \right)^2 = x^2 - 2\left(x \cdot \dfrac{x^3}{3!} \right) + 2\left(x \cdot \dfrac{x^5}{5!} \right) - 2\left(\dfrac{x^3}{3!} \cdot \dfrac{x^5}{5!} + x \cdot \dfrac{x^7}{7!} \right) + \ldots$$

It would be easier to use $\sin^2 x = \dfrac{1 - \cos 2x}{2}$ to write $\sin^2 x = \dfrac{1}{2} - \sum\limits_{n=0}^{\infty} \dfrac{(-1)^n (2x)^n}{(2n)!}.$

(7) Determine if $f(x) = e^x = 1 + x + \dfrac{x^2}{2!} + \dfrac{x^3}{3!} + \ldots$ is concave upward at x = 2.

Answer: The tangent line is the linear part of the series: y = 1 + x. The second derivative term , 1/2!, is positive (showing that f(x) lies *above* its tangent line at x = 2.) So this function is concave upward.

(8) Find an approximation for e.

Answer: Use $e^x = 1 + x + \dfrac{x^2}{2!} + \dfrac{x^3}{3!} + \ldots$ with x = 1.

SECTION 9.10

" Mathematics...is sublimely pure, and capable of a stern perfection such as only the greatest art can show"...Bertrand Russell

BINOMIAL SERIES

The term $\binom{s}{n}$ is called the "binomial coefficient". It is defined as : $\binom{s}{n} = \dfrac{s!}{n!(s-n)!}$ (Note:

the term inside the parentheses is not a fraction.)

A WORD ABOUT PROBABILITY

The binomial coefficient plays a large role in probability and statistics. The quantity $\binom{s}{n}$ is the number of ways we can pick a group of n things from a pool of s distinct things. E.g., the number of ways we can pick a committee of 3 people from 7 is given by:

$$\binom{7}{3} = \frac{7!}{3!4!} = \frac{7 \cdot 6 \cdot 5 \cdot 4!}{3! \quad 4!} = 35$$

the 4! terms cancel

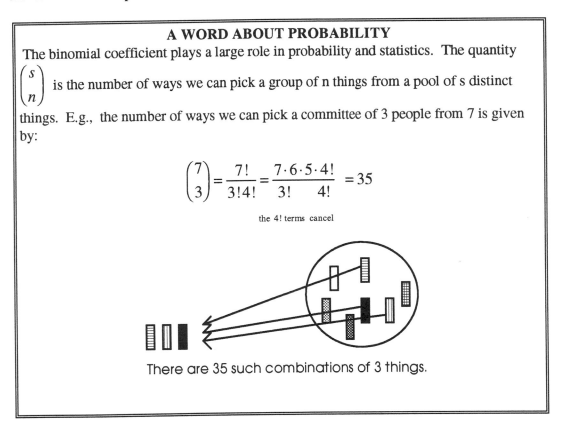

There are 35 such combinations of 3 things.

We can define the binomial coefficient in the following way for fractions, negative numbers, etc. (The term, n, in $\binom{s}{n}$ is always positive.)

$$\binom{-7}{3} = \frac{(-7)!}{3!(-7-3)!} = \frac{(-7)(-8)(-9)(-10)!}{3! \quad (-10)!} = -84$$

the terms go " down" in value with factorials

$$\binom{-1/2}{3} = \frac{(-1/2)!}{3!(-7-1/2)!} = \frac{(-1/2)(-3/2)(-5/2)(-9/2)...(-15/2)!}{3! \quad (-15/2)!} = 175.957...$$

Formulas to Memorize:
(Summation begins at n = 0)

$$(1+x)^s = \sum_{n=0}^{\infty} \binom{s}{n} x^n$$

$$(1-x)^s = \sum_{n=0}^{\infty} (-1)^n \binom{s}{n} x^n$$

EXAMPLE: Find $(1+x)^{1/2}$. By the formula :

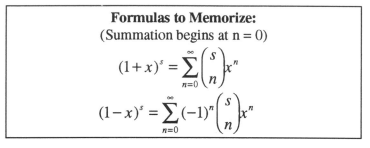

$$(1+x)^{1/2} = \sum_{n=0}^{\infty} \binom{1/2}{n} x^n = \binom{1/2}{0} + \binom{1/2}{1} x + \binom{1/2}{2} x^2 + \binom{1/2}{3} x^3 + ... \text{ where}$$

$$\binom{1/2}{0} = 1$$

$$\binom{1/2}{1} = \frac{(1/2)!}{1!(-1/2)!} = \frac{(1/2)(-1/2)!}{(-1/2)!} = \frac{1}{2}$$

$$\binom{1/2}{2} = \frac{(1/2)!}{2!(-3/2)!} = \frac{(1/2)(-1/2)(-3/2)!}{2!(-3/2)!} = -\frac{1}{2!}\left(\frac{1}{2}\right)\left(\frac{3}{2}\right)$$

$$\binom{1/2}{3} = \frac{(1/2)!}{3!(-5/2)!} = \frac{(1/2)(-1/2)(-3/2)(-5/2)!}{3!(-5/2)!} = +\frac{1}{3!}\left(\frac{1}{2}\right)\left(\frac{3}{2}\right)\left(\frac{5}{2}\right)$$

There is a Pattern.
COUNT DOWN IN NUMERATOR STARTING AT 1 / 2 - TERM

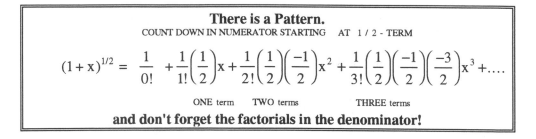

$$(1+x)^{1/2} = \frac{1}{0!} + \frac{1}{1!}\left(\frac{1}{2}\right)x + \frac{1}{2!}\left(\frac{1}{2}\right)\left(\frac{-1}{2}\right)x^2 + \frac{1}{3!}\left(\frac{1}{2}\right)\left(\frac{-1}{2}\right)\left(\frac{-3}{2}\right)x^3 + ...$$

ONE term   TWO terms   THREE terms

and don't forget the factorials in the denominator!

EXAMPLE: Approximate $\sqrt[4]{18}$ with an error less than .001.

$$\sqrt[4]{18} = \sqrt[4]{16+2} = \sqrt[4]{16\left(1+\frac{2}{16}\right)} = 2\sqrt[4]{1+\frac{2}{16}} = 2\left(1+\frac{1}{8}\right)^{1/4} \tag{1}$$

where

$$(1+x)^{1/4} = 1 + \frac{1}{4}x + \frac{1}{2!}\left(\frac{1}{4}\right)\left(\frac{-3}{4}\right)x^2 + \frac{1}{3!}\left(\frac{1}{4}\right)\left(\frac{-3}{4}\right)\left(\frac{-5}{4}\right)x^3 + \ldots$$

TWO terms   THREE   terms

So we can plug $x = 1/8$ into this series above. How many terms do we need? We need

$$|r_n| \le |s|\left(\frac{|x|^{n+1}}{1-|x|}\right) = \frac{1}{4}\left(\frac{(1/8)^{n+1}}{1-1/8}\right) = \frac{1}{4}\left(\frac{(1/8)^{n+1}}{7/8}\right) = \frac{1}{28(8)^n} < \frac{.001}{2} \text{ in order to make}$$

the error in (1) less than .001. So by trial and error, we find if $n = 3$, using up through the third derivative terms, the error is less than 6.98 E-5. We have with this approximation:

$$\sqrt[4]{18} \approx 2\{1 + \frac{1}{4}\left(\frac{1}{8}\right) + \frac{1}{2!}\left(\frac{1}{4}\right)\left(\frac{-3}{4}\right)\left(\frac{1}{8}\right)^2 + \left(\frac{1}{31}\right)\left(\frac{1}{4}\right)\left(\frac{-3}{4}\right)\left(\frac{-5}{4}\right)\left(\frac{1}{8}\right)^3 \} = 2.0597229..$$

Peeking at the calculator's approximation, $\sqrt[4]{18} \approx 2.059767...$

<hr>

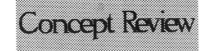

Concept Review

1. What is a power series? Are they functions?   How can you find where they converge? What types of convergence are possible? Can you begin to predict which ones converge and which don't? When convergence takes place in an interval, what about the endpoints? What about differentiating and integrating power series? Will we be sure the new results converge? Where does the derivative of a power series converge?

2. What is a Taylor Series? What's the point of expressing a function as a series? What benefits are there? Why do we use different values of "a" for expanding about? Why not always use $x = 0$? How can you tell from looking at a Taylor Series, what "a" has been used? How can you tell where a Taylor Series converges? Can you remember some importantTaylor Series expansions of functions? What does the Taylor Series of a polynomial look like? Can you find the Taylor Series for a given function, with given "a"? When does the Taylor Series actually represent the function?

3. What is a binomial series? What is the binomial coefficient? How do you estimate roots using the binomial series, with bounds on the error?

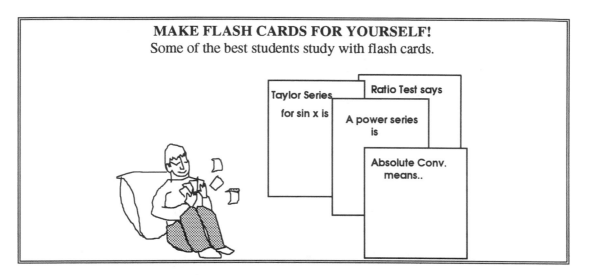

MAKE FLASH CARDS FOR YOURSELF!
Some of the best students study with flash cards.

Taylor Series for sin x is

Ratio Test says

A power series is

Absolute Conv. means..

Assessment

What skills from this chapter do you feel confident about? Reply in column 2.
After a test or quiz, come back to this table and fill in the last column with a comment.

TYPE OF PROBLEM	do you feel in control?	after test comments
Finding if Sequence Converges		
Theory about bounded sequences		
Proofs (if required) on sequence convergence		
Convergence of series: Partial sums, Comparison Test, Limit Comparison Test, Integral Test, Ratio Test, Root Test, Alternating Series Test		
Absolute and Conditional Convergence		
Power series, interval of convergence		
Taylor series, finding one		
Using Taylor series in approximations		
Binomial series		
RATE YOURSELF ON OVERALL UNDERSTANDING of the material in this chapter		

CHAPTER 10

" If you do not think about the future, you cannot have one"...John Galsworthy

CHAPTER OVERVIEW

The origin of the word "parameter" is "*through* the measure". A parameter is a variable, such as time, which is used to express other variables. At first this concept may seem confusing: why would we want to muddy the waters with yet another variable? But parametrizing allows for more flexibility, and adds great usefulness and beauty to the mathematics. Applied mathematicians, physicists, and engineers commonly use parametrizations of variables. And parametrization will allow us to describe the plane in a new coordinate system--polar coordinates.   Bring out your graphing calculator here! Finally, we will study in depth some curves you are probably already familiar with: the conic sections.

SECTION 10.1

"Life's a tough proposition, and the first 100 years are the hardest"...Wilson Mizner

PARAMETRIZING

We choose to express both x and y in terms of another variable, usually denoted t. When t represents time, we can think of points on the curve as being traced as a function of time. This provides more information about the graph than formerly. Before, the graph might have been regarded as mere "footprints in the snow"....but we can now say *when* the curve was *where.*

In the curve below, y is not a function of x, but the points on the curve (x, y) are functions of the time at which the curve was traced.

EXAMPLE: Sketch the curve given by x = 5 - t, y = 4t + 1, for t in [0,1].
Answer: First solve for t. We have t = 5 - x, and t = (y - 1)/4. Setting values of t equal, $\frac{y-1}{4} = (5-x)$ and $y = -4x + 21$.

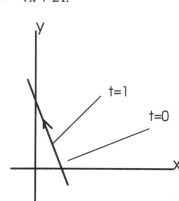

This is the equation of a straight line. When t = 0, (x, y) = (5, 1) and when t = 1, (x,y) = (4,5). In general, as t increases, x decreases and y increases. This helps explain how we determine (with an arrow) in which direction the line is traversed.

EXAMPLE: Sketch the curve parametrized by x = 2 cos t, y = 2 sin t for t in [0,2π].

Answer: We recognize this parametrization as giving a circle with radius 2, since x and y satisfy the equation: $x^2 + y^2 = 2^2$. We want to know how the circle is traversed. A chart helps us see that the circle is traversed counterclockwise. The "amplitude" of the circle (of radius 2) is 2, and the frequency is 1 (a point travels around the circle one time as t goes from 0 to 2π.)

t	(x, y)
0	$(2, 0)$
$\dfrac{\pi}{2}$	$(0, 2)$
π	$(-2, 0)$
$\dfrac{3\pi}{2}$	$(0, -2)$

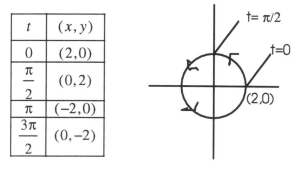

EXAMPLE: Find some other parametrizations of a circle:
Answer:
(a) x = cos 3 t, y = sin 3 t. This parametrization describes a circle of radius one, since $x^2 + y^2 = 1$, but the circle is traversed 3 times faster than the circle parametrized with x = cos t, y = sin t. The frequency (number of revolutions, as t goes from 0 to 2π) is 3, and one period is 2π/3 .(I.e., the circle will make a complete revolution in the time interval [0, 2π/3].) We conclude by plugging in points, that this circle is traversed in a counterclockwise direction.

(b) $x = \sin t$, $y = \cos t$. This parametrization gives a circle of radius one, but the circle is traversed *clockwise*. When $t = 0$, we are at the point $(0, 1)$.

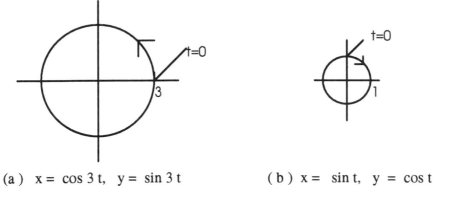

(a) $x = \cos 3t$, $y = \sin 3t$ (b) $x = \sin t$, $y = \cos t$

EXAMPLE: Parametrize an ellipse $\dfrac{x^2}{a^2} + \dfrac{y^2}{b^2} = 1$.

Answer: One parametrization of the ellipse is given by setting $\dfrac{x}{a} = \cos t$, $\dfrac{y}{b} = \sin t$, or

$x = a\cos t$, $y = b\sin t$. One revolution occurs as t goes from 0 to 2π.

SECTION 10.2

"The most beautiful thing we can experience is the mysterious. It is the source of all true art and science"...Albert Einstein

LENGTH OF CURVES AND SURFACE AREA

EXAMPLE: Find the length of the curve parametrized by $x = 3t$, $y = 2t^{3/2}$, for $0 \le t \le 5$.

Answer: We use the formula

$$L = \int_{a}^{b} \sqrt{(dx/dt)^2 + (dy/dt)^2}\, dt.$$

with $dx/dt = 3$, and $dy/dt = 3t^{1/2}$. The formula becomes:

$$L = \int_{a}^{b} \sqrt{(dx/dt)^2 + (dy/dt)^2}\, dt = \int_{0}^{5} \sqrt{3^2 + (3t^{1/2})^2}\, dt = 3\int_{0}^{5} \sqrt{1+t}\, dt \text{ , which can be evaluated,}$$

letting u $= 1 + t$.

Invent a Problem

As a good review on the section on parametrization, invent some problems for yourself:
For example:
1. Parametrize a circle of radius 4 in such a way that when t = 0, we are at the point (-2,0), then the circle is traversed 3 times clockwise, for t in [0, 2π]. Try out some other parametrizations of the circle and ellipse.
2. Find some parametrizations of a straight line or a line segment. For example, you might let $x = t^2$ and $y = ?$ How is this line traversed, with your various parametrizations?

SURFACE AREA

Recall from a previous chapter that when a curve is rotated, the resulting surface has a surface area element given by:

$$\Delta S = 2\pi \ (\text{radius}) \ \Delta(\text{arc length}).$$

Now that we are working with parametrized curves, we express all variables in terms of t.

♦ In the case of rotating a solid about the **x-axis:** $S = \int_a^b 2\pi \ y(t) \ \sqrt{\frac{dx}{dt}^2 + \frac{dy}{dt}^2} \ dt$.

♦ In the case of rotating a solid about the **y-axis:** $S = \int_a^b 2\pi \ x(t) \ \sqrt{\frac{dx}{dt}^2 + \frac{dy}{dt}^2} \ dt$.

EXAMPLE: Write the integral that expresses the surface area obtained when the curve parametrized by $x = 1 - t$, $y = t^2 + 3t$, with t in $[0, 4]$ is
(a) rotated about the y-axis, (b) rotated about the x-axis.
Answer:

(a) $S = \int_a^b 2\pi \ x(t) \ \sqrt{\frac{dx}{dt}^2 + \frac{dy}{dt}^2} \ dt = \int_0^4 2\pi \ (1-t) \ \sqrt{(-1)^2 + (2t+3)^2} \ dt =$

$2\pi \int_0^4 \ (1-t) \ \sqrt{4t^2 + 6t + 10} \ dt$.

(b) $S = \int_a^b 2\pi \ y(t) \ \sqrt{\frac{dx}{dt}^2 + \frac{dy}{dt}^2} \ dt = \int_0^4 2\pi \ (t^2 + 3t)\sqrt{(-1)^2 + (2t+3)^2} \ dt =$

$2\pi \int_0^4 \ (t^2 + 3t) \ \sqrt{4t^2 + 6t + 10} \ dt$.

NOTE: In many cases, the integral that expresses the arc length or the surface area or the volume of a solid of rotation is much more easily expressed using parametrization.

REVIEW Find equations for what you want to rotate, in PARAMETRIZED FORM , then produce the integral , in parametrized form, which expresses the mathematical construct in the table.

Integral?

Surface Area of a Sphere	
Surface Area of a Cone	
Circumference of a Circle	
Circumference of an Ellipse	
Surface Area of an Ellipsoid	

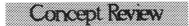 **Concept Review**

Explain what is meant by the parametrization of a curve; advantages and disadvantages. What is the formula for the arc length of a circle in *Cartesian coordinates*? In parametrized form? Find a parametrization of $y = \sqrt{x}$.

SECTION 10.3

" Literary intellectuals at one pole--at the other scientists....Between the two a gulf of mutual incomprehension."...C.P. Snow

POLAR COORDINATES

In this section, we replace the conventional Cartesian rectangular coordinate system with a new coordinate system. Although we are still on a plane, not a sphere, it might occasionally help to think of the "origin" or "pole" in this new system as the North Pole with only the zero-degree longitude line for an axis.

CHANGING FROM ONE COORDINATE SYSTEM TO ANOTHER:

From Polar to Rectangular: (If you know the polar coordinates os a point, you can find Cartesian coordinates.)

$$x = r \cos \theta$$
$$y = r \sin \theta$$

From Rectangular to Polar: If you know the Cartesian coordinates, you can find the polar coordinates. (Notice that in this direction, the description of coordinates is vaguer than above. The reason is that a point on the Cartesian plane can have many descriptions in polar coordinates.)

$$r^2 = x^2 + y^2$$

(This leaves open the possibility that r can be negative and positive   .)

$$\tan \theta = \frac{y}{x}$$

(where θ is in the appropriate quadrant corresponding to x and y   )

NOTE: We could not use $\theta = \tan^{-1} \frac{y}{x}$, because then values of θ could fall only in Quadrants I and II.

EXAMPLE: Write (x,y) = (1,1) in polar coordinates.

Answer : Using our eye, $\theta = \pi/4$ and $r = \sqrt{2}$ is one representation, but there are others. The complete answer is $(r, \theta) = (\sqrt{2}, \pi/4 + 2n\pi)$ or $(-\sqrt{2}, 5\pi/4 + 2n\pi)$, $n = 0, \pm 1, \pm 2, ...$

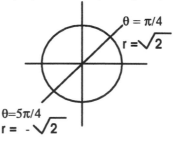

EXAMPLE: Write in Cartesian coordinates the point given in polar coordinates as $(2,\pi/3)$.
Answer: We have $x = 2 \cos \pi/3 = 1$, and $y = 2 \sin \pi/3 = 2\sqrt{3}/2 = \sqrt{3}$, so $(x, y) = (1, \sqrt{3})$.

NOTE: The pole, or origin, has the representation r = 0, where θ can be anything.

EXAMPLE: Write in Cartesian coordinates the equation: $\cot \theta = 3$.
Answer: We have $3 = \dfrac{\cos \theta}{\sin \theta} = \dfrac{r \cos \theta}{r \sin \theta} = \dfrac{x}{y}$, so $y = \dfrac{1}{3}(x)$. This is a line through the origin.

EXAMPLE: Write in polar coordinates, the curve: $x^2 + 2x + 1 + y^2 = 1$.
Answer:

Writing this equation as $(x + 1)^2 + y^2 = 1$, we have :

$(r \cos \theta + 1)^2 + (r \sin \theta)^2 = 1$, or $r^2 (\cos^2 \theta + \sin^2 \theta) + 2r \cos \theta + 1 = 1$, or $r^2 + 2r \cos \theta = 0$.

Thus $r = 0$ or $r = -2 \cos\theta$.

We use r = -2cos θ because this describes the curve, on which r = 0 does lie. Notice this is a circle centered at (-1,0) of radius 1.

EXAMPLE: Write in Cartesian coordinates: $r \sin\theta = 7$.
Answer: This is the line $y = 7$.

GRAPHING IN POLAR COORDINATES

CLASSIFYING CURVES IN POLAR COORDINATES
SEE THE DICTIONARY OF GRAPHS IN YOUR TEXT

NOTE: We use only one axis in polar coordinate--however, we will occasionally put a polar curve on a Cartesian grid, for clarity.

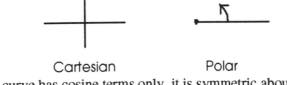

Cartesian   Polar

Basic ideas: If the curve has cosine terms only, it is symmetric about the x-axis; with sine terms only, it is symmetric about the y-axis. If you know the basic shape, find where r is largest--this will help put the curve "on the map".

CIRCLES:

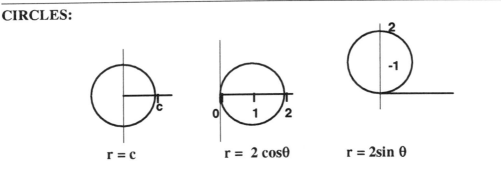

$r = c$ $r = 2 \cos\theta$ $r = 2\sin\theta$

CARDIOIDS : Heart-shaped curves. To graph, find where r is largest. Also notice the symmetry about the x- and y-axis is determined by the "cos" or "sin" term.

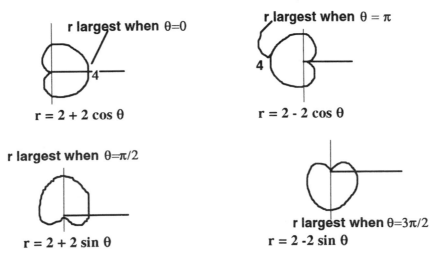

LIMACONS : Snail shaped cuves--corkscrews and "lima beans". These are of the form r = a ±b cos θ or r = a ± b sin θ. If a < b, we have a corkscrew spiral because r can be negative; if a > b, we have a lima bean because r is always positive; if a = b, we have a cardioid. The pictures below illustrate a situation in which a and b are both positive.

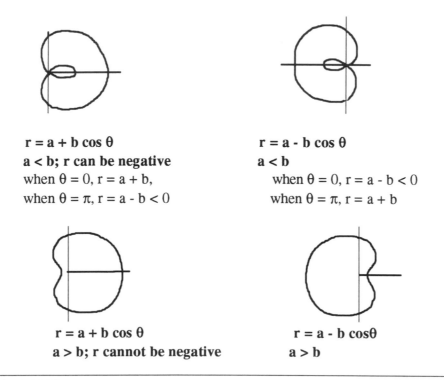

r = a + b cos θ
a < b; r can be negative
when θ = 0, r = a + b,
when θ = π, r = a - b < 0

r = a - b cos θ
a < b
  when θ = 0, r = a - b < 0
  when θ = π, r = a + b

r = a + b cos θ
a > b; r cannot be negative

r = a - b cosθ
a > b

ROSE CURVES: If angle is nθ with n even, there are 2n petals. If n is odd -- n petals. To determine the radial axis of the first petal, find the angle for which cos nθ or sin nθ is largest. Then petals are distributed 2π/n radians apart. With cosine, symmetry about the x-axis; with sine, symmetry about the y-axis.

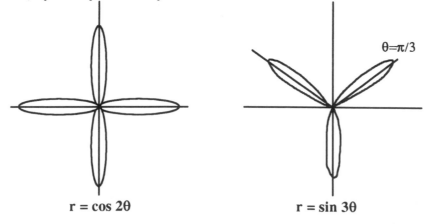

θ=π/3

r = cos 2θ   r = sin 3θ

LEMNISCATES: **Because r is squared, curve expands in 2 directions at once: dumb-bell shaped.**

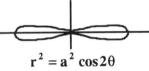

$$r^2 = a^2 \cos 2\theta$$

EXAMPLE: Describe the graph of $r = 1 + \cos\theta$.

Answer: This is a cardioid, symmetric about the x-axis. The largest value of r occurs at $\theta = 0$.

EXAMPLE: Describe the graph of $r = 1/2 - \sin\theta$.

Answer: This is a limacon, of the corkscrew variety, because $1/2 < 1$. The curve is aligned on the y-axis because of the sine term, and the largest value of r is $3/2$, when $\theta = 3\pi/2$.

EXAMPLE: Describe the graph of $r = 2 - \cos\theta$.

Answer: This is a "lima bean" limacon $(2 > 1)$. The curve is aligned on the x-axis, because of the cosine term, and the largest value of r is 3, occurring at $\theta = \pi$.

EXAMPLE: Describe the graph of $r = 5\sin 4\theta$.

Answer: This is a rose curve, with symmetry about the y-axis. There are 8 petals, since 4 is even. The "first" petal occurs where r is maximal or where $\sin 4\theta = 1$, namely, where $4\theta = \pi/2$, or $\theta = \pi/8$. The "radius" of each petal (maximal length) is 5 and the petals are distributed at angles $\pi/8 + k\,(2\pi/8)$, or $\pi/8 + k\,\pi/4$, where $k = 0, 1, 2, 3, 4, 5, 6, 7$.

NOTE ON GRAPHING: With your graphing calculator, you can see how the curve is traced. This will become very important when we find areas.

SECTION 10.4

" Go sir, gallop, and don't forget that the world was made in six days. You can ask me for anything you like, except time."...Napoleon Bonaparte

LENGTH OF CURVES IN POLAR COORDINATES

Using the formula for parametrized curves, we may obtain the formula for arc length, using θ as parameter:

$$L = \int_a^b \sqrt{\left(\frac{dr}{d\theta}\right)^2 + r^2}\, d\theta$$

EXAMPLE: Find the length of the curve $r = \sin^2 \frac{\theta}{2}$, for θ in $[0, \pi]$.

Answer: Note that $r' = (2 \sin\frac{\theta}{2}\cos\frac{\theta}{2})(\frac{1}{2}) = \sin\frac{\theta}{2}\cos\frac{\theta}{2}$. Now:

$$L = \int_a^b \sqrt{[f'(\theta)]^2 + [f(\theta)]^2}\, d\theta = \int_0^\pi \sqrt{\left(\sin\frac{\theta}{2}\cos\frac{\theta}{2}\right)^2 + \sin^4\frac{\theta}{2}}\ d\theta =$$

$$\int_0^\pi \sqrt{(\sin\frac{\theta}{2})^2\ (\cos^2\frac{\theta}{2} + \sin^2\frac{\theta}{2})}\ d\theta = \int_0^\pi \sin\frac{\theta}{2}\sqrt{1}\, d\theta = \int_0^\pi \sin\frac{\theta}{2}\ d\theta = -2\cos\frac{\theta}{2}\Big|_0^\pi = 2.$$

AREA IN POLAR COORDINATES

The formula for area in polar coordinates derives from the area of a wedge of a circle of radius r: The area of a wedge is $A = \frac{1}{2}r^2\theta$ where θ is the angle subtended by the wedge. This is obtained by noting:

Area of Wedge "is to" Area of Circle as Angle subtended by Wedge "is to" Angle of Circle. I.e.,

$$\frac{\text{Area of Wedge}}{\text{Area of Circle}} = \frac{\text{Angle subtended by Wedge}}{\text{Angle of Circle}}\ ,\ \text{or}$$

$$\frac{\text{Area of Wedge}}{\pi r^2} = \frac{\theta}{2\pi}.\ \text{So Area of Wedge} = \frac{1}{2}r^2\,\theta.$$

The formula for the area INSIDE a curve in polar coordinates is

$$A = \frac{1}{2}\int_a^b (f(\theta))^2 d\theta$$

THINGS TO WATCH FOR IN FINDING AREA:

(1) Area is measured **INSIDE** the curve, not **UNDER** the curve. Think of yourself again, as standing at the North Pole, "scooping in" area *toward the pole*.

Area is measured in pi-shaded regions, not rectangular regions. This requires a break in our usual ways of thinking about area.

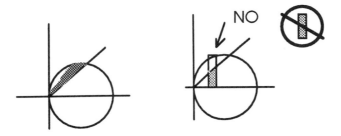

(2) Limits of integration in polar coordinates are not always what they "look like" on the graph. We have to use the equation in order to know *where* the curve was, *when* . In other words, finding the angle and radius of a point from the graph may be tricky.

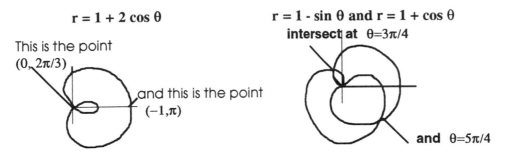

(3) When finding the area common to two curves (or inside of one and outside of the other), you will need to find where the curves intersect. Don't forget that they may intersect when r = 0--this may not show up when you set the two equations equal. Use the graph to guide you.

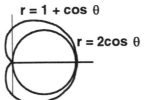

An intersection occurs when r = 0, although the angles on the 2 curves are different at the point of intersection . On the cardioid, r = 0 when θ = π and on the circle, when θ = π /2.

(4) Take advantage of symmetry wherever possible: find the area of a small piece, and double it, or triple it, etc. to avoid errors.

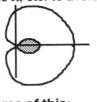   

Area of this: **is twice area of this.**

(5) Be sure to integrate in the positive direction of θ (the direction of increasing θ); the integrand is always positive, and this will be necessary to assure a positive area.

(6) You will be seeing a lot of these integrals. Memorize the formulas inside the integral sign.

$$\int_a^b \cos^2\theta \; d\theta = \int_a^b \frac{1 + \cos\; 2\theta}{2} \; d\theta$$

$$\int_a^b \sin^2\theta \; d\theta = \int_a^b \frac{1 - \cos\; 2\theta}{2} \; d\theta$$

EXAMPLE: Find the integral that represents the area inside the inner loop of the limacon given by $r = \frac{1}{2} + \cos\theta$.

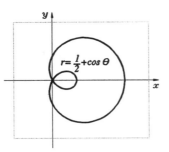

$$r = \frac{1}{2} + \cos\theta$$

Answer: We use symmetry to find half this area, then double it. Note that $r = 0$ when $\frac{1}{2} + \cos\;\theta = 0$, or $\theta = \frac{2\pi}{3}$. The area is then given by the integral:

$$A = 2\left(\frac{1}{2}\right)\int_{2\pi/3}^{\pi} \left(\frac{1}{2} + \cos\theta\right)^2 d\theta = \int_{2\pi/3}^{\pi} \frac{1}{4} + \cos\theta + \cos^2\theta \; d\theta \; .$$

EXAMPLE: Find the integral that represents the area common to the two circles: $r = \sin\theta$ and $r = \cos\theta$.

Answer: First we find that these curves intersect when $\sin\theta = \cos\theta$ or for $\theta = \pi/4$. We find the area INSIDE $r = \cos\theta$ (which is *shaded*) as θ goes from $\pi/4$ to $\pi/2$, and double this area. The area is $2\left(\frac{1}{2}\right)\int_{\pi/4}^{\pi/2} \cos^2\theta \; d\theta.$

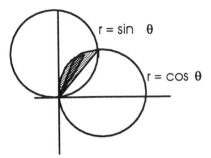

EXAMPLE: Find the integral that represents the shaded area below in Fig.(a):

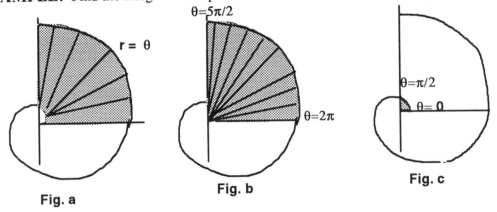

Fig. a   **Fig. b**   **Fig. c**

Don't forget: area is measured in wedges emanating from the origin

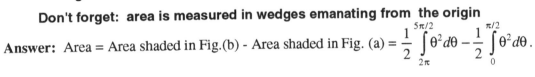

Answer: Area = Area shaded in Fig.(b) - Area shaded in Fig. (a) $= \dfrac{1}{2} \int\limits_{2\pi}^{5\pi/2} \theta^2 d\theta - \dfrac{1}{2} \int\limits_{0}^{\pi/2} \theta^2 d\theta$.

EXAMPLE: Find the integral that represents the area inside the curve r = 4 and outside the curve r = 2(1 + cos θ).

Answer: Graph the curves, find the intersection points: 4 = 2(1 + cos θ) when θ = 0. We see the cardioid lies inside the circle, and so the area is found by subtracting from the area of the circle, the area "scooped in" by the cardioid .

$$A = \frac{1}{2} \int\limits_{0}^{2\pi} 4^2 d\theta - \frac{1}{2} \int\limits_{0}^{2\pi} (2(1 + \cos\theta))^2 d\theta$$

EXAMPLE: Find the area outside the graph of r = 1 - sin θ and inside that of r = 1 + cos θ.

Answer: Graph the curves, find the intersection points. First note the curves intersect when r = 0 (for whatever angles), and also when 1 + sin θ = 1 + cos θ ; i.e., when θ = -π/4 or 3π/4.

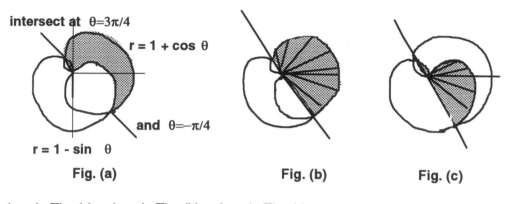

Fig. (a)   **Fig. (b)**   **Fig. (c)**

So Area in Fig. (a) = Area in Fig. (b) - Area in Fig. (c) , or

$$\text{Area} = \frac{1}{2} \int_{-\pi/4}^{3\pi/4} (1 + \cos\theta)^2 d\theta - \frac{1}{2} \int_{-\pi/4}^{3\pi/4} (1 - \sin\theta)^2 d\theta$$

....Summarizing..

Polar coordinates coordinatize the plane in terms of the two variables --------. Advantages of this system are -------------. However, certain disadvantages arise such as ------------. To change from Cartesian to polar coordinates, we use the equations --------------, and to change from polar to Cartesian we use -------------------. Problems arise with the latter translation, such as --------------. The arc length of a circle of radius r in polar coordinates would be given by the intetgral----------whereas in Cartesian coordinates, this integral would be -------. In sketching curves in polar coordinates, the following ideas are useful ----------------. A brief run-down on the types of curves and their equations, that occur in the world of polar coordinates is :------------------. In finding areas inside a polar coordinate curve, we use the formula -------------- (did you remember the 1/2-term?). In finding areas inside curves in the polar coordinate system, we must be careful about the following:------------------.

SECTION 10.5

"It is better to know some of the questions than all of the answers"...James Thurber

CONIC SECTIONS

All conic sections can be expressed in the form

$$\boxed{Ax^2 + Bxy + Cy^2 + Dx + Ey + F = 0}$$

The presence of the "B" term signals that there is a rotation--the curve is "skewed" like the hyperbola xy = 1; we will study these curves in the next section. In this section, we will not see the "B" term. The presence of the "C" or "D" term tells us the conic section has been translated from its standard position at the origin.

Ruling out degeneracies (where there is no solution, or the solution is a point, a pair of straight lines, etc.), the curve is

(1) a **PARABOLA** if there is one and only one squared term

(2) an **ELLIPSE** if there are two squared terms having coefficients of like sign

(3) a **HYPERBOLA** if there are two squared terms with coefficients of opposite sign

(4) a **CIRCLE** (special case of an ellipse) if the two squared terms have equal coefficients.

THE PARABOLA

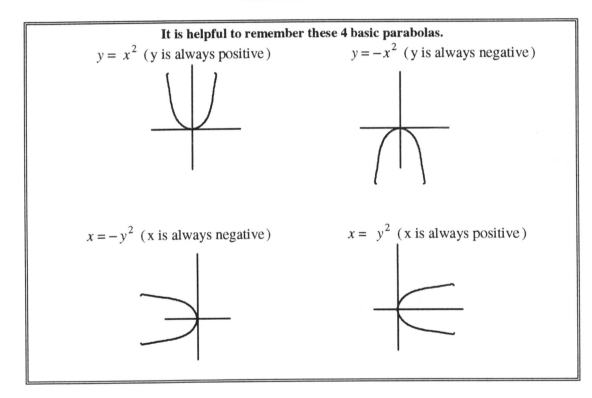

It is helpful to remember these 4 basic parabolas.

$y = x^2$ (y is always positive)

$y = -x^2$ (y is always negative)

$x = -y^2$ (x is always negative)

$x = y^2$ (x is always positive)

NOTE: In using the chart, if we are given, for instance, $-4c(y+4) = (x-2)^2$, we are reminded of the $y = -x^2$ graph, and will think : "*y is basically negative*". Therefore, this curve opens downward.

NOTE: The parabola comes with **a VERTEX, an AXIS OF SYMMETRY, a FOCUS, and a DIRECTRIX.**

EXAMPLE: Find the parabola whose focus is (-2,1) and whose directrix is y = 3.
Answer:
1.) We know the directrix is perpendicular to the axis of symmetry of the curve.
2.) This, and the fact that the focus lies *below* the directrix tells us the parabola turns downward.
3.) The basic equation of a turned-down parabola is (consult 4 basic types): $-4cy = x^2$.
4.) " c " is half the distance from focus to directrix: c = (3-1)/2 = 1.
5.) The vertex of the parabola is (-2,2). (We know x = -2, because the vertex will lie on the same axis as the focus. And y = 2 because the vertex is halfway between the focus and the directrix.)
6.) The basic equation, given in (3) with vertex at (-2,2) is now $-4c(y-2) = (x+2)^2$.
7.) The equation with c = 1, is finally: $\boxed{-4(y-2) = (x+2)^2}$.

EXAMPLE: Find all characteristics of the conic section given by $2x^2 + 4x - 3y + 8 = 0$.
Answer:
1.) We recognize this as a parabola, because of the single squared term. In fact, because this "looks like" $x^2 - y = 0$ or $x^2 = y$ (y is always positive), this parabola opens upward.

2.) Complete the square (See Chapter 1 of this Guide.)
$$2x^2 + 4x - 3y + 8 = 0$$
$$2(x^2 + 2x \qquad) - 3y + 8 = 0$$
$$2(x^2 + 2x + 1) - 3y + 8 - 2 = 0$$
$$2(x+1)^2 = 3y - 6$$
$$(x+1)^2 = \frac{3}{2}(y-2)$$

3.) The vertex of the parabola is (-1,2).
4.) The focus can be found by setting $4c = 3/2$. Thus c = 3/8. The focus and vertex are on the axis of symmetry, to find the focus, we add 3/8 to the y-coordinate of the vertex. The focus is (-1, 2+3/8) = (-1, 19/8).
5.) The axis of symmetry is found (by looking at vertex) to be x = -1.
6.) The directrix is the same distance from the vertex as the focus, and is perpendicular to the axis of symmetry. Therefore, the directrix is y = 2-(3/8) , or y = 13/8.

EXAMPLE: Find the number d such that x + y = d is tangent to the parabola $2y^2 = x$.

Answer: Differentiating, we have $4y\dfrac{dy}{dx} = 1$, while the slope of the line x + y = d is -1.

Therefore, we need $\dfrac{dy}{dx} = -1$, or 4y = -1. Thus y = - 1/4, and from the equation of the

parabola, x = 1/8. The point (1/8, - 1/4) must lie on the line x + y = d. So   1/8-1/4 = d,
and d = -1/8.

EXAMPLE: Find the point on the parabola $x^2 = 4y$ closest to (0,1).

Answer: From the graph, we might guess that this point is the origin . The "distance-squared" , from the unknown point on the parabola (x, y) to the point (0,1) is

$D^2 = (x)^2 + (y-1)^2$. Differentiating, $2D\dfrac{dD}{dx} = 2x + 2(y-1)\dfrac{dy}{dx}$, and since we want

$\dfrac{dD}{dx} = 0$, the right side of the equation above must be 0. Thus, at the point (x, y) closest

to (0,1), we have $\dfrac{dy}{dx} = -\dfrac{x}{(y-1)}$. Now differentiating the parabola equation, $2x = 4\dfrac{dy}{dx} =$

or $\dfrac{dy}{dx} = \dfrac{x}{2} = -\dfrac{x}{(y-1)}$. Solving this equation, one solution is (0,0), a point whose

distance from (0,1) is 1. If x $\neq$ 0, then y-1 = -2, or y = -1, which is not possible. Thus,
as we suspected, the origin is the point on the parabola closes to (0,1) .

THE ELLIPSE

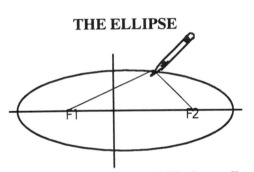

To draw an ellipse, attach a string to nails at Fl and F2, then pull a pencil tightly inside the
string, attempting to draw a circle. The result will be an ellipse with foci at Fl and F2.

NOTE: The ellipse comes with a **CENTER (not the vertex), 4 VERTICES (the
"corners"), 2 FOCI, and a MAJOR AND MINOR AXES OF SYMMETRY (note:
these are line segments.)**

NOTE: From the equation $\dfrac{x^2}{a^2} + \dfrac{y^2}{b^2} = 1$, mentally "blot out" the x-term (x = 0). Then

y = ±b; so (0, ±b) are the two y-intercepts. Similarly, (± a, 0) are the two x-intercepts.
These points are the vertices of the ellipse in standard position. This technique helps line up
any ellipse quickly. As for finding the foci,

$$c^2 = (Bigger \text{ of a or b})^2 - (Smaller \text{ of a or b})^2 \quad .$$

You needn't get locked into thinking of formulas involving "a" and "b".

EXAMPLE: Find the equation of this ellipse in standard form: $4x^2 + 9y^2 = 1$.

Answer: This one can stump you at first: $\dfrac{x^2}{1/4} + \dfrac{y^2}{1/9} = 1$.

EXAMPLE: If the foci of an ellipse are (3,0) and (-3,0), and the vertices are (4,0) and (-4,0), find the ellipse.

Answer: The foci lie on the longer, major axis of the ellipse . Thus c = 3, a = 4, and b = $\sqrt{7}$. This ellipse is centered at (0,0) , the point halfway between foci. The equation of the ellipse becomes $\dfrac{x^2}{16} + \dfrac{y^2}{7} = 1$.

EXAMPLE: Find the equation of an ellipse passing through (1 , -1) and (-2 , 1/2) in standard position.

Hint: The equation of an ellipse in standard position is: $\dfrac{x^2}{a^2} + \dfrac{y^2}{b^2} = 1$. Using (1,-1) in this equation, $\dfrac{1}{a^2} + \dfrac{1}{b^2} = 1$ or $a^2 + b^2 = a^2 b^2$.   (Equ. 1)

Using (-1, 1/2), we obtain $4b^2 + (1/4)a^2 = a^2 b^2$.   (Equ. 2)

Solve the equations together for a and b and use in the equation $\dfrac{x^2}{a^2} + \dfrac{y^2}{b^2} = 1$.

EXAMPLE: Find all characteristics of the conic section given by $2x^2 + 5y^2 + 4x + 30y + 37 = 0$. (Note this is an ellipse because AC > 0.)

Answer:

$2x^2 + 5y^2 + 4x + 30y + 37 = 0$, or

$2(x^2 + 2x + \quad) + 5(y^2 + 6y + \quad) = -37$ and

$2(x^2 + 2x + 1) + 5(y^2 + 6y + 9) = -37 + 2 + 45 = 10$ (completed the square) . So

$2(x+1)^2 + 5(y+3)^2 = 10$ and

$\dfrac{(x+1)^2}{5} + \dfrac{(y+3)^2}{2} = 1$

NOTE: Here is a technique that is useful in translating a conic section:

Imagine the ellipse centered at the origin: $\dfrac{(x - 0)^2}{5} + \dfrac{(y - 0)^2}{2} = 1$

We have

1.) Center at (0,0) (of course)

2.) Vertices at $(\pm\sqrt{5},0)$ and $(0,\pm\sqrt{2})$

3.) Foci at $(\pm\sqrt{3},0)$

4.) Major axis, the line segment between $(+\sqrt{5},0)$ and $(-\sqrt{5},0)$

5.) Minor axis, the line segment between $(0,+\sqrt{2})$ and $(0,-\sqrt{2})$

HOWEVER, our ellipse is centered at (-1,-3). So translate the information above. Now we have an ellipse with

1.) Center at (-1,-3)

2.) Vertices at $(\pm\sqrt{5},0)+(-1,-3)$ and $(0,\pm\sqrt{2})+(-1,-3)$, or $(-1\pm\sqrt{5},-3)$ and $(-1, -3\pm\sqrt{2})$

3.) Foci at $(\pm\sqrt{3},0) + (-1,-3) = (-1\pm\sqrt{3},-3)$

4.) Major axis the line segment between the two vertices $(-1\pm\sqrt{5},-3)$

5.) Minor axis the line segment between the two vertices $(-1, -3\pm\sqrt{2})$

THE HYPERBOLA

NOTE: The hyperbola comes with a **CENTER, 2 VERTICES, 2 FOCI, a PRINCIPAL AXIS OF SYMMETRY, and 2 ASYMPTOTES**

NOTE: It might be convenient to remember two types of hyperbolas ("hyperbolae" is the formal term):

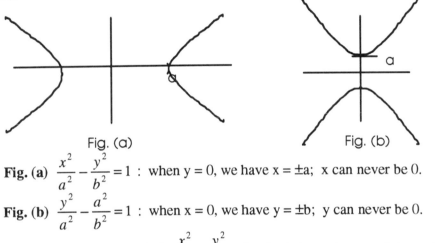

Fig. (a)   Fig. (b)

Fig. (a) $\dfrac{x^2}{a^2} - \dfrac{y^2}{b^2} = 1$: when y = 0, we have x = $\pm$a; x can never be 0.

Fig. (b) $\dfrac{y^2}{a^2} - \dfrac{a^2}{b^2} = 1$: when x = 0, we have y = $\pm$b; y can never be 0.

NOTE: Below, for the hyperbola $\dfrac{x^2}{a^2} - \dfrac{y^2}{b^2} = 1$, the intercepts are x-intercepts: (a,0) and (-a,0). Draw the box with horizontal sides of length 2a and vertical sides of length 2b. Then the asymptotes of the hyperbola are y = $\pm$ (b/a) x; which means they pass through the corners of the box.

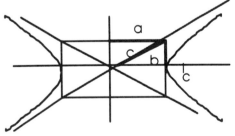

A useful way to find asymptotes is to rewrite the hyperbola equation $\dfrac{x^2}{a^2} - \dfrac{y^2}{b^2} = 1$ as $\dfrac{x^2}{a^2} - \dfrac{y^2}{b^2} = 0$. (Can you explain a rationale for doing this ?) From this equation the asymptotes are seen to be $y = \pm\dfrac{b}{a}x$.

EXAMPLE: Sketch the conic section: $x^2 - 2x - 4y^2 + 12y = -8$.

Answer: This is a hyperbola because AC < 0.

$x^2 - 2x - 4y^2 + 12y = -8$

$x^2 - 2x + \quad - 4(y^2 - 3y + \quad) = -8$

$x^2 - 2x + 1 - 4(y^2 - 3y + \dfrac{9}{4}) = -8 + 1 - 9 = -16$ (completed the square)

$(x-1)^2 - 4(y - \dfrac{3}{2})^2 = -16$

$4(y - \dfrac{3}{2})^2 - (x-1)^2 = 16$

$\dfrac{(y - \dfrac{3}{2})^2}{2^2} - \dfrac{(x-1)^2}{4^2} = 1$

We see immediately that this hyperbola opens up and down. So the principal axis of symmetry is vertical.

If the center were the origin (think of $\dfrac{y^2}{2^2} - \dfrac{x^2}{4^2} = 1$), we would have:

1.) Vertices $(0, \pm 2)$.

2.) Principal axis of symmetry, the y-axis, x = 0.

3.) $c^2 = a^2 + b^2$, and $c^2 = 2^2 + 4^2 = 20$, $c = \sqrt{20} = 2\sqrt{5}$, so the two foci are $(0, \ \pm 2\sqrt{5})$

4.) Asymptotes occur where $\dfrac{y^2}{2^2} - \dfrac{x^2}{4^2} = 0$ or $y = \pm \dfrac{1}{2}x$.

But now--siince the center is actually (1, 3/2), we translate all information using this new center. *All of these expressions can be simplifed.*

1.) New vertices $(0, \pm 2) + (1, 3/2)$, or $(1, \ 3/2 \pm 2)$

2.) Principal axis of symmetry is the line x = 0 + 1, or x = 1.

3.) Foci are $(0, \pm 2\sqrt{5}) + (1, 3/2) = (1, \ 3/2 \pm 2\sqrt{5})$

4.) Asymptotes are $y - \dfrac{3}{2} = \pm \dfrac{1}{2}(x-1)$.

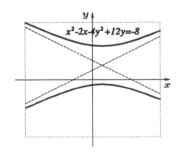

$$x^2 - 2x - 4y^2 + 12y = -8$$

CONIC SECTIONS REVIEW

CONIC SECTION	DEFINING PROPERTY	EQUATION(S) IN STANDARD FORM	SPECIAL CHARACTER-ISTICS
Circle	locus of all points equidistant from a fixed point	$x^2 + y^2 = r^2$	center, radius
Ellipse			center, foci , major and minor axes. Foci given using this formula for finding c: (**Write formula here:**)
Parabola			
Hyperbola		$\dfrac{x^2}{a^2} - \dfrac{y^2}{b^2} = 1$	

SECTION 10.6

"...turn to a place where you come out right"....Shaker hymn

ROTATING CONIC SECTIONS

Here we examine those conic sections that have an "xy" term.

NOTE: Given the conic section $Ax^2 + Bxy + Cy^2 + Dx + Ey + F = 0$, the "discriminant" term $B^2 - 4AC$ tells us what type of conic section to expect (ruling out degeneracies).

$B^2 - 4AC > 0$ tells us the curve is a **hyperbola**,

$B^2 - 4AC < 0$ tells us the curve is an **ellipse**,

$B^2 - 4AC = 0$ tells us the curve is a **parabola**.

EXAMPLE: Remove the xy - term for $2x^2 - 72xy + 23y^2 + 100x - 50y = 0$.

Answer: The discriminant is positive, so we have a hyperbola. Now

$\tan 2\theta = \dfrac{B}{A - C} = \dfrac{-72}{2 - 23} = \dfrac{72}{21} = \dfrac{24}{7}$, and $0 < 2\theta < \dfrac{\pi}{2}$ (tan 2θ is positive) so sin 2θ and

$\cos 2\theta$ are both positive. We use the 7-24-25 Pythagorean triple to find $\cos 2\theta = 7/25$. We use:

$$\cos\theta = \sqrt{\frac{1+\cos 2\theta}{2}} \quad \text{and} \quad \sin\theta = \sqrt{\frac{1-\cos 2\theta}{2}}$$

to find that $\cos\theta = 4/5$ and $\sin\theta = 3/5$. Now $x = \dfrac{4X-3Y}{5}$ and $y = \dfrac{3X+4Y}{5}$, so

$$\frac{2}{25}(4X-3Y)^2 - \frac{72}{25}(4X-3Y)(3X+4Y) + \frac{23}{25}(3X+4Y)^2 + \frac{100}{5}(4X-3Y) - \frac{50}{5}(3X+4Y) = 0,$$

or $-\dfrac{625}{225}X^2 + \dfrac{1250}{25}Y^2 + 50X - 100Y = 0$, or $-X^2 + 2Y^2 + 2X - 4Y = 0$, which gives:

$$\frac{(Y-1)^2}{(1/\sqrt{2})^2} - \frac{(X-1)^2}{(1)^2} = 1 , \text{ in the new coordinate system.}$$

SECTION 10.7

" The wave of the future is coming, and there's no fighting it"...Anne Morrow Lindbergh

CONIC SECTIONS REVISITED

ECCENTRICITY , e, is a term which describes, in some sense, the "width" of the curve:
$$e = c/a$$

EXAMPLE: Find the eccentricity of the conic section $\dfrac{x^2}{25} + \dfrac{y^2}{9} = 1$.

Answer: Since $c = \sqrt{25-9} = 4$, the eccentricity e is given as e = 4/5.

EXAMPLE: Find the conic section with foci (0,5) and (0,-5) and eccentricity 5/3.

Answer: We have c = 5 and c/a = 5/3 so a = 3. We cannot have the relationship $c^2 = a^2 - b^2$ with c = 5 and a = 3. Therefore we do not have an ellipse, but rather, a hyperbola, with $c^2 = a^2 + b^2$. (The general argument goes: if e > 1, we have a hyperbola.) Now b = 4. The hyperbola is in standard position (observing the foci) so the equation is $\dfrac{y^2}{16} - \dfrac{x^2}{9} = 1$.

NOTE: Conic sections with vertex at the origin can be expressed with two parameters (e.g., a and b) . In polar form, with focus at the origin, they can be expressed with eccentricity e = c/a and directrix x = ± k or y = ± k , where for an ellipse or hyperbola, k = a^2/c , and for a parabola, k = c.

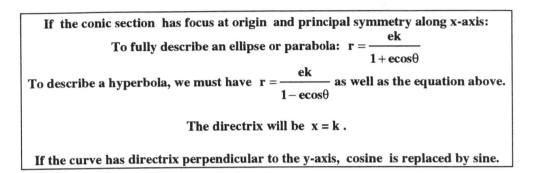

NOTE: The directrix is perpendicular to the "long" direction of the curve.

EXAMPLE: What is the conic section defined by $r = \dfrac{1}{1 + \frac{1}{2}\sin\theta}$?

Answer: This is an ellipse, with e = 1/2 and ek = 1 (so k = 2). This tells us that the directrix of the ellipse (perpendicular to the major axis) is y = 2. Plotting points in polar coordinates, we can find r when θ = π/2 and -.π/2 . The points given (vertices along the major axis) are (0,2/3) and (0,-2). The center of the ellipse is midway between, at (0,-2/3) . Then from the fact that a = 4/3 and c = 2/3 , we can find b, and the equation for the ellipse.

Here is another way. We have r + (r/2) sin θ = 1, and since y = r sinθ, r + y/2 = 1, or

r = 1 - y/2. Squaring, $r^2 = (1 - y/2)^2$ or $x^2 + y^2 = 1 - y + y^2/4$, or $x^2 + \dfrac{3}{4}y^2 + y = 1$.

After completing the square, we have $\dfrac{x^2}{4/3} + \dfrac{(y + 2/3)^2}{16/9} = 1$.

CHAPTER REVIEW

Consider the ways in which the material in this chapter "hangs together". We studied parametrization, which is ------ A common parametrization of the circle is -----and a common parametrization of an ellipse is ------- . The arc length (length of the curve) formula for a parametrized curve becomes------How this formula comes about is explained by ------------With polar coordinates, we change to two parameters, ---------, and the conversion formulas that bring us back and forth from polar to Cartesian coordinates are ---------. Sometimes in the conversion, points may not be given unique representations. For example, ---------. Some examples of polar curves are ----------, and their representative equations are ----------. Some useful tips for graphing are ----------. The arc length formula in polar coordinates is---and the area formula, --------. The rationale for these formulas is ----------. To find the area between two curves given in polar coordinates, we do the following:------------. The general conic section is given by the equation----------. We can tell an ellipse, parabola, hyperbola, or circle from the formula as follows: --------- Characteristic properties of each of these conic sections are --------. Examples of types of problems we have seen with each conic section are the following : ----------To eliminate the xy-term, we perform a rotation; the angle given by-------. The formulas for conversion (if you must learn them) are---------A unified approach to conic sections involves eccentricity, defined as -----and the directrix, whose use is --------. The general formula for a conic section in polar coordinates is (are)-------------The defining properties of the conic sections are -------, the reflection properties are ------------------.

CHAPTER 11

"Push on ...keep moving"...Thomas Morton

In this chapter, we develop a foundation for understanding vectors. If you have not had physics or engineering courses, where these concepts are generally used, you may need to adjust to the new ideas and notation. The mathematical constructs seen here will be the core of the entire course, so work especially hard at this beginning material. You must develop such familiarity with vectors in this chapter that when you are reviewing, you will think the problems are *obvious*. Try to do more than the homework assigned--every problem has a special facet that will help you understand vectors more instinctively. Do not confine yourself to "types" of problems--stretch yourself. Also, since this material is the foundation for so much to come, MEMORIZE and MEMORIZE some more! You will need to have automatic recall of such things as dot products, projections, cross-products, equations of planes, lines, parametric representations of the same, etc.

SECTION 11.1

"Mathematics ...is changing. The field shows a renewed emphasis on applications, a return to concrete images, and an increasing role for mathematical experiments."...Ivars Peterson

CARTESIAN (RECTANGULAR) COORDINATES IN SPACE

EXAMPLE: Find the distance D between the point (4,1, 3) and (-2,05).
Answer:

$$D = \sqrt{(\text{change in x})^2 + (\text{change in y})^2 + (\text{change in z})^2} = \sqrt{(4-(-2))^2 + (1-0)^2 + (3-5)^2} = \sqrt{41}$$

EXAMPLE: Find the following:
(a) the equation of a sphere of radius 2 with center (1,-4,5):
Answer: $(x-1)^2 + (y+4)^2 + (z-5)^2 = 2^2$.
(b) the locus of points in an open ball of radius 2 with center (1,-4,5):
Answer: $\{(x,y,z)$ such that $(x-1)^2 + (y+4)^2 + (z-5)^2 < 2^2\}$.
(c) the locus of points exterior to a sphere of radius 2 with center (1,-4,5):
Answer: $\{(x,y,z)$ such that $(x-1)^2 + (y+4)^2 + (z-5)^2 > 2^2\}$.

EXAMPLE: Find the center and radius of the sphere : $x^2 + 6x + y^2 - 12y + z^2 + 2z = 3$.
Answer: Complete the square on x, y, and z at the same time:

$x^2 + 6x + y^2 - 12y + z^2 + 2z = 3$

$(x^2 + 6x \quad) + (y^2 - 12y \quad) + (z^2 + 2z \quad) = 3$ (Factor out leading coefficients)

$(x^2 + 6x + 9) + (y^2 - 12y + 36) + (z^2 + 2z + 1) = 3 + 9 + 36 + 1$

Add [half of the middle coefficient squared] to the parentheses on the left side of the
equation and add this quantity to the right side of the equation as well .

$(x + 3)^2 + (y - 6)^2 + (z + 1)^2 = 49 = 7^2$

This is the equation of a sphere with radius 7 and center (-3,6,-1).

SECTION 11.2

*"The truth is, the science of Nature has been already too long made only a work of the brain and
the fancy: It is now high time that it should return to the plainness and soundness of observations
on material and obvious things"..Robert Hooke*

VECTORS IN SPACE

A point in 3-space will be denoted $P = (x, y, z)$. When we think of this point as a vector,
we write it as $\mathbf{P} = x\,\mathbf{i} + y\,\mathbf{j} + z\,\mathbf{k}.$ Vectors are written in boldface in the text and in this
guide. Thus, a is a number, and **a (written in boldface)** is a vector. To contrast with
"vectors", real numbers are often called "scalars".

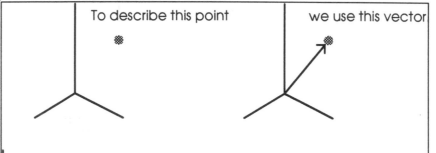

We say that x, y, z are **coordinates of the point** P and x , y, and z are **components of the
vector P.** (Sometimes the vector **P** is written $\vec{\mathbf{P}}$.)

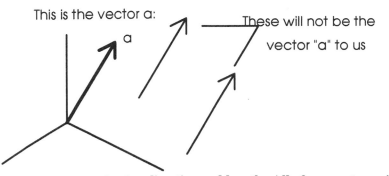

This is the vector a: These will not be the
                                     vector "a" to us

It is useful to think of vectors as having direction and length. All of our vectors will originate at the origin, unlike the "free" vectors one encounters in physics courses, for example.

ADDING AND SUBTRACTING VECTORS: It is possible to add and subtract vectors **a** and **b** by adding "pointwise"; that is, by adding their components.

EXAMPLE: Find the sum   **a + b** where **a** = 3**i** + 2**j** and **b** = **i** - **j** + **k**.
Answer: a + b = (3 +1) **i** + (2 - 1) **j** + (0 + 1) **k** = 4**i** + **j** + **k**.

Adding Vectors:

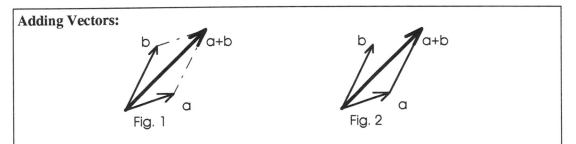

In Fig. 1, we think of adding **a** and **b** by the parallelogram rule: **a** + **b** is the vector on the main diagonal of the parallelogram with sides **a** and **b**. In Fig. 2, we think of adding **a** and **b** by the "tip-to-toe" method, where we add the "toe" of **b** to the "tip" of **a**, or vice versa.

Keep a Diary of Mistakes: Reserve a section of your notebook for a record of the types of mistakes you tend to make--not only algebraic errors, but also confusions that prevented you from solving exercises from the homework, quizzes and exams. (Be sure to review this before taking another quiz or exam!)

Subtracting Vectors:

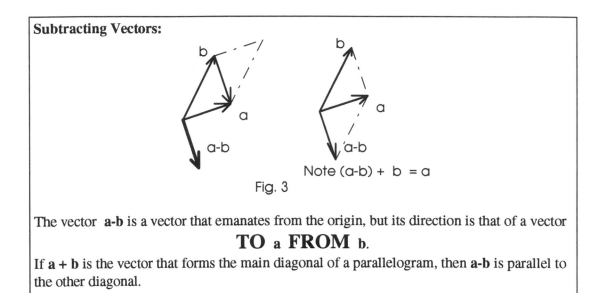

Fig. 3

Note (a-b) + b = a

The vector **a-b** is a vector that emanates from the origin, but its direction is that of a vector

TO a FROM b.

If **a + b** is the vector that forms the main diagonal of a parallelogram, then **a-b** is parallel to the other diagonal.

NOTE: If we are only computing with vectors , the diagrams we draw in this Guide may not be to scale, or even accurate. (Inaccurate drawings can still be useful.)

EXAMPLE: If **a = 2i + 5j -k** and **b = i - j** , find **a - b**.

Answer: **a - b = (2-1) i + (5-(-1))j + (-1-0)k = i + 6j - k.**

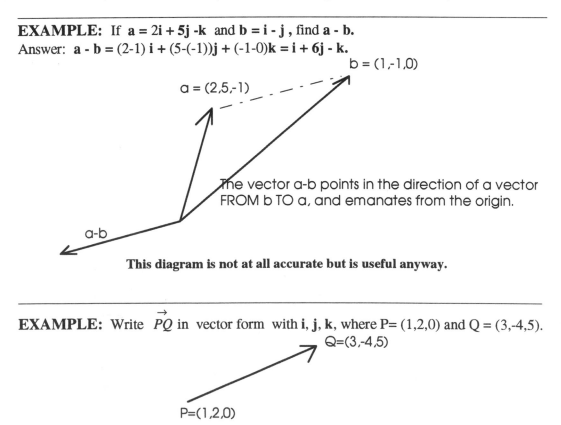

The vector a-b points in the direction of a vector FROM b TO a, and emanates from the origin.

This diagram is not at all accurate but is useful anyway.

EXAMPLE: Write $\overrightarrow{PQ}$ in vector form with **i, j, k**, where P= (1,2,0) and Q = (3,-4,5).

Answer: To find the vector $\overrightarrow{PQ}$ from P to Q subtract P coordinates from Q coordinates. Thus: change in x = (3-1) = **2,** change in y = (-4-2) = **-6,** change in z = (5-0) = **5.** And

$\overrightarrow{PQ} = 2\,\mathbf{i} - 6\,\mathbf{j} + 5\,\mathbf{k}$.

NOTE: When we write "0", we mean the number 0, but boldface "**0**" denotes the **0**-vector, or (0,0,0). Namely, **0** is the origin in 3-space.

NOTE: We speak a lot about "parallel" vectors; these are vectors that point in the same direction. We are trying to combine visual ease (where we may imagine vectors such as **a - b** as free vectors, that float about in space) with the fact that vectors must originate at the origin.

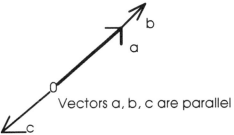

Vectors a, b, c are parallel

NOTE: The length of a vector **a** is denoted ‖a‖. A "unit" vector is a vector of length 1. We can obtain a unit vector parallel to (or in the same direction as) any given vector **a** by dividing a by its length. Thus, $\dfrac{\mathbf{a}}{\|\mathbf{a}\|}$ is recognizable as a unit vector.

EXAMPLE:
(a) Find the unit vector **u** having the same direction as (or parallel to) **a** = 3 **i** + 4**j** - **k**.

Answer: The length of this vector is: $\|\mathbf{a}\| = \sqrt{3^2 + 4^2 + (-1)^2} = \sqrt{26}$. Thus $\mathbf{u} = \dfrac{\mathbf{a}}{\|\mathbf{a}\|} =$

$\dfrac{3}{\sqrt{26}}\mathbf{i} + \dfrac{4}{\sqrt{26}}\mathbf{j} - \dfrac{1}{\sqrt{26}}\mathbf{k}$.

(b) Find a vector in the same direction as **a**, but having length 6.

Answer: Just multiply the unit vector above by 6: $\dfrac{18}{\sqrt{26}}\mathbf{i} + \dfrac{24}{\sqrt{26}}\mathbf{j} - \dfrac{6}{\sqrt{26}}\mathbf{k}$.

(c) Find the midpoint of the vector **a** above.
Answer: This is found by halving every component. We write this point as (3/2, 4/2, -1/2) or (3/2, 2, -1/2). As a vector, this is **v** = (3/2) **i** + 2**j** - (1/2) **k**.

(d) Find a vector pointing in the opposite direction of the vector **a**, and having the same length.
Answer: This vector would be -**a** , or -3 **i** - 4**j** + **k**.

EXAMPLE: Show graphically that the vector **a** in the plane can be represented as
a = (‖a‖ cos θ) **i** + (‖a‖ sin θ) **j** where θ is the angle between the x-axis and the vector **a**.

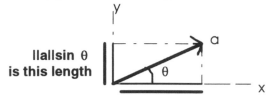

So vector **a** = ‖a‖cos θ **i** + ‖a‖sin θ **j**

SECTION 11.3

" Thou hast seen nothing yet"...Miguel Cervantes

THE DOT PRODUCT

The dot product of two vectors is a **real number, not a vector**. The dot product has many uses.

NOTE: The dot product makes it possible to find the angle between two vectors. (See examples below). Note: $\mathbf{a} \cdot \mathbf{a} = \|a\|^2$. Can you prove this? Can you prove $\mathbf{a} \cdot \mathbf{b} = \|a\|\ \|b\|\cos\theta$ from the Law of Cosines?

EXAMPLE: Let **a** = **i** - **j** and **b** = 2 **i** + 3**j** + 5**k**. Find (i) **a** · **b** , (ii) the cosine of the angle between the vectors, and (iii) the angle between the vectors.
Answer:
(i) $\mathbf{a} \cdot \mathbf{b} = (1\cdot2) + (-1\cdot3) + (0\cdot5) = -1$.

(ii) $\|a\| = \sqrt{1^2 + (-1)^2} = \sqrt{2}$ and $\|b\| = \sqrt{2^2 + 3^2 + 5^2} = \sqrt{38}$. And $\mathbf{a} \cdot \mathbf{b} = \|a\|\ \|b\|\cos\theta$.
Thus $-1 = \sqrt{2}\sqrt{38}\cos\theta = \sqrt{2\cdot38}\cos\theta = \sqrt{2\cdot2\cdot19}\cos\theta = 2\sqrt{19}\cos\theta$ and $\cos\theta = -\dfrac{1}{2\sqrt{19}}$.

(iii) $\theta = \cos^{-1}\left(-\dfrac{1}{2\sqrt{19}}\right) = 1.6857$ radians (obtained from a calculator).

NOTE: **a** and **b** are perpendicular if and only if $\mathbf{a} \cdot \mathbf{b} = 0$.

EXAMPLE: Are the vectors $\mathbf{a} = 2\mathbf{i} + 3\mathbf{j} - \mathbf{k}$ and $\mathbf{b} = -\mathbf{i} + \mathbf{k}$ perpendicular?
Answer: Since $\mathbf{a} \cdot \mathbf{b} = (2)(-1) + (3)(0) + (-1)(1) = -3$, not 0, the vectors are not perpendicular.

EXAMPLE: Use the definition of dot product to prove $\mathbf{a} \cdot \mathbf{b} = \mathbf{b} \cdot \mathbf{a}$.
Answer: Write $\mathbf{a} = a_1\mathbf{i} + a_2\mathbf{j} + a_3\mathbf{k}$ and $\mathbf{b} = b_1\mathbf{i} + b_2\mathbf{j} + b_3\mathbf{k}$. Then
$\mathbf{a} \cdot \mathbf{b} = (a_1 b_1) + (a_2 b_2) + (a_3 b_3) = (b_1 a_1) + (b_2 a_2) + (b_3 a_3) = \mathbf{b} \cdot \mathbf{a}$.

NOTE: We are not usually able to visualize the dot product.

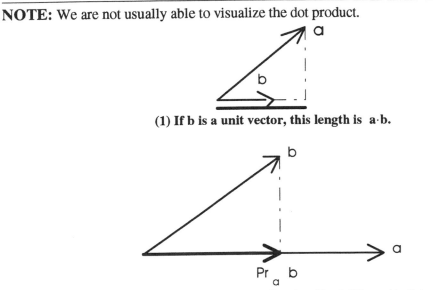

(1) If b is a unit vector, this length is a·b.

(2) The projection of b onto a is more easily visualized. We write this vector as $\mathbf{Pr_a b}$

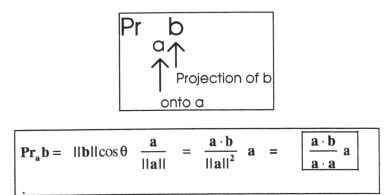

$$\mathbf{Pr_a b} = \|\mathbf{b}\|\cos\theta \, \frac{\mathbf{a}}{\|\mathbf{a}\|} = \frac{\mathbf{a} \cdot \mathbf{b}}{\|\mathbf{a}\|^2}\mathbf{a} = \boxed{\frac{\mathbf{a} \cdot \mathbf{b}}{\mathbf{a} \cdot \mathbf{a}}\mathbf{a}}$$

(3) The first equality says $\mathbf{Pr_a b}$ equals (length of b) times (the unit vector in the direction of a). The second equality writes this differently, as does the third equality, with 4 a-terms and 1 b-term. Remember that the direction of this vector is a!

EXAMPLE: Find $\mathbf{Pr_a b}$ for $\mathbf{a} = 2\mathbf{i} + 3\mathbf{j} - \mathbf{k}$ and $\mathbf{b} = \mathbf{i} - \mathbf{j}$.

Answer: Note $\|\mathbf{a}\| = \sqrt{2^2 + 3^2 + (-1)^2} = \sqrt{14}$. $\mathbf{Pr_a b} = \dfrac{\mathbf{a} \cdot \mathbf{b}}{\|\mathbf{a}\|^2}\mathbf{a} = \dfrac{-1}{14}(2\mathbf{i} + 3\mathbf{j} - \mathbf{k}) =$

$(-1/7)\,\mathbf{i} - (3/14)\,\mathbf{j} + (1/14)\,\mathbf{k}$. The Projection of **b** onto **a** is a **vector** pointing in the direction of **a.**

NOTE: In general, when we are asked to find projections of vectors, the vectors do not come labelled as "**a**" and "**b**" , so be sure you understand how to project a vector onto another.

NOTE: If **u** is a unit vector, then $\mathbf{Pr_u b} = (\mathbf{u \cdot b})\mathbf{u}$. Why?

EXAMPLE: Determine whether $\overrightarrow{PQ}$ and $\overrightarrow{PR}$ are perpendicular, where $P = (0,1,2)$, . $Q = (3,1,-4)$, and $R = (2,0,3)$.

Answer: Since $\overrightarrow{PQ} = 3\mathbf{i} + 0\,\mathbf{j} - 6\mathbf{k}$ and $\overrightarrow{PR} = 2\mathbf{i} - \mathbf{j} + \mathbf{k}$,

$\overrightarrow{PQ} \cdot \overrightarrow{PR} = (3)(2) + 0(-1) + (-6)(1) = 0$ and the vectors are perpendicular.

EXAMPLE: Let $\mathbf{a} = \mathbf{i} + \mathbf{j} - \mathbf{k}$, $\mathbf{a'} = 2\mathbf{i} - 3\mathbf{j} - \mathbf{k}$, and $\mathbf{b} = 4\mathbf{i} - \mathbf{j} - 3\mathbf{k}$. (It is given that these vectors lie on the same plane.) Resolve vector **b** into vectors parallel to **a** and **a'**

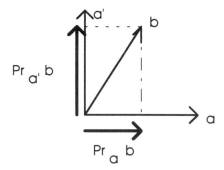

Answer: (Hint) We have $\mathbf{b} = \mathbf{Pr_a b} + \mathbf{Pr_{a'} b}$ where $\mathbf{Pr_a b} = \dfrac{\mathbf{a \cdot b}}{\mathbf{a \cdot a}}\,\mathbf{a}$ and $\mathbf{Pr_{a'} b} = \dfrac{\mathbf{a' \cdot b}}{\mathbf{a' \cdot a'}}\,\mathbf{a'}$.

EXAMPLE: Direction angles. Find the direction angles of $\mathbf{a} = 3\mathbf{i} + 4\mathbf{j} - \mathbf{k}$.

Answer: When the vector $\mathbf{a} = a_1\mathbf{i} + a_2\mathbf{j} + a_3\mathbf{k}$ is a unit vector, the **coordinates** of **i, j,** and **k** are **direction cosines**; these are the cosines that the vector **a** makes with the x, y, and z - axes, respectively. I.e., $a_1 = \cos\alpha$, where α is the angle between the x-axis and the vector **a**, etc. The angles can then be found from the cosines.

We write $\mathbf{a} = 3\mathbf{i} + 4\mathbf{j} - \mathbf{k} = \sqrt{26}\left(\dfrac{3}{\sqrt{26}}\mathbf{i} + \dfrac{4}{\sqrt{26}}\mathbf{j} - \dfrac{1}{\sqrt{26}}\mathbf{k}\right)$ where we have multiplied

and divided by the vector's length. Now $\left(\dfrac{3}{\sqrt{26}}\mathbf{i} + \dfrac{4}{\sqrt{26}}\mathbf{j} - \dfrac{1}{\sqrt{26}}\mathbf{k}\right)$ is a unit vector. Thus

$\cos\alpha = 3/\sqrt{26}$, $\cos\beta = 4/\sqrt{26}$ and $\cos\gamma = -1/\sqrt{26}$, where α, β, γ are the angles between

a and the x- , y- , and z-axes. These angles can be found by using $\alpha = \cos^{-1}\dfrac{3}{\sqrt{26}}$, etc.

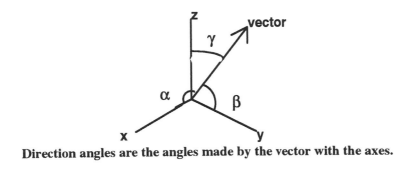

Direction angles are the angles made by the vector with the axes.

THE CROSS PRODUCT

The cross-product is a product of vectors whose result gives a **vector**. It can be used only in 3-dimensions, and does not generalize as the dot product does, to higher dimensions. The cross-product of **a** and **b** is denoted **a x b** where

(1) **a x b** is a **vector** pointing in a direction perpendicular to **a** and **b** and given by the right-hand rule.

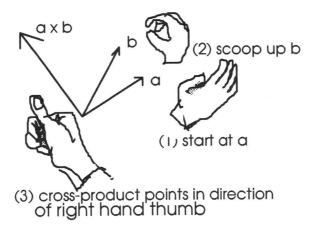

(2) **scoop up b**

(1) start at a

(3) cross-product points in direction of right hand thumb

(2) the **length of a x b,** or ‖**a x b**‖ is given by ‖**a x b**‖= ‖**a**‖ ‖**b**‖ sin θ where θ is the angle between **a** and **b**. (Note that sin θ ≥ 0 if θ lies in [0, π].)

The cross-product is also a way to find angle between two vectors, in 3-space. But it is not as useful as the dot-product for this purpose.

(3) the **length of a x b,** or ‖**a x b**‖ is in fact equal to the **area of the parallelogram** with sides **a** and **b** **So the cross-product gives a useful way to find areas.** (Do not get confused! The cross-product is a vector, but its length is a real numberwhich can be interpreted as an area.)

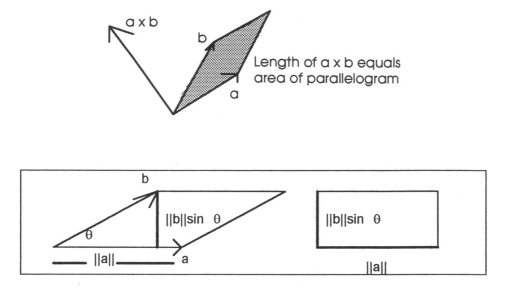

Length of a x b equals
area of parallelogram

Area of parallelogram is base times height. thus, area of parallelogram equals area of rectangle with the same base and height as the parallelogram.

(4) Two non-zero vectors **a** and **b** are collinear (point along the same line) if and only if **a x b = 0** , the **0**-vector. (The area of the parallelogram is 0).

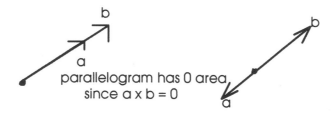

(5) To find **a x b** use determinants (see Example below, after comment #8.)

(6) The cross-product can tell us not only about area, but about volume, as follows. If we have 3 vectors in 3-space, **a**, **b**, and **c**, then |**c**·(**a x b**)| represents the volume of the parallelepiped with sides **a, b, c**. In the case that **a, b, c** are all co-planar, or lie on the same plane, the volume of the parallelepiped they span is 0. Notice we can use a, b, c in any order in the expression |**c**·(**a x b**)| to find the volume of the parallelepiped.

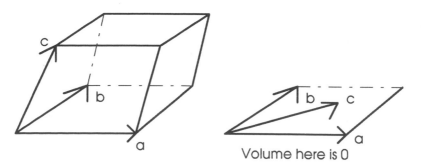

(7) A not-so-wonderful aspect of the cross product is that **a x b** ≠ **b x a** in general. In fact, (**a x b**) = - (**b x a**). **The cross-product is not commutative.** (Some pairs of vectors may commute, but this is not the general rule.)

The commutative property ab= ba is simliar to commuters who go back and forth. The cross product is not commutative.

Nor is the cross product associative (although some triples of vectors may have the associative property.) It is not true, for example, that (**i x i**) **x j** = **i x** (**i x j**). Since cross product is defined only between two vectors, not three, we cannot take a cross-product of three vectors such as **a x b x c**, without specifying where the parentheses should go.

In the associative property (axb) x c = a x (bxc). But the cross prduct is not associative.

(8) It's useful to remember that **i** x **i** = **j** x **j**= **k** x **k** = 0. Think of the multiplication of the vectors **i**, **j**, **k** taking place on a circle. Multiplying in a counterclockwise direction, the cross product of two consecutive vectors is the next vector (Fig. a) Multiplying in a clockwise direction, the cross product is the negative of the next vector (Fig. b).

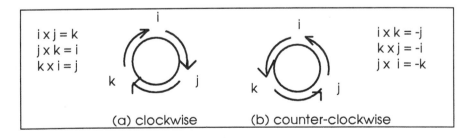

(9) When it's easy to do, we can find cross products without the use of determinants--in the same way that we would multiply polynomial factors--but we must remember the multiplication rules for **i, j, k** above. Thus we can find (**i + 2j**) × (**j-k**) as
(**i** ×**j**) +(2**j**×**j**)) + (**i**×**k**) -(2**j**×**k**) = **k** + 0 + (-**j**) - 2(**i**) = -2**i** - **j** + **k.**

NOTE: Finding determinants:

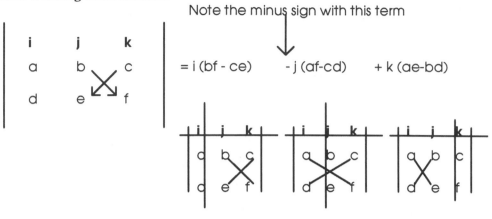

Note the minus sign with this term

$$\begin{vmatrix} i & j & k \\ a & b & c \\ d & e & f \end{vmatrix} = i\,(bf - ce) \quad - j\,(af - cd) \quad + k\,(ae - bd)$$

EXAMPLE: Let **a** = **j** + **k** ,   **b** = 2**i** - **j**+ 3**k**, **c** = -3 **i** - **j** -**k**. Find (i) **a** x **b** (ii) **c**·(**a** x **b**) (iii) the volume of the paralellepiped with sides **a, b,** and **c,** (iv) the area of the parallelogram with sides **a** and **b**   and (v) a vector perpendicular to both **a** and **b**.

Answers:

(i) $\mathbf{a} \times \mathbf{b} = \begin{vmatrix} i & j & k \\ 0 & 1 & 1 \\ 2 & -1 & 3 \end{vmatrix} = i\,(3 - (-1)) - j\,(0 - 2) + k\,(0 - 2) = \ 4\mathbf{i} + 2\mathbf{j} - 2\mathbf{k}$

(ii) **c**·(**a** x **b**) = (-3**i** - **j** -**k**) · (4**i** + 2**j** - 2**k**) = -12-2+2 = -10.

(iii) The volume of the parallelepiped with sides **a, b,** and **c** is | **c**·(**a** x **b**) | = |-10| = 10.

(iv) The area of the parallelogram with sides **a** and **b** is ‖ **a** x **b** ‖ = $\sqrt{4^2 + 2^2 + 2^2}$ =2$\sqrt{3}$.

(v)   A vector perpendicular to both **a** and **b** is given by **a** x **b** in (i) above.

EXAMPLE: Find the volume of the parallelepiped with 4 vertices given by P = (0,0,0), Q = (1,2,3), R = (-1,3,5), and S = (2,0,4).

Answer: Hint: Find the vectors $\overrightarrow{PQ}$, $\overrightarrow{PR}$, and $\overrightarrow{PS}$ and find $|\overrightarrow{PQ} \cdot (\overrightarrow{PR} \times \overrightarrow{PS})|$.

NOTE: If you are finding it difficult to keep track of what's a vector and what's a scalar (real number)--that's normal! But keep at it; this is important! What may complicate the issue is that your teacher, in lecture, will probably not be using boldface, and will denote vectors as $\vec{a}$, $\vec{b}$, $\vec{P}$. In fact, many teachers are accustomed to making no distinction between the **point** P = (x, y, z) and the **vector,** $\vec{P}$ = x **i** + y **j** +z **k**, and it may seem that they are taking dot and cross products of (what seem to you) points. Before it's too late, be clear about this with your teacher. It is not a dumb question to ask what types of notation are going to be used!

TRUE or FALSE

1. Given points P and Q, $\overrightarrow{PQ} = -\overrightarrow{QP}$.
2. The dot product has the property that $| \mathbf{a} \cdot \mathbf{b} | = \|\mathbf{a}\| \, \|\mathbf{b}\| \sin \theta$
3. The cross product has the property that $\mathbf{a} \times \mathbf{b} = \|\mathbf{a}\| \, \|\mathbf{b}\| \sin \theta$
4. $\mathbf{a} \times \mathbf{b} = - \mathbf{b} \times \mathbf{a}$
5. $\mathbf{a} \times \mathbf{b} = 0$ if and only if the vectors $\mathbf{a}$ and $\mathbf{b}$ are perpendicular.
6. $\mathbf{i} \times \mathbf{i} = 0$.
7. $\mathbf{a} \cdot (\mathbf{b} \times \mathbf{c})$ is the volume of a parallelepiped with sides $\mathbf{a}, \mathbf{b}, \mathbf{c}$.
8. "Parallelepiped" is how you spell parallelepiped.
9. The area of a parallelogram with sides given by vectors $\mathbf{a}, \mathbf{b}$ is $\mathbf{a} \times \mathbf{b}$.
10. We can divide vectors.
11. You cannot add a scalar to a vector.
12. You can divide a vector by a scalar.
13. The vector $-\mathbf{a}$ is parallel to the vector $\mathbf{a}$.
14. The dot product is commutative.
15. When working with dot product, $\mathbf{a} \cdot (\mathbf{b} \cdot \mathbf{c}) = (\mathbf{a} \cdot \mathbf{b}) \cdot \mathbf{c}$.
16. It is true that $(\mathbf{a} \times \mathbf{b}) \times \mathbf{c} = \mathbf{a} \times (\mathbf{b} \times \mathbf{c})$. (The cross product is associative.)
17. If a is resolved into two components, then the second component is
$$\sqrt{\|\mathbf{a}\|^2 - (\text{first component})^2}\,.$$

Answers: (1) T (2) F (3) F (although $\|\mathbf{a} \times \mathbf{b}\| = \|\mathbf{a}\| \, \|\mathbf{b}\| \, |\sin\theta|$) (4) T (5) F (parallel) (6) T (7) F -- need absolute value (8) T (9) F --need length, or norm (10) F (11) T (12) T (13) T (14) T (15) doesn't make sense; we don't take dot product of a vector and scalar, so F (16) F--try an example with $\mathbf{i}, \mathbf{i}$ and $\mathbf{j}$ (17) F--true only if the two vectors on which we are taking components are perpendicular.

Concept Review

1. How would you find the length of a given vector? A unit vector parallel to a given vector? The angle between 2 vectors? The projection of a vector onto another vector? How would you know if the angle between two vectors is acute? How would you know if two vectors are perpendicular? How do you find the direction angles of a vector? How do you resolve a vector into components? All of these techniques use dot product.

2. How would you find a vector perpendicular to 2 given vectors? The area of the parallelogram spanned by 2 vectors? The area of a triangle in 3-space? (Hint: this is 1/2 the area of the parallelogram). How would you know if 3 vectors lie on the same plane? How would you find the volume of a parallelepiped whose sides are given by 3 vectors? -- All of these techniques use the cross product.

SECTION 11.5

" Ne'er look for birds of this year in the nests of the last."...Miguel Cervantes

LINES IN SPACE

When it is understood that we are working in 3-dimensional space, or 3-space, the equation $y = mx + b$ no longer describes a line in 3-space. In fact it describes a plane. **One** equation will describe a **plane**; **two** equations will be required to describe a **line**. This has to do with equations pinning down "degrees of freedom". The plane "wants" freedom in 2 dimensions; one equation is sufficient to pin it down. The line "wants" freedom in 1 dimension; two equations are now necessary to define it.

We can describe a straight line in three ways.

(1) **VECTOR EQUATION**. A line is given by a point and a slope (ask Euclid!) Suppose we want the straight line that passes through $P = (2,-1,3)$ and has direction (slope) vector given by $4\mathbf{i} + 2\mathbf{j} - \mathbf{k}.$

> **[Note: in 3-space it takes a vector to give direction, or slope.]**

Then if $x\,\mathbf{i} + y\,\mathbf{j} + z\,\mathbf{k}$ is *any* vector on the line, the equation of this line is:

$$x\,\mathbf{i} + y\,\mathbf{j} + z\,\mathbf{k} = t\,(4\mathbf{i} + 2\mathbf{j} - \mathbf{k}) + (2\mathbf{i} - \mathbf{j} + 3\mathbf{k})$$
generic (x,y,z)-point on line = t (slope vector) + (given point on line)

This formula translates a line through the origin , $x\,\mathbf{i} + y\,\mathbf{j} + z\,\mathbf{k} = t\,(4\mathbf{i} + 2\mathbf{j} - \mathbf{k})$ to pass through the point $(2\mathbf{i} - \mathbf{j} + 3\mathbf{k})$ when $t = 0$.

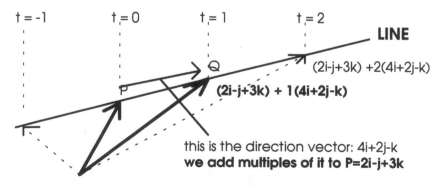

As t varies, the point obtained slides along the line. Note the vectors on the line for various values of t.
Note that the line is travelled in a certain direction, as t increases.

(2) **PARAMETRIC EQUATION:** Here we express the coordinates x, y, and z in terms of t, to obtain 3 separate equations. In (1) we could have written:

$$x\,\mathbf{i} + y\,\mathbf{j} + z\,\mathbf{k} = (2\mathbf{i} - \mathbf{j} + 3\mathbf{k}) + t\,(4\mathbf{i} + 2\mathbf{j} - \mathbf{k}) = (2 + 4t)\,\mathbf{i} + (-1+2t)\,\mathbf{j} + (3-t)\,\mathbf{k}.$$

As a result, the parametric equations are

$$x = \ 2 + 4t$$
$$y = \ -1 + 2t$$
$$z = \ \ 3 - t$$

with t any real number.

Note that the components of the direction (slope) vector 4i + 2j - k appear with the "t" terms, and that the point P = (2, -1, 3) appears as the constant terms.

This is another description of the same line.

(3) **SYMMETRIC EQUATIONS:** Here we set all t's in (2) equal to each other . For example, solving for t, we find $t = \dfrac{x-2}{4}$ and $t = \dfrac{y+1}{2}$, etc. Setting these values equal we have the symmetric equation of a line:

$$\frac{x-2}{4} = \frac{y+1}{2} = \frac{z-3}{-1}.$$

Note that the direction (slope) vector, 4i + 2j - k, appears in the denominator, and the point P = (2, -1, 3) appears in the numerator. NOTE THAT IT TAKES TWO EQUATIONS IN x AND y TO DESCRIBE A LINE. (Convince yourself that the two "equal signs" amount to two equations.)

NOTE: If the equation of a line has (for example) the appearance $\dfrac{x-1}{2} = \dfrac{y+3}{1};\quad z = 4$, we have our two equations. The direction (slope) vector in this case is $(2,1,\mathbf{0})$, because there is no change in the z-coordinate. Sometimes it may be a bit tricky to find the direction (slope) vector. See the next example.

EXAMPLE: Find the direction (slope) vector of the line given in symmetric form as:
$$\frac{1-3x}{2} = 2+5z;\quad y = 0.$$

Answer: We work with the terms to write:
$$\frac{(-1/3)}{(-1/3)}\left(\frac{1-3x}{2}\right) = \frac{(1/5)}{(1/5)}(2+5z); y = 0, \text{ or } \frac{x-(1/3)}{(-2/3)} = \frac{z+(2/5)}{1/5}; y = 0, \text{ and the slope}$$
vector can now be seen to be (-2/3, 0, 1/5).

EXAMPLE: Working with equations of lines
(1) **Find parametric equations for the line going through 2j + 4k and parallel to -4i + 5j + 6k. .**

Answer: A point on the line is (0,2,4) and the direction vector is -4**i** + 5**j** + 6**k**. So the parametric equations are: x = 0 - 4t, y = 2 + 5t, z = 4 + 6t, for t any real number.

(2) **Find parametric equations for the line containing P = (2,3,4) and Q = (-5,6,4).**
Anwer: We need a point on the line: let's use P = (2,3,4) . We also need a direction

(slope) vector, say $\overrightarrow{PQ} = (-5-2)\mathbf{i} + (6-3)\mathbf{j} + (4-4)\mathbf{k} = -7\mathbf{i} + 3\mathbf{j}$. Thus x = 2 -7t,
y = 3+3t, z = 4 + 0t = 4 are our parametric equations.

(3) **Find symmetric equations for the line above.**
Answer: Having found a point on the line, (2,3,4) and the direction vector we simply equate all t's:

Since t = $\dfrac{x-2}{7} = \dfrac{y-3}{3}$ and z has nothing to do with t , and is constantly 4, we have

$$\frac{x-2}{7} = \frac{y-3}{3} \; ; z \; = \; 4$$

(4) **Show that the line containing (1, 17, 5) and (3,2,-1) is parallel to the line containing (2,-2,5) and (-2,8,17).**

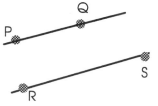

Answer: Let P = (1, 7, 5), Q = (3,2,-1), so
$\overrightarrow{PQ} = (3-1)\mathbf{i} + (2-7)\mathbf{j} + (-1-5)\mathbf{k} = 2\mathbf{i} - 5\mathbf{j} - 6\mathbf{k}$

Let R = (2,-2,5) and S = (-2, 8, 17), so
$\overrightarrow{RS} = (-2-2)\mathbf{i} + (8+2)\mathbf{j} + (17-5)\mathbf{k} = (-4)\mathbf{i} + (10)\mathbf{j} + (12)\mathbf{k}$

Then $\overrightarrow{RS} = (-2)\overrightarrow{PQ}$ and the direction vectors are parallel. This says the lines are parallel.

(5) **The symmetric equations of a line are :** $\dfrac{z-3}{2} = 3(1-y)$; x = 5. **Find the vector equation of the line.**

Answer: Write $\dfrac{(y-1)}{-(1/3)} = \dfrac{z-3}{2}$;   x = 5, a point on the line (reading from numerators,
and the fact that x = 5) is (5, 1, 3) and the slope vector (reading from denominators) is
(0, -1/3, 2). The equation of the line is:

$$x\,\mathbf{i} + y\,\mathbf{j} + z\,\mathbf{k} = t(\,-(1/3)\mathbf{j} + 2\mathbf{k}) + (5\mathbf{i} + \mathbf{j} + 3\mathbf{k})$$

Hints:

(1) To show that a line containing P and Q is perpendicular to a line containing R and S, show $\overrightarrow{PQ} \cdot \overrightarrow{RS} = 0$.

(2) To show that the line containing P and Q is parallel to the line with equation: $\dfrac{x - A}{a} = \dfrac{y - B}{b}; z = C$, use that the slope vector of the line is (a, b, 0). Now show that vector $\overrightarrow{PQ}$ is a scalar multiple of this vector.

EXAMPLE:

(1) Find parametric equations for the line that goes through (2,3,4) when t = 0 and through (1, 6,-7) when t = 1.

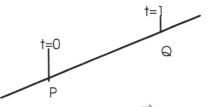

Answer: Let P = (2,3,4) and Q = (1, 6, -7) , so that $\overrightarrow{PQ}$ = -**i** + 3 **j** -11**k** is the direction (slope) vector. We can use that P = (2,3,4) is a point on the curve. So the parametrization of this line is given by:

$$x = 2 + t\,(-1)\,,\quad y = 3 + t\,(3)\,,\quad z = 4 + t\,(-11)\,,\quad \text{or}$$
$$x = 2 - t,\ y = 3 + 3t\,,\ z = 4 - 11t.$$

Notice, when t = 0, we are at P and when t = 1, we are at Q.

(2) Find a parametrization that is at P when t = 0 and at Q when t = 2.
Answer: Use the above but replace t with t/2. This line takes twice as long to cover .

(3) Find the equation of the line segment between P and Q.
Answer: The line segment is merely the piece of line between P and Q. We use the parametrization in (1) but restrict t to $0 \le t \le 1$. Thus this line segment "begins" at P and "ends" at Q.

(4) Find the midpoint of the line segment above.
Answer: The midpoint of the line segment is the point on the segment where t = 1/2 . The midpoint is (1/2) ((2,3,4)+(1, 6, -7)] = (3/2, 9/2, 3/2).

THE INTERSECTION OF TWO LINES

To find where two lines in 3-space intersect (if indeed they do), it is not enough to set the vector equations equal, and solve for t. Since the parameter t is like a "time" value, it is possible for the two lines to intersect, but at different times.

EXAMPLE: Where (if ever) do the lines (1) and (2) given below, intersect?
(1) $x\mathbf{i} + y\mathbf{j} + z\mathbf{k} = t(\mathbf{i} - 3\mathbf{j} + 4\mathbf{k}) + (2\mathbf{i} + \mathbf{j} + \mathbf{k})$ and
(2) $x\mathbf{i} + y\mathbf{j} + z\mathbf{k} = t(2\mathbf{i} - 6\mathbf{j} + 6\mathbf{k}) + (2\mathbf{i} + \mathbf{j})$

Answer: Although both equations are presented with parameter "t", we change one parameter to "s". (If we were to set the equations equal to each other, we would find no intersection.) We have:

(1) $x\mathbf{i} + y\mathbf{j} + z\mathbf{k} = t(\mathbf{i} - 3\mathbf{j} + 4\mathbf{k}) + (2\mathbf{i} + \mathbf{j} + \mathbf{k}) = (t+2)\mathbf{i} + (-3t + 1)\mathbf{j} + (4t + 1)\mathbf{k}$
(2) $x\mathbf{i} + y\mathbf{j} + z\mathbf{k} = s(2\mathbf{i} - 6\mathbf{j} + 6\mathbf{k}) + (2\mathbf{i} + \mathbf{j}) = (2s + 2)\mathbf{i} + (-6s + 1)\mathbf{j} + 6s\mathbf{k}.$

Setting components equal, we have three simultaneous equations:
$t + 2 = 2s + 2,$
$-3t + 1 = -6s + 1$
$4t + 1 = 6s.$

These equations can be solved if $t = 2s$. The last equation gives $8s + 1 = 6s$, or $s = -1/2$ on line (2) and thus, $t = -1$ on line (1). We will find the point of intersection by substituting $t = -1$ into line(1) (or $s = -1/2$ into line (2)). The point of intersection is (3, -2, 5).

SECTION 11.6

"Scientists have it within them to know what a future-directed society feels like, for science itself, in its human aspect, is just that."...C.P.Snow

PLANES IN SPACE

NOTE: Remember that points on a plane, such as (x,y,z) or (x_0, y_0, z_0), are identified as vectors from the origin. The vector $x\mathbf{i} + y\mathbf{j} + z\mathbf{k}$ will describe a "generic" variable point, while a vector such as $x_0\mathbf{i} + y_0\mathbf{j} + z_0\mathbf{k}$ will describe a specific point. In the equation of a plane we will need both. Vectors of the form $(x - x_0)\mathbf{i} + (y - y_0)\mathbf{j} + (z - z_0)\mathbf{k}$ may seem to reside "flat" on the plane; but actually, since vectors emanate from the origin, these vectors are only parallel to those on the plane.

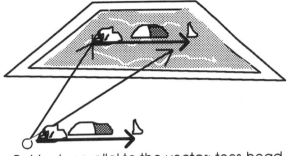

His head and toes are given by 2 different vectors

But he is parallel to the vector: toes-head.
(All vectors actually emanate from the origin.)

It takes one equation in 3-space to determine a plane. (this plane is also called a "hyperplane". A "hyperplane" is a space that is one dimension less than that of the space we are in -- the "home space".)

NOTE: An axiom states that a plane is determined by a given point on a plane and the normal to the plane .

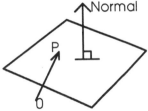

Here:
(1) (x,y,z) is an arbitrary point on the plane
(1) (x_0, y_0, z_0) is a given point on the plane
(2) The vector $a\,\mathbf{i} + b\,\mathbf{j} + c\,\mathbf{k}$ is normal (perpendicular) to the plane.

EQUATION OF A PLANE

(vector parallel to the plane) dot (normal to plane) = 0, or
$[(x - x_0)\mathbf{i} + (y - y_0)\mathbf{j} + (z - z_0)\mathbf{k}] \cdot (a\mathbf{i} + b\mathbf{j} + c\mathbf{k}) = 0$, or
$a\,x + b\,y + c\,z = e$, where e is a constant.

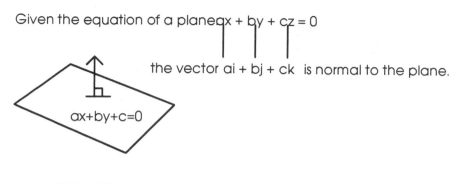

Given the equation of a plane $ax + by + cz = 0$

the vector $ai + bj + ck$ is normal to the plane.

$ax+by+c=0$

IN FACT, RETURNING TO THE 2-DIMENSIONAL CASE:

Given the equation of a LINE: $ax + by = c$

the vector $ai + bj$ is perpendicular to the line.

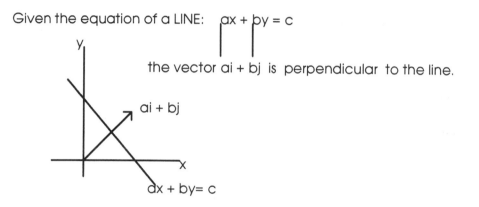

$ai + bj$

$ax + by = c$

EXAMPLE: **Find the normal** to the plane $2x - 3y + 4z = 3$.
Answer: Reading from the coefficients, the normal is $2i - 3j + 4k$.

EXAMPLE: Finding the equation of a plane, given a point and the normal vector.
Find an equation of a plane that contains $(1,-2,3)$ and has $N = -2i + 4j$ as a normal vector.

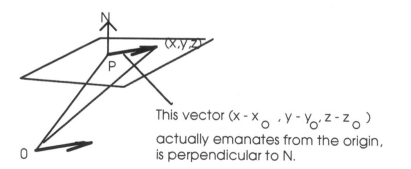

This vector $(x - x_o, y - y_o, z - z_o)$
actually emanates from the origin,
is perpendicular to N.

Answer: We write $a(x - x_o) + b(y - y_0) + c(z - z_0) = 0$, or
$-2(x-1) + 4(y - (-2)) + 0(z - 3) = 0$ (This can be simplified as $-x + 2y = -3$

NOTE: Planes of the form $ax + by + cz = 0$ are the plane which pass through the origin.
Note: $(0,0,0)$ must be a point on the plane.

NOTE: The normal vector is useful only for giving direction. Any scalar multiple would do as a normal vector, so there is an infinite supply of normal vectors. Also, there are "many" points on a plane! You may be wondering if all the different points and different normal vectors will really lead us to the same equation--they will.

EXAMPLE: Finding the equation of a plane from three given points.

Find an equation of a plane that contains the three points P = (0,1,2), Q = (-4, 3,6), and R = (2, 1, -2).

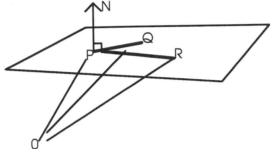

Answer:
(1) First, we find two vectors that lie on the plane:

$$\overrightarrow{PQ} = -4\mathbf{i} + 2\mathbf{j} + 4\mathbf{k} \text{ and } \overrightarrow{PR} = 2\mathbf{i} + 0\mathbf{j} - 4\mathbf{k}.$$

(2) Now we find $\mathbf{N} = \overrightarrow{PQ} \times \overrightarrow{PR} =$

$$\begin{vmatrix} \mathbf{i} & \mathbf{j} & \mathbf{k} \\ -4 & 2 & 4 \\ 2 & 0 & -4 \end{vmatrix} = -8\mathbf{i} - 8\mathbf{j} + -4\mathbf{k}.$$

(3) Now that we have a point on the plane (we can use P) and the normal we can determine the plane. The normal vector is -8i-8j-4k, but we can use instead, 2i + 2j + k. (Note: these vectors are not the same; we only need the direction.) The equation becomes:

$$a(x - x_o) + b(y - y_0) + c(z - z_0) = 0 \text{ or } 2(x - 0) + 2(y - 1) + (z - 2) = 0$$
or 2x + 2y + z = 4.

EXAMPLE: Finding the equation of a plane containing two parallel lines. Find the equation of the plane containing the two parallel lines;

$$\frac{x-3}{2} = \frac{y-4}{-1}; z = 3 \text{ and } \frac{x+5}{4} = \frac{y-4}{-2}; z = 5.$$

Answer: Note: going by the denominators, the slope vectors of the lines are 2i - j and 4i -2j (which is a scalar multiple of the first vector). The lines are indeed parallel.

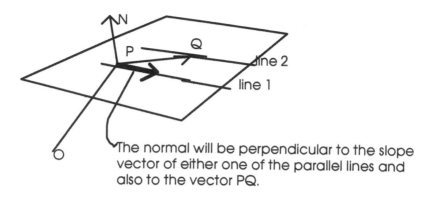

The normal will be perpendicular to the slope vector of either one of the parallel lines and also to the vector PQ.

(1) We need the normal, and for this we need two vectors on the plane for a cross product. We can use the direction (slope) vector as one of these vectors: $2\mathbf{i} - \mathbf{j}$. For another vector, use $\overrightarrow{PQ}$ where P is a point on one line, and Q is a point on the other. Pick any point P = (x, y, z) that satisfies $\dfrac{x-3}{2} = \dfrac{y-4}{-1}$; $z = 3$; let's choose (3, 4, 3). Now pick any point Q that satisfies $\dfrac{x+5}{4} = \dfrac{y-4}{-2}$; $z = 5$; e.g. (-5, 4, 5).

Now $\overrightarrow{PQ}$ = (-5-3) $\mathbf{i}$ + (4-4)$\mathbf{j}$ + (5-3) $\mathbf{k}$ = -8$\mathbf{i}$ + 2 $\mathbf{k}$.

(2) Form $\mathbf{N} = \overrightarrow{PQ} \times (2\mathbf{i} - \mathbf{j}) = \begin{vmatrix} \mathbf{i} & \mathbf{j} & \mathbf{k} \\ -8 & 0 & 2 \\ 2 & -1 & 0 \end{vmatrix} = 2\mathbf{i} + 4\mathbf{j} + 8\mathbf{k}$, and since direction is all we need, we'll simplify, and use instead: $\mathbf{N} = \mathbf{i} + 2\mathbf{j} + 4\mathbf{k}.$

(3) Now recall the formula for a plane : $a(x - x_o) + b(y - y_0) + c(z - z_0) = 0$. Here, using for the point on the plane, (3,4,3), we have: (x-3) + 2(y-4) + 4(z-3) = 0, which can be simplified.

EXAMPLE: **Find the equation of a plane passing through the point P and perpendicular to a given line.** (Hint:) Find the direction (slope) vector of the line. This will be a normal to the plane. Then we are back to the problem of constructing the equation for a plane from a normal and a point on the plane.

THE INTERSECTION OF PLANES

EXAMPLE: Find the line that is the intersection of the planes x + 3y - z = 1 and x + y + z = 2.

Answer: Solve the equations simultaneously:

x + 3y - z = 1
x + y + z = 2

We can eliminate the "x"-term by subtracting: $2y - 2z = -1$. Now eliminate "z " by adding: $2x + 4y = 3$. Since "y" is common to both of these equations, we may write:

$y = (-1+2z)/2$ and $y = (3-2x)/4$. Equating these: $(-1+2z)/2 = (3-2x)/4 = y$. To get this into the usual form for a line in 3-space:

$$ y \;=\; z-(1/2) \;=\; \frac{-(x-3/2)}{2} $$

EXAMPLE: Find the symmetric equations of a line passing through $(1,3,0)$ and perpendicular to the plane $x - y + 4z = 1$.

Answer: A normal vector to the plane is $\mathbf{i} - \mathbf{j} + 4\mathbf{k}$. This will be the direction (slope) vector of our line. Now we have the symmetric equations:
$$ \frac{x-1}{1} = \frac{y-3}{-1} = \frac{z-0}{4} \,. $$
Parametric equations for the line are $x - 1 = t,\ y - 3 = -t,\ z = 4t$, or rewritten,

$$ x = 1 + t \,,\ y = 3 - t \,,\ z = 4t, \text{ for all } t. $$

EXAMPLE: Sketch the plane $x - y + 2z = 3$.

Answer: Planes are a challenge t to sketch! However, we know the intercepts with the axes are the points $(3, 0, 0)$, $(0, -3, 0)$, and $(0,0, 3/2)$, and 3 points determine a plane. In this case a normal to the plane is given by $1\mathbf{i} - \mathbf{j} + 2\mathbf{k}$ (taken from the coefficients of x, y, z in the equation of the plane)

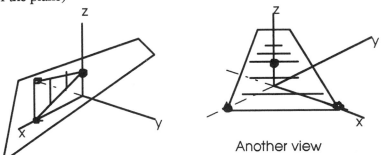

Another view

IF YOU DON'T DO WELL ON TESTS

Everyone is entitled to a "bad day", but if not doing well on tests is a chronic problem, you may need to look closely at the way you study. Chances are, you are learning *passively* rather than *actively*. This type of learning stresses recognizing rather than constructing and is probably encouraged by traditional approaches to teaching and learning. In mathematics we read and are lectured to in a "foreign language", but tests ask us to *speak* this language fluently. In order to make learning more of an active process, try writing summaries and explanations for yourself. Rewrite your notes: teachers write tests based on the material they cover, and they remember. And above all, keep up with the material--cramming never works!

REVIEW   COULD YOU.....

1. Find a line in 3-space given 2 points?
2. Find a line in 3-space given a point on the line and a vector giving its direction ?
3. Find a line in 3-space parallel to a given line and going through a given point?
4. Find a plane in 3-space given 3 points?
5. Find a plane in 3-space given 2 parallel lines?
6. Find a plane in 3-space given 2 intersecting lines?
7. Find a plane in 3-space given a line in the plane and a point on the plane, not on the line?
8. Find the distance from a point to a point?
9. Find the distance from a point to a line?
10. Find the distance from a point to a plane?
11. Find the distance from a line to a plane?
12. Find the vector, parametric, or symmetric equations for a line?
13. Find the intersection of 2 planes?

(Practice these problems with friends at a study session!)

❧STUDY SESSION❧

In this chapter, we first discussed 3-space with Cartesian coordinates. The distance between two points is found in this way ----. The equation of a sphere in 3-space is -----, and the points in its interior are expressed by the inequality ------. We introduced vectors in 3-space. Vectors could be added and subtracted in this way-----. The geometrical interpretation of these operations on vectors is ------. Vectors could be multiplied by scalars, which are ----. The geometrical effect of multiplying a vector by a scalar is ----. We discussed the dot product of two vectors. This could be found in two ways: ------. We could find if two vectors were orthogonal by -----. We could find the angle between two vectors ----. A formula for the length of a vector in terms of the dot product is ----. A unit vector in the direction of a given vector could be found as follows ---- The Projection of one vector onto another is given by the formula -----and the geometrical interpretation of this vector is -----.The direction cosines of a vector are ----. Resolving a vector into components means ----.The cross product of two vectors is a ----- which points in the direction ---- and whose length is -----. The interpretation of this length relates to the area of a parallelogram in this way---. A way to find if vectors are coplanar is ----. The cross product is found by taking a "determinant" in this way -----. The quantity that represents the volume of a parallelepiped spanned by 3 vectors is ----. The equation of a line in space can be achieved in 3 ways---. (Find an example, and go through all three ways). The equation of a plane in space can be determined by a point and a normal as follows.... Some of this theory dealt with finding the intersection of lines and planes in 3-space; recall as much as you can ----.

I Keep Making Careless Mistakes on Tests

Not all mistakes that students think are careless really are. And careless mistakes are nevertheless important. Bring your tests to your teacher for some analysis of what you can do to avoid making those types of mistakes. One suggestion is to take care writing down homework problems. Perhaps you are taking shortcuts that are disguising areas where you need more practice.

Chapter 12

" I reject the cynical notion that nation after nation must spiral down a militaristic stairway into the hell of nuclear destruction. I believe that unarmed truth and unconditional love will have the final word in reality"...Martin Luther King

In this chapter many ideas from first-year calculus are reworked in 3-dimensions: concepts such as limits and continuity, for example. Keep looking back in your text to see how these ideas were explained in first-year calculus: what is the same, what has to change? The subject we are studying here is actually the beginning of differential geometry, a subject that Einstein used extensively in relativity theory.

SECTION 12.1

"You will find it a very good practice always to verify your references, sir"...Martin Routh

VECTOR-VALUED FUNCTIONS

Now that we have studied vectors (Chapter 11), we will investigate *functions* which send real numbers to vectors, and later-- vectors to vectors.

$F(t) = f_1(t)\mathbf{i} + f_2(t)\mathbf{j} + f_3(t)\mathbf{k}$ gives a map from the **reals** (where t lives) to **3-space** (where $\mathbf{i}, \mathbf{j}, \mathbf{k}$ live). We can form sums, differences, dot products and cross products of functions like these.

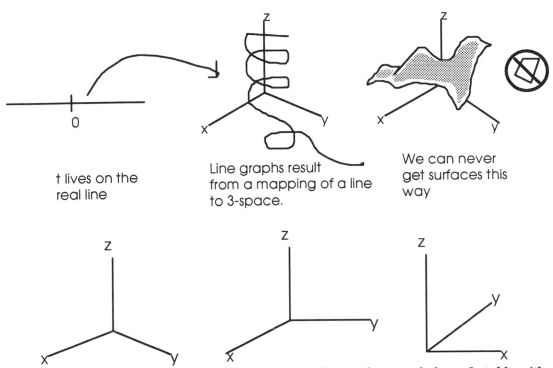

t lives on the
real line

Line graphs result
from a mapping of a line
to 3-space.

We can never
get surfaces this
way

**There are many ways to draw coordinate systems. Draw what you feel comfortable with;
you may want to change the look of the coordinate frame for different views. But be careful
of orientation, and observe the "right hand rule" with respect to the x, y, and z - axes.**

EXAMPLE: Determine the domain and the component functions of
$F(t) = t^2\mathbf{i} + (\ln t)\mathbf{j} + \sqrt{t-1}\,\mathbf{k}$.

Answer: The component functions are $x(t) = t^2$, $y(t) = \ln t$, $z(t) = \sqrt{t-1}$. The
domain is those t for which x, y, and z are simultaneously defined. The definition of y
requires that $t > 0$, and the definition of z, that $t \geq 1$. So the most restrictive set of t's
constitutes the domain and this is t such that $t \geq 1$.

NOTE: Tips on Graphing.

1. Look for circular relationships   when you see components with cos and sin terms.
    Recall terms like frequency and amplitude. You have seen parametrization in calculus I.
    For example, points of the form (cos **2**t, sin **2**t) describe a circle with frequency **2** and
    radius 1. This circle has period π; it completes one loop as t goes from 0 to π . On the
    other hand points (**2** cos t, **2** sin t) describe a circle with radius **2**, and period 2π.

2. Look for cross-sections: what happens when x = 0, y = 0, z = 0? Remember that x = 0
    describes the yz plane, etc.

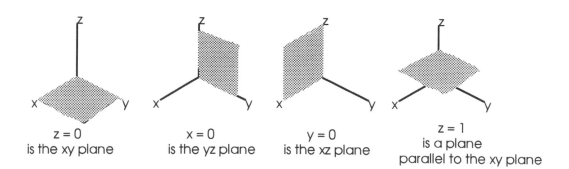

z = 0
is the xy plane

x = 0
is the yz plane

y = 0
is the xz plane

z = 1
is a plane
parallel to the xy plane

3. Restrictions on t (for example, t ε [a, b]) generally mean we are using only a segment of a curve. Recall how to find a **line segment** (Chapter 11).

4. **Linear** expressions in t , or terms in the form "at + b" with a, b real numbers, imply the curve will be a line.

5. Two useful things to know: (1) the equation $x^2 + y^2 + z^2 = r^2$ describes a **sphere** and (2) the equation $x^2 + y^2 = z^2$ describes a **cone**.

EXAMPLE: Sketch the curve traced out by the vector-valued function, indicating the direction in which the curve is traced.

1. Graph of $F(t) = (2t - 1)\mathbf{i} + (3t)\mathbf{j} + (t + 1)\mathbf{k}$

Answer: We seek a relation between x, y, and z.

$x = 2t - 1 \Rightarrow t = (x+1)/2$

$y = 3t \quad \Rightarrow t = y/3$

$z = t + 1 \Rightarrow t = z - 1.$

This is the line, $\dfrac{x+1}{2} = \dfrac{y}{3} = z - 1$, where t is any real. Notice the direction (slope) vector of this line is (2,3,1), and the line goes through the point (-1,0,1).

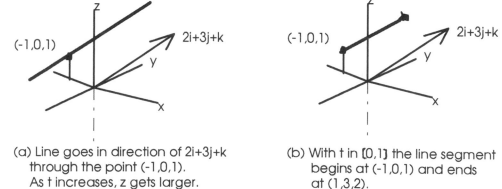

(a) Line goes in direction of 2i+3j+k
through the point (-1,0,1).
As t increases, z gets larger.

(b) With t in [0,1] the line segment
begins at (-1,0,1) and ends
at (1,3,2).

2. Graph of $F(t) = (2t - 1)\mathbf{i} + (3t)\mathbf{j} + (t + 1)\mathbf{k}$ with t ε [0,1].

Answer: This is the line segment in Fig. (b) above. The endpoints of the segment are the points on the line for which t = 0, and t = 1.

3. Graph of F(t) = $(\cos t)$ **i** + $(\sin t)$ **j** + 2**k** .

Answer: Since $x^2 + y^2 = 1$, x and y are points on a circle of radius 1. Meanwhile z is fixed at z = 2. The circle is being traced clockwise in the plane z = 2. (See Fig. (c) below.)

4. Graph F(t) = $(\cos t)$ **i** + $(\sin t)$ **j** + t^2**k** .

Answer: This is a spiral, acounterclockwise helix. Looking down the z-axis, we see the graph of a circle of radius 1, since $x^2 + y^2 = 1$. We will see how the curve is traced for t in $(-\infty, \infty)$. When t comes from $-\infty$, the spiral is traced downward, and at t = 0, lands at (1,0,0) -- then the spiral is traced upward as t goes to $+\infty$.

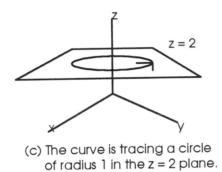

(c) The curve is tracing a circle
of radius 1 in the z = 2 plane.

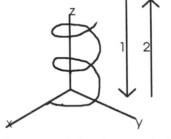

(d) The curve is tracing a spiral
on the z-axis

5. Graph F(t) = t **i** + $(\cos t)$**j** + $(\sin t)$**k** , where t ε [0, 2π] .

Answer: A circle is taking place in the yz plane, and the fact that z = t tells us the spiral is moving along the x-axis. The limits of t tell us we achieve one circle as x = t goes from 0 to 2π . (See Fig. (e) below.)

6. Graph F(t) = $t(\cos 2t)$**i** + $t(\sin 2t)$**j** + t**k** .

Answer: Seeking a relationship between x, y, and z, we find $x^2 + y^2 = t^2$ or $x^2 + y^2 = z^2$. There is a circular relationship between x and y. Looking down the z-axis we see circles of radius t, having a frequency of 2. The radius of the circles is changing with time, as well. In fact, from the equation $x^2 + y^2 = z^2$, we note that in the yz plane, where x = 0, we have $z^2 = y^2 + 0$, or z = $\pm y$, a straight line. Similarly, in the xz plane, when y = 0, we have z = $\pm x$. In the xy plane, when z = 0, we have circles. We are seeing counterclockwise spirals on a cone.

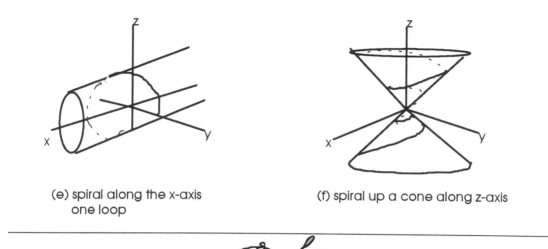

(e) spiral along the x-axis
one loop

(f) spiral up a cone along z-axis

The right- handed bindweed and the left-handed honeysuckle spiral in different directions. "'Maybe our luck'll take a 'turn' for the better,' said the bindweed to the honeysuckle." ..Flanders and Swann. Spirals that spiral in a counterclockwise direction and those that spiral clockwise are two different things entirely-- What are some examples from real life where the left-handedness or right-handedness of a spiral matters?

SECTION 12.2

"I make it a point to think of at least ten impossible things before breakfast" ..Lewis Carroll

LIMITS AND CONTINUITY OF VECTOR-VALUED FUNCTIONS

NOTE: Compare the definitions of limits and continuity with definitions you saw in first year calculus. Read your text thoroughly. You will want to absorb this theory carefully.

LIMITS OF A VECTOR-VALUED FUNCTION

The vector-valued function **F** has a limit **L** (also a vector) as t approaches t_0, or

$$\lim_{t \to t_0} \mathbf{F}(t) = \mathbf{L}$$

If:

given any ε > 0   (given any arbitrarily small ball around **L**; ε is its radius)

there exists a δ > 0   (there is a small interval about t_0; δ is its radius)

such that

|t - t_0| < δ   (whenever t is so close to t_0, it falls in that interval)

implies ||F(t) - L|| < ε   (the **vectors** **F**(t) will lie in the ε-ball).

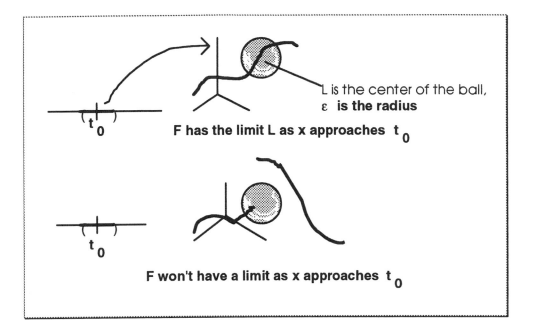

CONTINUITY OF A VECTOR-VALUED FUNCTION

The vector-valued function $\mathbf{F}$ is continuous at t_0 if $\lim\limits_{t \to t_0} \mathbf{F}(t) = \mathbf{F}(t_0)$.

This seemingly simple definition says 3 things: (1) the left hand side exists, (2) the right hand side exists, and (3) the left hand side equals the right hand side. Refresh your first year calculus for the meaning of this definition.

NOTE: In defining F to be continuous, we used the definition of LIMIT and added the expectation that $\lim\limits_{t \to t_0} \mathbf{F}(t)$ equal F($\mathbf{t_0}$). In the expression:

$$\lim\limits_{t \to t_0} \mathbf{F}(t) = \mathbf{F}(t_0)$$

we are requiring that

> **What we PREDICT** (the left side) **equals**
> **What Actually HAPPENS** (the right side)

A function that is continuous has "no surprises".

NOTE: We will not often run into non-continuous functions in this course. The graphs of such functions would have punctures or rips. A function that does not have a limit at a given point is rare, too.

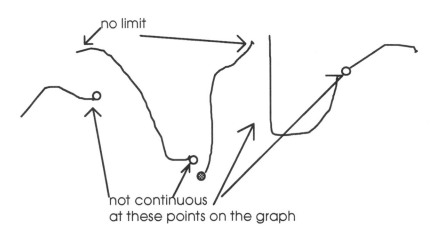

no limit

not continuous
at these points on the graph

A Wonderful Result

A vector valued function F(t) = x(t) i + y(t) j + z(t) k satisfies properties of limits, continuity, differentiability, and integrability if and only if the component functions x(t), y(t), and z(t) satisfies those properties.

(As a result, we reduce all issues involving these concepts to issues of single-variable calculus.)

EXAMPLE: Compute

$$\lim_{t \to 0} F(t) \text{ where } F(t) = \left(\frac{\sin t}{t} \right) i + \sqrt{t^2 + 1} \, j + (\cos t) \, k .$$

Answer: We merely compute all the limits of the component functions.

$$\lim_{t \to 0} F(t) = \lim_{t \to 0} \left(\frac{\sin t}{t} \right) i + \lim_{t \to 0} \sqrt{t^2 + 1} \, j + \lim_{t \to 0} (\cos t) \, k = i + j + k .$$

EXAMPLE: Compute $\lim_{t \to 0} F \cdot G \,(t)$ where

$$F(t) = e^{-t} i + |t| \, j - t \, k \text{ and } G(t) = 3i + \frac{1}{t} j - \frac{1}{t} k .$$

Answer: $F \cdot G(t) = 3e^{-t} i + \frac{|t|}{t} j - \frac{t}{t} k$. So $\lim_{t \to 0} F \cdot G(t)$ does not exist, because

$\lim_{t \to 0} \frac{|t|}{t}$ does not exist.

Study: College Grads earn $1,039 more per month

The Star Tribune (Minneapolis) reports on Jan. 28, 1993 that on average, persons with a bachelor's degree earn $2,116 a month, and high school graduates earn an average of $1,077 a month; the gap in earning grows with further education. Forty-three percent of adults earned their degrees within 4 years of high school graduation. Females were more likely than males to finish college within 4 years. The article concludes that the payoff comes several years after graduation from college, as college graduates are promoted past their less-educated colleagues.

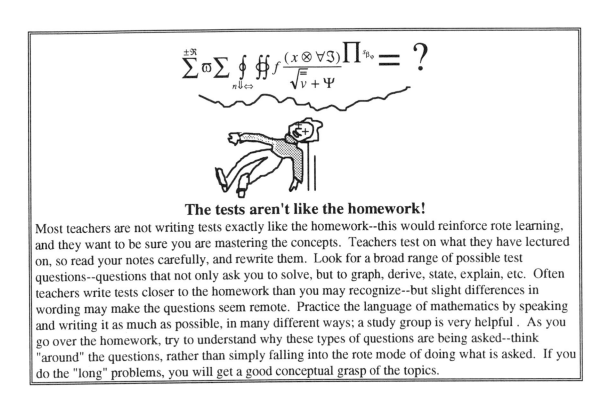

The tests aren't like the homework!

Most teachers are not writing tests exactly like the homework--this would reinforce rote learning, and they want to be sure you are mastering the concepts. Teachers test on what they have lectured on, so read your notes carefully, and rewrite them. Look for a broad range of possible test questions--questions that not only ask you to solve, but to graph, derive, state, explain, etc. Often teachers write tests closer to the homework than you may recognize--but slight differences in wording may make the questions seem remote. Practice the language of mathematics by speaking and writing it as much as possible, in many different ways; a study group is very helpful . As you go over the homework, try to understand why these types of questions are being asked--think "around" the questions, rather than simply falling into the rote mode of doing what is asked. If you do the "long" problems, you will get a good conceptual grasp of the topics.

SECTION 12.3

"I agree with you that there is a natural aristocracy among men. The grounds of this are virtue and talents."..Thomas Jefferson

DERIVATIVES AND INTEGRALS OF VECTOR-VALUED FUNCTIONS

Here are two important definitions . First we will write

$$\mathbf{F}(t) = f_1(t)\mathbf{i} + f_2(t)\mathbf{j} + f_3(t)\mathbf{k}$$

instead of $\mathbf{F}(t) = x(t)\mathbf{i} + y(t)\mathbf{j} + z(t)\mathbf{k}$. Then the derivative and integral of F(t) are defined as:

> **DERIVATIVE:** $\mathbf{F}'(t_0) = \lim\limits_{t \to t_0} \dfrac{\mathbf{F}(t) - \mathbf{F}(t_0)}{t - t_0}$, if this limit exists
>
> **INTEGRAL:** $\int \mathbf{F}(t)dt = \left(\int f_1(t)dt\right)\mathbf{i} + \left(\int f_2(t)dt\right)\mathbf{j} + \left(\int f_3(t)dt\right)\mathbf{k}$,
> where the individual components are integrable.

EXAMPLE: Find the derivative of $\mathbf{F}(t) = (e^t\cos t)\,\mathbf{i} + (\cosh t)\,\mathbf{j} + (\tan^{-1}t)\,\mathbf{k}$.

Answer: $\mathbf{F}'\,(t) = (e^t\cos t - e^t\sin t)\,\mathbf{i} + (\sinh t)\,\mathbf{j} + \dfrac{1}{1+t^2}\,\mathbf{k}$

EXAMPLE: Find the derivative of $\mathbf{F} \times \mathbf{G}\ (t)$ where $\mathbf{F}(t) = 3\,t\,\mathbf{i} + t^2\,\mathbf{k}$ and $\mathbf{G}(t) = (\cos t)\,\mathbf{i} + (\sin t)\,\mathbf{j} + (\tan t)\,\mathbf{k}$.

Answer: First find $\mathbf{F} \times \mathbf{G}\ (t) =$

$\begin{vmatrix} \mathbf{i} & \mathbf{j} & \mathbf{k} \\ 3t & 0 & t^2 \\ \cos t & \sin t & \tan t \end{vmatrix} = -(t^2\sin t)\,\mathbf{i} - (3t\,\tan t - t^2\cos t)\,\mathbf{j} + (3t\sin t)\,\mathbf{k},$

Then $(\mathbf{F} \times \mathbf{G})'(t) =$

$-(2t\sin t + t^2\cos t)\,\mathbf{i} - (3\tan t + 3t\sec^2 t - 2t\cos t + t^2\sin t)\,\mathbf{j} + (3\sin t + 3t\cos t)\,\mathbf{k}$

EXAMPLE: Evaluate the integral $\int (t^2\,\mathbf{i} + \cos t\ \mathbf{j} + \sinh t\ \mathbf{k})\,dt$.

Answer:

$\int (t^2\,\mathbf{i} + \cos t\ \mathbf{j} + \sinh t\ \mathbf{k})\,dt = \left(\dfrac{t^3}{3} + C_1\right)\mathbf{i} + (\sin t + C_2)\mathbf{j} + (\cosh t + C_3)\mathbf{k}$, which we

can write as $\left(\left(\dfrac{t^3}{3}\right)\mathbf{i} + (\sin t)\,\mathbf{j} + (\cosh t)\ \mathbf{k}\right) + (C_1\mathbf{i} + C_1\mathbf{j} + C_1\mathbf{k})$, which would

be like adding a "constant" vector to an antiderivative vector.

EXAMPLE: Evaluate the integral $\displaystyle\int_0^1 (t^2\,\mathbf{i} + (\cos t)\ \mathbf{j} + (\sinh t)\ \mathbf{k})\,dt$.

Answer:

$\displaystyle\int_0^1 (t^2\,\mathbf{i} + \cos t\ \mathbf{j} + (\sinh t)\ \mathbf{k})\,dt = \left(\dfrac{t^3}{3}\,\mathbf{i} + \sin t\ \mathbf{j} + (\cosh t)\ \mathbf{k}\right)\Bigg|_0^1 =$

$\dfrac{1}{3}\,\mathbf{i} + (\sin 1)\ \mathbf{j} + (\cosh 1)\ \mathbf{k} - (0\,\mathbf{i} + (\sin 0)\ \mathbf{j} + (\cosh 0)\ \mathbf{k}) =$

$\dfrac{1}{3}\,\mathbf{i} + (\sin 1)\ \mathbf{j} + ((\cosh 1) - 1)\ \mathbf{k}.$ $(\cosh 0 = 1.)$

> **POSITION, VELOCITY AND ACCELERATION ARE NOW VECTORS:**
> **Derivative of Position is Velocity**
> **Derivative of Velocity is Acceleration**

EXAMPLE: Find the position, velocity, and speed of an object having acceleration $\mathbf{a}(t) = \cos t\,\mathbf{i} + \sin t\,\mathbf{j}$, where $\mathbf{v_o} = \mathbf{k}$ and $\mathbf{r_0} = \mathbf{i}$.

Answer:

(1) The derivative of velocity is acceleration, $\mathbf{a}(t)$. So velocity is the integral of acceleration. Or $\mathbf{v}(t) = \int (\cos t)\ \mathbf{i} + (\sin t)\ \mathbf{j}\ dt = (\sin t)\ \mathbf{i} - (\cos t)\ \mathbf{j} + C$.

(2) Use initial conditions: $\mathbf{v}_0 = \mathbf{k}$. Thus, $\mathbf{v}(0) = \mathbf{v}_0 = \mathbf{k}$ and
$(\sin 0)\ \mathbf{i} - (\cos 0)\ \mathbf{j} + C = \mathbf{k}$ or $-\mathbf{j} + C = \mathbf{k}$, and $C = \mathbf{j} + \mathbf{k}$. So
$\mathbf{v}(t) = (\sin t)\ \mathbf{i} - (\cos t)\ \mathbf{j} + (\mathbf{j} + \mathbf{k}) = (\sin t)\ \mathbf{i} + (1 - \cos t)\ \mathbf{j} + \mathbf{k}$.

(3) The speed is
$$\| \mathbf{v}(t) \| = \sqrt{\sin^2 t + (1 - \cos t)^2 + 1} = \sqrt{\sin^2 t + 2 - 2\cos t + \cos^2 t} = \sqrt{3 - 2\cos t} .$$

(4) The derivative of position is velocity. So position is the integral of velocity given in (3):
$$\mathbf{r}(t) = \int (\sin t)\mathbf{i} - (\cos t - 1)\mathbf{j} + \mathbf{k}\ dt = -\cos t\ \mathbf{i} - (\sin t - t)\mathbf{j} + t\ \mathbf{k} + C$$

(5) Use initial conditions $\mathbf{r}_0 = \mathbf{i}$. Plugging into the equation in (4),
$\mathbf{r}_0 = \mathbf{r}(0) = (-\cos 0)\ \mathbf{i} - (\sin 0 - 0)\mathbf{j} + 0\ \mathbf{k} + C = -\mathbf{i} + C = \mathbf{i}$, so $C = 2\mathbf{i}$. Thus
$\mathbf{r}(t) = -\cos t\ \mathbf{i} - (\sin t - t)\mathbf{j} + t\ \mathbf{k} + 2\mathbf{i} = (2 - \cos t)\ \mathbf{i} + (t - \sin t)\ \mathbf{j} + t\ \mathbf{k}$.

SECTION 12.4

*"[Bishop Berkeley wrote that] Newton's fluxions were as 'obscure, repugnant and precarious' as any point in divinity.'...' I shall claim the privilege of a Free-thinker', wrote the Bishop, 'and take the liberty to inquire into the object, principles, and method of demonstration admitted by the mathematicians of the present date, with the same freedom that you presume to treat the principles and mysteries of Religion'...[Berkeley felt that to consider 32 + 16dt to be the same as 32 was unintelligible.] 'Nor will it avail,' he wrote, 'to say that the term negalected is a quantity exceedingly small; since we are told that [errors no matter how small, cannot be neglected]' ",
...Philip Davis and Reuben Hersh in "The Mathematical Experience", p.243*

SPACE CURVES AND THEIR LENGTHS

This is similar to finding arc length of a curve as we did in first year calculus. Read your text carefully, and refresh your memory from your calculus 1 text.

$$L = \int_a^b \| \mathbf{r}'(t) \| dt \quad \text{or} \quad L = \int_a^b \sqrt{\left(\frac{dx}{dt}\right)^2 + \left(\frac{dy}{dt}\right)^2 + \left(\frac{dz}{dt}\right)^2}\ dt$$

Arc length is found by taking a limit of the sum of lengths of secant lines.

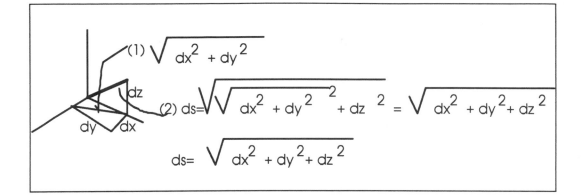

ABOUT POSITION AND VELOCITY

 Suppose an object moves with its position given by r(t) = x(t) **i** + y(t) **j** + z(t) **k**. We will think of that "object" as being a point on the curve. Thus, we are thinking of the description of a curve as having an implicit velocity, **r**' (t), the velocity with which the curve is being traced. When this is the case, the speed with which the curve is traced is the instantaneous rate of change of arc length with respect to time, namely, $\dfrac{ds}{dt}$, and this is given by

$$\frac{ds}{dt} = \| \mathbf{r}' (t) \|$$

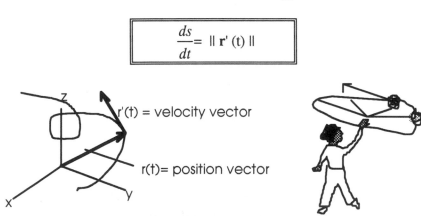

The velocity vector points in the direction of the instantaneous change in position. Speed is the length of the velocity vector. Note that the velocity vector looks tangent to the curve traced out by the position (ray) vector.

CURVES CAN SPEED UP AND SLOW DOWN: AN EXAMPLE FROM THE PLANE
(The same thing happens in 3-space)

Consider the parametrization of the ellipse $\dfrac{x^2}{3^2} + \dfrac{y^2}{4^2} = 1$, given by x = 3 cos t , y = 4 sin t.

With this particular parametrization , points on the ellipse are not traced at the same speed as t goes from 0 to 2π.

The position vector is $\mathbf{r}(t) = x(t)\,\mathbf{i} + y(t)\,\mathbf{j}$, or $\mathbf{r}(t) = (3\cos t)\,\mathbf{i} + (4\,\sin t)\,\mathbf{j}$. Thus, the velocity vector is $\mathbf{r}'(t) = -(3\sin t)\,\mathbf{i} + (4\,\cos t)\,\mathbf{j}$, and the speed is $ds/dt = \sqrt{(3\sin t)^2 + (4\cos t)^2}$.

As t goes from 0 to 2π, the speed is fastest when t = 0 and π, and slowest when t = $\pi/2$ and 3 $\pi/2$.

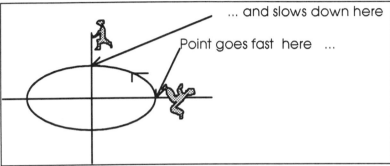

NOTE: The velocity and the speed depend on the parametrization.

The circle given by $x(t) = \cos t$, $y(t) = \sin t$ has position vector $\mathbf{r}(t) = (\cos t)\,\mathbf{i} + (\sin t)\,\mathbf{j}$ and velocity vector $\mathbf{r}'(t) = -(\sin t)\,\mathbf{i} + (\cos t)\,\mathbf{j}$. The speed is constantly 1:

$$\frac{ds}{dt} = \|\mathbf{r}'(t)\| = \sqrt{(-\sin t)^2 + (\cos t)^2} = 1$$

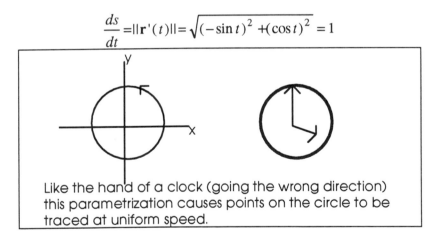

But if the circle above is given a different parametrization, say, $x(t) = \cos 2t$, $y(t) = \sin 2t$ (where the frequency is 2), then the position vector is $\mathbf{r}(t) = (\cos 2t)\,\mathbf{i} + (\sin 2t)\,\mathbf{j}$, and $\mathbf{r}'(t) = -(2\sin 2t)\,\mathbf{i} + (2\cos 2t)\,\mathbf{j}$. Now the speed is

$$\frac{ds}{dt} = \|\mathbf{r}'(t)\| = \sqrt{(-2\sin t)^2 + (2\cos t)^2} = 2.$$

The circle will "look" the same, but is traced twice as fast.

NOTE: The function $\mathbf{r}$ is considered smooth on an interval I if $\mathbf{r}$ has a continuous derivative on an interval I and $\mathbf{r}'(t) \neq \mathbf{0}$. Situations where $\mathbf{r}'(t) = \mathbf{0}$ are signals to **watch out!**

The planar curve $\mathbf{r}(t) = t\,\mathbf{i} + |t|\,\mathbf{j} + 0\,\mathbf{k}$, which corresponds to $y = |x|$, is considered only piecewise smooth, because $\mathbf{r}'(0)$ does not exist. Points on a curve where $\mathbf{r}'(t)$ does not exist or $\mathbf{r}'(t) = \mathbf{0}$ are points where the curve may have a cusp or be "sharp ". You can think of the curve as temporarily slowing down to a halt to "make up its mind" which way to go.

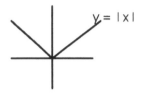

EXAMPLE: Determine which of the parametrizations are smooth, piecewise smooth, and neither.

(a) $\mathbf{r}(t) = t\,\mathbf{i} + (\sin t)\,\mathbf{j} + t^{1/3}\,\mathbf{k}$.

Answer: Since $\mathbf{r}'(0)$ does not exist , $\mathbf{r}$ is only piecewise smooth.

(b) $\mathbf{r}(t) = \cos t\,\mathbf{i} + \sin t\,\mathbf{j} + t\,\mathbf{k}$.

Answer: Since $\mathbf{r}'$ always exists and is never $\mathbf{0}$, $\mathbf{r}$ is smooth.

(c) $\mathbf{r}(t) = (e^t - t)\,\mathbf{i} + \cos t\,\mathbf{j} + t^2\,\mathbf{k}$.

Answer: Since $\mathbf{r}'(t) = (e^t - 1)\,\mathbf{i} - (\sin t)\,\mathbf{j} + 2t\,\mathbf{k}$ and $\mathbf{r}'(t) = \mathbf{0}$ when $t = 0$, the curve is piecewise smooth.

HERE ARE SOME USEFUL PARAMETRIZATIONS TO KNOW

EXAMPLE: A parametrization of a line. Find a smooth parametrization of a straight line from (1,0,-2) to (3, 4, 6).

Answer: The line between two points P and Q is given by $\mathbf{r}(t) = t\,\vec{P} + (1 - t)\,\vec{Q}$. or here, $\mathbf{r}(t) = t\,(\mathbf{i} + 0\,\mathbf{j} - 2\mathbf{k}) + (1 - t)\,(3\mathbf{i} + 4\mathbf{j} + 6\mathbf{k}) = (3 - 2t)\,\mathbf{i} + (4 - 4t)\,\mathbf{j} + (6 - 8t)\,\mathbf{k}.$

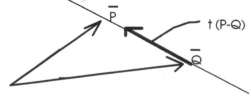

The line between vectors P and Q is given by
$r(t) = Q + t (P - Q)$ or $r(t) = t P + (1-t) Q$.

EXAMPLE: A parametrization of a line segment. Find a parametrization of the line segment between (1,0,-2) and (3,4,6) .

Answer: We use the same position vector as above, but restrict t to lie in [0, 1].

$\mathbf{r}(t) = (3 - 2t)\,\mathbf{i} + (4 - 4t)\,\mathbf{j} + (6 - 8t)\,\mathbf{k},$ **with t in [0,1].**

EXAMPLE: A parametrization of a circle in a plane. Find a smooth parametrization of a circle in the plane z = 3, centered at (1, 2, 3) and having radius 2.

Answer: We think of $(x-1)^2 + (y-2)^2 = 2^2$ for the circle.

So, we take x - 1 = 2 cos t, y - 2 = 2 sin t, and z = 3. A smooth parametrization is then

r(t) = (2cost + 1) **i** + (2 + 2sin t) **j** + 3 **k**.

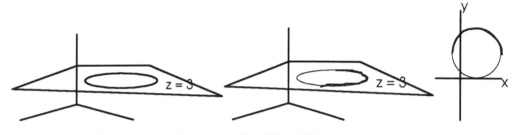

EXAMPLE: A parametrization of a piece of that circle. Find a parametrization of the "top half" looking down the z-axis, of the circle above.

Answer: We use the parametrization above, but restrict t to lie in [0, π].

EXAMPLE: Arc length. Find the arc length of the curve given by

r(t) = (sinh t) **i** + (cosh t) **j** + t **k** for $0 \le t \le 1$.

Answer:

$$L = \int_a^b \|\mathbf{r}'(t)\| \, dt = \int_0^1 \sqrt{(dx/dt)^2 + (dy/dt)^2 + (dz/dt)^2} \, dt =$$

$$\int_0^1 \sqrt{(\sinh t)^2 + (\cosh t)^2 + (\cosh t)^2 - (\sinh t)^2} \, dt \quad (\text{using } \cosh^2 t - \sinh^2 t = 1) =$$

$$\int_0^1 \sqrt{2(\cosh t)^2} \, dt = \sqrt{2} \int_0^1 \cosh t \, dt = \sqrt{2} \, \sinh t \Big|_0^1 = \sqrt{2}(\sinh 1 - \sinh 0) = \sqrt{2}(\sinh 1)$$

EXAMPLE: If **r** (t) = cos t **i** + sin t **j** + t **k**, find ds/dt for values of t such that $0 \le t \le \pi$.

Answer: ds/dt = ‖ **r**'(t) ‖ =

$$\sqrt{(dx/dt)^2 + (dy/dt)^2 + (dz/dt)^2} = \sqrt{(-\sin t)^2 + (\cos t)^2 + 1} = \sqrt{2}.$$

This is the "speed" with which this curve (a spiral) is traced out; it is constant.

EXAMPLE: Find s(t) for **r** (t) = cos t **i** + sin t **j** + t **k** where $0 \leq t$.

Answer: This is the arc length *as a function of time, t.* Notice "t" is the upper limit.

$$s(t) = \int_0^t \frac{ds}{dt} dt = \int_0^t \sqrt{2} dt = \sqrt{2}t .$$

ds/dt gives the rate at which
the curve is spinning out
length; s(t) is the arc length
spun out at time t.

t=0

SOLVE A PROBLEM

*Find the **tangent line** at the point t = 0 to the graph of the vector-valued function*
given by **r**(t) = (cos t)**i** + (sin t) **j** + 3t **k** .

Divide a sheet of paper into two columns . In the left column write out the steps that solve this problem. Simultaneously, in the right column, write out an explanation of what you have done at every step. Many times we think we understand an idea until we try to put it in writing. Writing mathematics continually helps you develop skill with this new language.

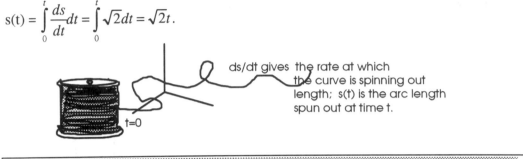

RIP VAN WINKLE has more or less lost track of the subject matter.

In fact, he is desperately behind. He wants the answers to some questions. Can you help him?

Rip: I'm confused about all these vector gadgets. Last I knew they stayed in one place. Now there's all this moving around, with time t. What is all this stuff with the t?

You say: (Your answer mentions the words "vector-valued functions".)

Rip: What's the domain of this here function? The range? Where do the vectors come out? How do you graph 'em?

You say:

Rip: Well now.. What use are they? And what's all this here stuff about limits and continuity? And differentiatin' and integratin' ? How's that accomplished, now? What do them terms mean anyhow? And what do these critters look like?

You say:

Rip: Last I heard, they were telling me a curve speeds up and slows down. That there's nonsense, right?

What's this parametrization idea anyhow? And arc length? Where's the curve if we're talkin' vectors? Whew--seems to me I gotta get rest!

You can't take it any more and leave...!

COMMON ERRORS : *List the common errors you make when working*

with this material. You should be able to come up with a good dozen.

SECTION 12.5

" I had to grow old to learn what I wanted to know, and I should need to be young to say well what I know"...William Drennan

TANGENTS AND NORMALS TO CURVES

Because the mathematics works out nicely, we define the tangent vector to the curve to be the **unit tangent vector**. With this assumption, the normal vector to the curve, defined below, is perpendicular to the tangent vector and points into the direction in which the curve is turning.

$\mathbf{v}(t) = \mathbf{r}'(t)$ **Velocity Vector,** appears tangent to curve but is not called the "tangent vector"	$\mathbf{T}(t) = \dfrac{\mathbf{r}'(t)}{\|\mathbf{r}'(t)\|}$ **(Unit) Tangent Vector** to curve
$\mathbf{a}(t) = \mathbf{v}'(t)$ **Acceleration Vector;** is not generally normal to the curve	$\mathbf{N}(t) = \dfrac{\mathbf{T}'(t)}{\|\mathbf{T}'(t)\|}$ **(Unit) Normal to Curve**

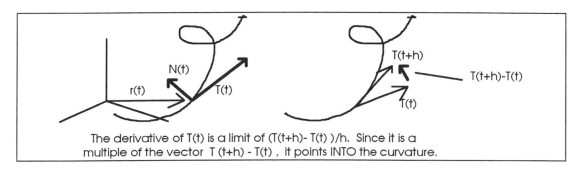

The derivative of T(t) is a limit of (T(t+h)- T(t))/h. Since it is a multiple of the vector T (t+h) - T(t) , it points INTO the curvature.

NOTE: It would be convenient if the acceleration vector (the derivative of the velocity vector) were normal to the curve, but it isn't, in general. The normal vector is obtained by differentiating the **unit tangent vector** and the acceleration vector, by differentiating the **tangent vector**. It is nice, however, that the acceleration vector lives on the plane determined by $\mathbf{T}(t)$ and $\mathbf{N}(t)$ and can be written in terms of these vectors.

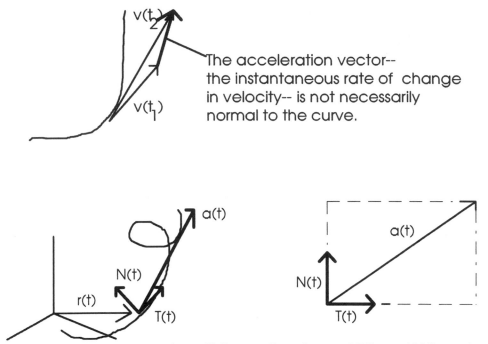

The acceleration vector--
the instantaneous rate of change
in velocity-- is not necessarily
normal to the curve.

But the acceleration vector a(t) lies on the plane of T(t) and N(t) and can be
expressed in terms of T(t) and N(t).

The component in the T(t) direction, is

$$\mathbf{a_T}(t) = \frac{d\|\mathbf{v}(t)\|}{dt}$$

(acceleration "along the curve" is the derivative of speed with respect to time)

We aim to write: $\mathbf{a} = \mathbf{a_T}(t)\mathbf{T(t)} + \mathbf{a_N}(t)\mathbf{N(t)}$

So the component in the N(t) direction :

$$\mathbf{a_N}(t) = \sqrt{\|\mathbf{a}\|^2 - (\mathbf{a_T}(t))^2}$$

(since we must have $\mathbf{a_T}^2 + \mathbf{a_N}^2 = \|\mathbf{a}\|^2$)

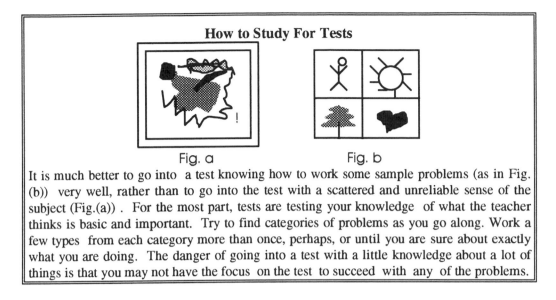

How to Study For Tests

Fig. a   Fig. b

It is much better to go into a test knowing how to work some sample problems (as in Fig. (b)) very well, rather than to go into the test with a scattered and unreliable sense of the subject (Fig.(a)) . For the most part, tests are testing your knowledge of what the teacher thinks is basic and important. Try to find categories of problems as you go along. Work a few types from each category more than once, perhaps, or until you are sure about exactly what you are doing. The danger of going into a test with a little knowledge about a lot of things is that you may not have the focus on the test to succeed with any of the problems.

There is another vector we speak of: the **Binormal**, denoted **B** and perpendicular to **T** and **N**

$$\mathbf{B} = \mathbf{T} \times \mathbf{N} = \frac{\mathbf{v} \times \mathbf{a}}{\|\mathbf{v} \times \mathbf{a}\|}$$

(Warning: Remember the order of cross-product is important! $\mathbf{T} \times \mathbf{N} \neq \mathbf{N} \times \mathbf{T}$)

NOTE: The vector B is a unit vector (why?) and with T and N forms a coordinate system that can chase around a curve. Mathematicians and scientists make use of this idea to study the motion of a particle in space.

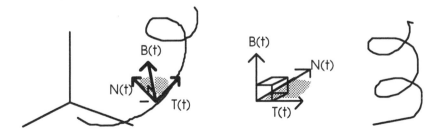

The binormal points up out of the plane of the tangent and normal to the curve. Where would it point in the spiral at right as you walk up the spiral?

EXAMPLE: A curve is parametrized by $\mathbf{r}(t) = \sin t \, \mathbf{i} + \sin t \, \mathbf{j} + \sqrt{2} \cos t \, \mathbf{k}$. Find

♦ **Velocity vector: Answer:** $\mathbf{v}(t) = \mathbf{r}'(t) = (\cos t) \, \mathbf{i} + (\cos t) \, \mathbf{j} - \sqrt{2}(\sin t) \, \mathbf{k}$.

♦ **Speed** (a scalar): **Answer:** $\|\mathbf{v}(t)\| = \sqrt{(\cos t)^2 + (\cos t)^2 + 2(-\sin t)^2} = \sqrt{2}$

♦ **Tangent vector** (unit) :

Answer: $T(t) = \dfrac{\mathbf{v}(t)}{\|\mathbf{v}(t)\|} = \dfrac{1}{\sqrt{2}}\big((\cos t)\ \mathbf{i}\ +\ (\cos t)\ \mathbf{j}\ -\ \sqrt{2}(\sin t)\ \mathbf{k}\ \big)$

♦ **Normal vector** (unit) :

Answer: $N(t) = \dfrac{d\mathbf{T}(t)/dt}{\|d\mathbf{T}(t)/dt\|} = \dfrac{1}{\sqrt{2}}\big((-\sin t)\ \mathbf{i}\ +\ (-\sin t)\ \mathbf{j}\ -\ \sqrt{2}\ (\cos t)\ \mathbf{k}\big).$

(Note $\|d\mathbf{T}/dt\| = 1$.) Check that **T** and **N** are perpendicular: $\mathbf{T} \cdot \mathbf{N} = 0$. (Check!)

♦ **Acceleration vector** : **Answer:** $\mathbf{a}(t) = d\mathbf{v}(t)/dt = (-\sin t)\ \mathbf{i}\ +\ (-\sin t)\ \mathbf{j}\ -\ \sqrt{2}(\cos t)\ \mathbf{k}$

(In this special case, **a**(t) is parallel to **N**.)

♦ **Tangential component of the acceleration:**

Answer: $\mathbf{a}_T(t) = \dfrac{d\|\mathbf{v}\|}{dt} = 0$ (We knew this, as **a**(t) is parallel to **N**; and speed is

constant.)

♦ **Normal component of the acceleration:**

Answer: $\mathbf{a}_N(t) = \sqrt{\|\mathbf{a}\|^2 - (\mathbf{a}_T(t))^2} = \|\mathbf{a}\| = \sqrt{2(-\sin t)^2 + (-\sqrt{2}\cos t)^2} = \sqrt{2}$

♦ **The binormal vector: Answer: B = T × N =**

$$\dfrac{(\cos t)\ \mathbf{i}\ +\ (\cos t)\ \mathbf{j}\ -\ \sqrt{2}(\sin t)\ \mathbf{k}}{\sqrt{2}} \times \dfrac{(-\sin t)\ \mathbf{i}\ +\ (-\sin t)\ \mathbf{j}\ -\ \sqrt{2}(\cos t)\ \mathbf{k}}{\sqrt{2}} =$$

$$\dfrac{1}{\sqrt{2}} \cdot \dfrac{1}{\sqrt{2}} \begin{vmatrix} \mathbf{i} & \mathbf{j} & \mathbf{k} \\ \cos t & \cos t & -\sqrt{2}\sin t \\ -\sin t & -\sin t & -\sqrt{2}\cos t \end{vmatrix} \text{(we have factored out the scalars)}$$

$$= \dfrac{1}{2}(-\sqrt{2}\mathbf{i} + \sqrt{2}\mathbf{j} + 0\mathbf{k}) = -\dfrac{1}{\sqrt{2}}\mathbf{i} + \dfrac{1}{\sqrt{2}}\mathbf{j}$$ (Since in the initial description of **r**(t),

the x component equals the y component, this curve lies in the plane given by y = x. Thus it is no surprise that the binormal is normal to this plane.)

EXAMPLE: Find a piecewise smooth parametrization of the semicircle in the xz plane that begins at (4,0,0), passes through (0,0,4) and ends at (-4,0,0).

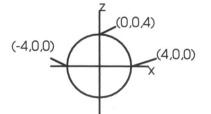

When graphing in the xz plane, ignore y, which is 0.

Answer: First guess: $r(t) = (\cos t)\,\mathbf{i} + 0\,\mathbf{j} + (\sin t)\,\mathbf{k}$ -- this gives a circle in the xz plane.

Second guess: $r(t) = (4\cos t)\,\mathbf{i} + (4\sin t)\,\mathbf{k}$ -- the circle has radius 4.

Final touch: $r(t) = (4\cos t)\,\mathbf{i} + (4\sin t)\,\mathbf{k}$, where $t \,\varepsilon\, [0, \pi]$. This half of the circle begins and ends at the appropriate points. The parametrization is smooth except at the endpoints, whre the derivative does not exist; thus, it is piecewise smooth.

EXAMPLE: Show that the curves parametrized by
$r_1(t) = t\mathbf{i} + t^2\mathbf{j} + 2t\mathbf{k}$ and $r_2(t) = t^2\mathbf{i} + (2 - t^2)\mathbf{j} + (1 - t)\mathbf{k}$ intersect at $(1,1,2)$ and that the vectors tangent to the two curves are perpendicular there.
Hint: The point in common, $(1,1,2)$ occurs on r_1 when $t = 1$ and on r_2 when
$t = -1$. Show $r_1{}'(1) \cdot r_2{}'(-1) = 0$.

SECTION 12.6

" That which has always been accepted by everyone, everywhere, is almost certain to be false"...Paul Valery

CURVATURE
The curvature of a curve at a point is a real number, denoted "kappa", and the radius of curvature, "rho".

The curvature at a point at t is $\kappa(t) = \left\|\dfrac{d\mathbf{T}}{ds}\right\| = \dfrac{\|\mathbf{T}'(t)\|}{\|\mathbf{r}'(t)\|}$

The radius of curvature at t is $\rho(t) = \dfrac{1}{\kappa(t)}$

Sometimes we compute $\kappa(t)$ as $\kappa(t) = \dfrac{\|\mathbf{v} \times \mathbf{a}\|}{\|\mathbf{v}\|^3}$

EXAMPLE (FROM THE PLANE): Consider the curve from first-year calculus,
$f(x) = x^2$, and parametrize it as $r(t) = t\mathbf{i} + t^2\mathbf{j} + 0\,\mathbf{k}$. Find the curvature.

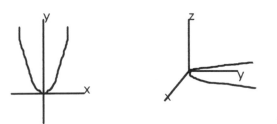

Answer: We use $\kappa(t) = \dfrac{\|\mathbf{v} \times \mathbf{a}\|}{\|\mathbf{v}\|^3}$ where $\mathbf{v}(t) = \mathbf{r}'(t) = \mathbf{i} + 2t\,\mathbf{j}$, and $\mathbf{a}(t) = \mathbf{v}'(t) = 2\mathbf{j}$. Then

$$\kappa(t) = \frac{\|\mathbf{v} \times \mathbf{a}\|}{\|\mathbf{v}\|^3} = \frac{\|(\mathbf{i} + 2\mathbf{j}) \times (2\mathbf{j})\|}{(1 + 4t^2)^{3/2}} = \frac{\|2\mathbf{k}\|}{(1 + 4t^2)^{3/2}} = \frac{2}{(1 + 4t^2)^{3/2}}.$$ At $t = 0$, for

example, the curvature is 2, and at $t = 1$, the curvature is $\dfrac{2}{5^{3/2}}$. The curvature becomes

closer to 0 (curve is flatter) as t becomes positively or negatively large. Notice that we are doing more than taking the second derivative of $f(x) = x^2$, which would be the constant 2. We would not expect the curvature of this curve in 3-space to be a constant.

ABOUT CURVATURE: A REVIEW. In Calculus I, the sign of the second derivative told us about curvature; i.e., it told us whether the curve was concave upward or downward. For us, its magnitude did not have a meaning. Curves seen in this chapter can be travelled at different rates of speed, depending on the parametrization. So taking the derivative of the tangent vector (finding acceleration) will not lead to a well-defined number giving curvature. Nevertheless, if a parametrization is chosen so that ds/dt = 1 (the curve travels at a uniform rate of speed) then the denominator in the expression above is 1, and the curvature is simply the magnitue of the rate of change in the unit tangent vector (like a second derivative). It can also be thought of as the magnitude of the rate of change in the unit tangent vector with respect to arc length; i.e. , as ||dT/ds||. Arc length is a more natural parameter than "time" when thinking about a curve geometrically. The *radius of curvature* is essentially the radius of the best-fitting circle that can be drawn tangent to the curve at t. This circle is called the "osculating" circle-- the "kissing" circle.

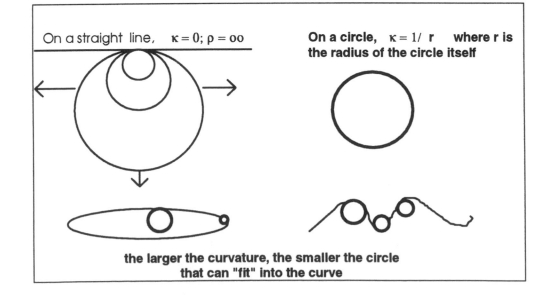

EXAMPLE: Let $\mathbf{r}(t) = \sin t\, \mathbf{i} + \sin t\, \mathbf{j} - \sqrt{2}\cos t\, \mathbf{k}$. Find the curvature of the curve traced out by $\mathbf{r}$:

Answer:

We use $\kappa(t) = \dfrac{\|\mathbf{T}'(t)\|}{\|\mathbf{r}'(t)\|}$. Now

(i) $\mathbf{T}(t) =$

$\dfrac{\mathbf{r}'(t)}{\|\mathbf{r}'(t)\|} = \dfrac{\cos t\ \mathbf{i} + \cos t\ \mathbf{j} + \sqrt{2}\sin t\ \mathbf{k}}{\sqrt{2(\cos t\ )^2 + (\sqrt{2}\sin t\ )^2}} = \dfrac{1}{\sqrt{2}}\left(\cos t\ \ \mathbf{i} + \cos t\ \ \mathbf{j} + \sqrt{2}\sin t\ \ \mathbf{k}\right),$

(ii) $\mathbf{T}'(t) = \dfrac{1}{\sqrt{2}}\left(-\sin t\ \ \mathbf{i} - \sin t\ \ \mathbf{j} + \sqrt{2}\cos t\ \ \mathbf{k}\right)$. So

(iii)

$\kappa(t) = \dfrac{\|\mathbf{T}'(t)\|}{\|\mathbf{r}'(t)\|} = \dfrac{\left\|\dfrac{1}{\sqrt{2}}\left(-\sin t\ \ \mathbf{i} - \sin t\ \ \mathbf{j} + \sqrt{2}\cos t\ \ \mathbf{k}\right)\right\|}{\|\sqrt{2}\|} = \dfrac{\sqrt{2(-\sin t)^2 + (\sqrt{2}\cos t)^2}\,/\,\sqrt{2}}{\sqrt{2}} = \dfrac{1}{\sqrt{2}}.$

(The curvature is constant.)

EXAMPLE: Compute the curvature and the radius of curvature for the curve traced by $\mathbf{r}(t) = t\sin t\, \mathbf{j} + t\cos t\, \mathbf{k}$ using the formula $\kappa(t) = \dfrac{\|\mathbf{v} \times \mathbf{a}\|}{\|\mathbf{v}\|^3}$ (because the derivative, $\mathbf{r}'(t)$

gets messier).

Answer:

(i) $\mathbf{v}(t) = \mathbf{r}'(t) = (t\cos t + \sin t\,)\mathbf{j} + (-t\sin t + \cos t)\mathbf{k}$ and

$\|\mathbf{v}(t)\| = \sqrt{(t\cos t + \sin t\,)^2 + (-t\sin t + \cos t)^2} = \sqrt{t^2 + 1}$ and

$\mathbf{a}(t) = \mathbf{v}'(t) = (\cos t - t\sin t + \cos t)\mathbf{j} + (-\sin t - t\cos t - \sin t)\mathbf{k} =$
$(2\cos t - t\sin t)\mathbf{j} - (2\sin t + t\cos t)\mathbf{k}$

(ii) $\mathbf{v} \times \mathbf{a} =$

$\begin{vmatrix} \mathbf{i} & \mathbf{j} & \mathbf{k} \\ 0 & t\cos t + \sin t & -t\sin t + \cos t \\ 0 & 2\cos t - t\sin t & -(2\sin t + t\cos t) \end{vmatrix} =$

$[-(t\cos t + \sin t)(2\sin t + t\cos t) - (-t\sin t + \cos t)(2\cos t - t\sin t)]\mathbf{i} - 0\mathbf{j} - 0\mathbf{k} = (-2 - t^2)\mathbf{i}$

(iii) $\kappa(t) = \dfrac{\|\mathbf{v} \times \mathbf{a}\|}{\|\mathbf{v}\|^3} = \dfrac{|-2 - t^2|}{(t^2 + 1)^{3/2}} = \dfrac{t^2 + 2}{(t^2 + 1)^{3/2}}.$

(iv) $\rho(t) = \dfrac{(t^2 + 1)^{3/2}}{(t^2 + 2)}$

Review of Concepts

How do you define a vector-valued function from the real line to 3-space?

How is the limit of such a function defined?

How is continuity of such a function defined?

How can you find the derivative of a vector-valued function, $\mathbf{r}(t)$?

What does the derivative represent, geometrically?

How do we obtain a unit tangent vector?

How do we obtain a unit vector normal to a curve? The unit binormal vector?

What is acceleration?

What is curvature? Give a geometrical explanation. What is meant by radius of curvature?

What's the Formula?

Can you give the formula for the following mathematical quantities? Can you give some idea of how these formulas are derived or why they make sense?

Construct	Formula?	Explain how the formula comes about	Give an example in which you find this construct
$\mathbf{T}(t)$			
$\mathbf{N}(t)$			
$\mathbf{B}(t)$			
velocity of a vector-valued function			
acceleration			
speed			
arc length			
curvature			
radius of curvature			
$\mathbf{a}_T$ and $\mathbf{a}_N$			

KEEP A DOUBLE ENTRY JOURNAL

This is a useful way of keeping track of what you know and how to learn. Keep a notebook with two columns. In the left column, daily or at least a few times a week, keep a record of your study patterns, what subjects are causing problems, questions you want to ask, what you do and don't feel "solid" with. A week later, in the right hand column, compare how you feel now about what you have said. Then look back at all of this before a test. You will have a record of your progress that will be encouraging and useful. It is well known that writing helps you pull the subject together and helps your retention. And what you have written will give you a reflection of how you learn that can be invaluable.

CHAPTER 13

"Philosophy is written in this grand book--I mean the universe--which stands continually open to our gaze, but it cannot be understood unless one first learns to comprehend the language and interpret the characters in which it is written. It is written in the language of mathematics"...Gallileo Gallilei

In this chapter, we enter a new realm of 3-dimensional imagery as we consider functions of several variables. The graphs of these functions will be in 3-space (and higher, but these we cannot visualize). While we will be imitating what has been done in first year calculus, there will be many new concepts to understand. For example, what will take the place of the first year concepts of derivative, or area under a curve ? A good ability to graph and "get around " in 3-space will help enormously, so keep reviewing from your first year text. There are many interesting computer algebra systems that have graphics packages allowing us to see the graphs of these surfaces from a variety of angles; ask about these.

SECTION 13.1

"God does not play dice with the universe"...Albert Einstein. "He just plays hide-and-seek"...Woody Allen

FUNCTIONS OF SEVERAL VARIABLES

In the previous chapters we studied mappings of the **line** to **3-space**, and the image was generally a curve. When we map a **plane** into a higher dimension the result of the graph can be a surface. The dimension of a line is one, that of a plane 2, etc. The dimension of the surface we get is always less than or equal to the dimension of the domain of our mapping because a function cannot expand the dimension of the domain space. (The definition of function requires that every element in the domain be sent to one and only one element in the range.)

There are many graphs in your text. Don't feel intimidated by drawing them--if *we can, you can!* In fact, it's useful to practice drawing these graphs; you will understand the functions better if you do.

NOTE: A **level surface** is a surface obtained by holding f(x,y) fixed. The intersection of a level surface with one of the planes x = 0, y = 0, or z = 0 is called the **trace** of the level surface; it is a "cross-section" of the surface.

TIPS ON GRAPHING : SOME PROTOTYPE GRAPHS
Learn these; then variations on them will be easy to learn.

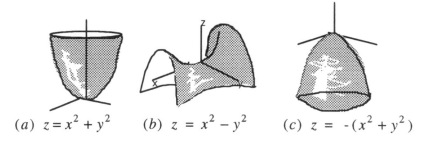

$(a)\ z = x^2 + y^2$ $(b)\ z = x^2 - y^2$ $(c)\ z = -(x^2 + y^2)$

(a) **Paraboloid**: z is always positive. Traces with z constant are the curves, $x^2 + y^2 = c^2$, circles of radius c. This surface goes through the origin.

(b) **Hyperbolic Paraboloid**. When y = 0 (the xz-plane) we have parabolas, $z = x^2$, opening upwards. When x = 0, in the yz-plane, we have parabolas, $z = -y^2$, opening downwards. Looking down on the xy-plane, when z is a constant, we have hyperbolas of 2 sheets:

$$x^2 - y^2 = c^2 \quad \text{or} \quad \frac{x^2}{c^2} - \frac{y^2}{c^2} = 1.$$

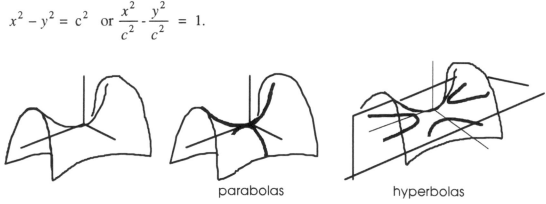

parabolas   hyperbolas

(c) **Paraboloid**: Like (a), but z is always negative.

MORE PROTOTYPE GRAPHS TO LEARN

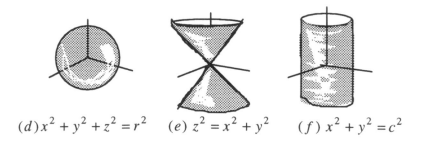

$(d)\,x^2 + y^2 + z^2 = r^2$ $(e)\ z^2 = x^2 + y^2$ $(f)\ x^2 + y^2 = c^2$

(d) **Sphere** of radius r, centered at origin. The top hemisphere might be expressed as $z = \sqrt{r^2 - x^2 - y^2}$. Given square roots, you can square them, but remember only positive values of z are wanted.

(e) **Cone**. Note that z can be positive or negative. When x = 0, we have the straight lines z = ±y. When y = 0, z = ±x. When z is constant we have circles of varying radius. This surface goes through the origin. Positive values of z; i.e., $z = \sqrt{x^2 + y^2}$, give the top half of the cone.

(f) **Cylinder** of radius r. Notice the absence of the "z"-term--z doesn't matter. This means we may translate the curve $x^2 + y^2 = c^2$ (a circle) up and down the z-axis; thereby obtaining a cylinder.

PLANES

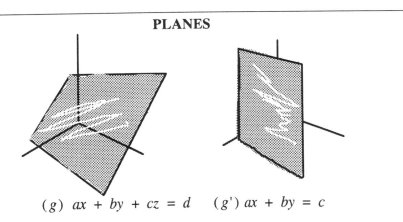

$(g)\ ax + by + cz = d \qquad (g')\ ax + by = c$

(g) A **Plane** is expressed in linear terms of x, y, and z. The typical equation is ax + by + cz = d. If a variable is missing, as in (g'), the plane is parallel to one of the coordinate planes. In the plane in (g'), z doesn't "matter". We might guess the equation of the surface in (g') is x = 5. The normal to the plane in (g) is **ai + bj + ck**, read from the coefficients of the equation.

SHEETS

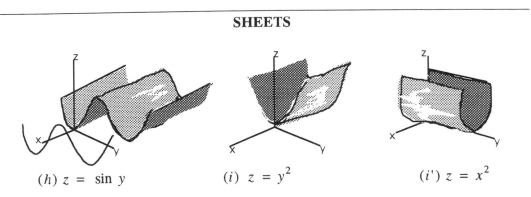

$(h)\ z = \sin y \qquad\qquad (i)\ z = y^2 \qquad\qquad (i')\ z = x^2$

(h) , (i) , (i'): **Surfaces with sheets**: a variable is missing; a curve has been translated along straight lines. In both (h) and (i) a curve has been translated along the x-axis ("x" is missing), and in (i'), along the y-axis. The graphs in (i) an (i') are **Parabolic Cylinders**.

SURFACES WITH LINEAR (ax + by) TERMS

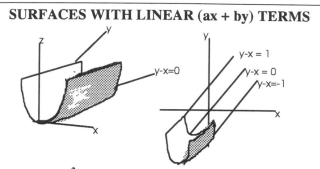

The surface above, $z = (y - x)^2$ has a linear term, "ax + by". Surfaces like these may have traces given by ax + by = constant; their traces are straight lines (In this case, the straight lines are skewed with respect to the axes.)

SURFACES WITH CIRCLES OR ELLIPSES AS TRACES

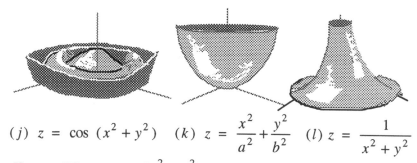

$$(j) \quad z = \cos(x^2 + y^2) \quad (k) \quad z = \frac{x^2}{a^2} + \frac{y^2}{b^2} \quad (l) \quad z = \frac{1}{x^2 + y^2}$$

(j) **Circular Traces.** When we see $(x^2 + y^2)$ in the expression for f(x,y), this usually means traces with z equal aconstant are circles. In (j), when $x^2 + y^2 = c$, we have $z = \cos(c)$.

(k) **Elliptic Traces: An elliptic paraboloid.**

(l) **Circular Hyperboloid:** Traces when y = 0 or x = 0 are the curves $z = 1/x^2$ and $z = 1/y^2$

THE ELLIPSOIDS

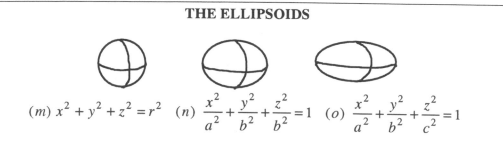

$$(m) \quad x^2 + y^2 + z^2 = r^2 \quad (n) \quad \frac{x^2}{a^2} + \frac{y^2}{b^2} + \frac{z^2}{b^2} = 1 \quad (o) \quad \frac{x^2}{a^2} + \frac{y^2}{b^2} + \frac{z^2}{c^2} = 1$$

(m) **Sphere**

(n) **Ellipsoid** in which a circle is taking place in the yz-plane, ellipses in the other coordinate planes (assuming a ≠ b).

(o) **Ellipsoid** in which ellipses are taking place in all the coordinate planes (assuming a, b, c are all unequal.)

GRAPHING ESOTERICA

Your text gives graphs of a number of hyperbolic surfaces. Here we have removed the "ellipses" from the equations and replaced them with circles, to analyze the essential shapes of the graph of $\dfrac{x^2}{a^2} + \dfrac{y^2}{b^2} - \dfrac{z^2}{c^2} = 1$ can be recognized from the graph of $x^2 - y^2 + z^2 = 1$.

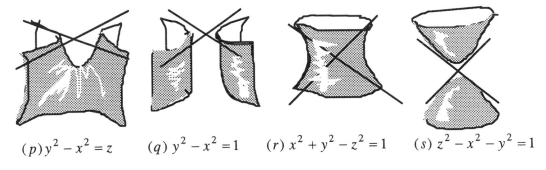

$(p)\, y^2 - x^2 = z$ $(q)\, y^2 - x^2 = 1$ $(r)\, x^2 + y^2 - z^2 = 1$ $(s)\, z^2 - x^2 - y^2 = 1$

(p) This is a hyperbolic paraboloid as we saw before, in (b). Notice the traces (when x, or y, or z is constant). The hyperbolas are in the xy plane and change direction depending on whether the traces are of the form $y^2 - x^2 = 1$ or $x^2 - y^2 = 1$.

(q) Two hyperbolic sheets. The absence of a "z"-term means we translate hyperbolas in the xy- plane up and down the z-axis.

(r) Hyperboloid of one sheet. When z = constant, we have circles, radius ≥ 1. Hyperbolas when x or y is constant, in the yz- or xz-planes, and these hyperbolas rotate around the z-axis. **This surface has one negative sign, is in one piece.**

(s) Hyperpoloid of two sheets. There are a range of z values that are forbidden: we require z ≥ 1. Notice z may not have values in the interval (-1,1). We have hyperbolas when x or y is constant, in the yz- or xz - planes. These hyperbolas rotate around the z-axis. Looking down the z-axis, we see circles. **This surface has 2 negative signs, is in 2 pieces.**

EXAMPLE: Find the domain of $g(x,y) = \sqrt{9 - x^2 - y^2}$.

Answer: We must have $9 - x^2 - y^2 \geq 0$, or $x^2 + y^2 \leq 3^2$ (See Fig. (a) below.)

Writing $z = \sqrt{9 - x^2 - y^2}$ or $z^2 = 9 - x^2 - y^2$ or $x^2 + y^2 + z^2 = 3^2$, this graph is easily recognizable as the top hemisphere of a sphere of radius 3.

EXAMPLE: Find the domain of $g(x,y,z) = \sqrt{9 - x^2 - y^2 - z^2}$.

Answer: The graph of this function is the "top" hemisphere of a sphere in 4 dimensions. We can't visualize the graph, but we can find the domain: this is all (x,y,z) such that $x^2 + y^2 + z^2 \leq 3^2$. See Fig. (b) below. This domain is called a **ball** of radius 3, in 3-space . A ball is filled, a sphere is hollow.

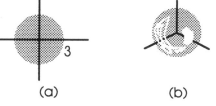

(a)   (b)

4-DIMENSIONS

While we can't visualize in 4-dimensions, we can get some insights into its complexities by generalizing from lower dimensions.

If function sends line to line: $y = f(x)$, the graph is in 2-space
If function sends line into plane: $r(t) = x(t)\,\mathbf{i} + y(t)\,\mathbf{j}$, the graph is in 2-space
If function sends plane to line: $z = f(x,y)$, the graph is in 3-space
If function sends 3-space to line: $w = f(x,y,z)$, the graph is in 4-space!

EXAMPLE: Sketch the graph (hints)

(a) $f(x,y) = 2x + 3y + 1$.
Hint: Write this as $z = 2x + 3y + 1$, or $2x + 3y - z = -1$. This is a plane with normal vetor $2\mathbf{i} + 2\mathbf{j} - \mathbf{k}$. You might graph it by finding intercepts.

(b) $f(x,y) = \sqrt{x^2 + 9y^2}$.
Hint: Write this as $z = \sqrt{x^2 + 9y^2}$. This is the top half of a cone with elliptic cross-sections.

(c) $f(x,y) = x^3$.
Hint: Write this as $z = x^3$. Since y is absent, we translate the curve $z = x^3$ along the y-axis. This is a "sheeted" surface.

(d) $y = x^2 + 1$.

Hint: The absence of z implies we translate the curve $y = x^2 + 1$ in the xy plane along the z-axis.

(e) $y = 4$.

Hint: This is a plane with parallel to the xz-plane with y = 4.

Quadric Surfaces: (polynomials in x and y and z)

(f) $y = \sqrt{1 - x^2 - z^2}$.

Hint: Temporarily square both sides. (Remember $y \geq 0$.) Then $y^2 = 1 - x^2 - z^2$, or $x^2 + y^2 + z^2 = 1$, which is a sphere of radius 1. But we want the hemisphere for which $y \geq 0$.

(g) $\dfrac{x^2}{2^2} + \dfrac{y^2}{3^2} + z^2 = 4$.

Hint: This is an ellipsoid--all traces parallel to planes x = 0, y = 0, and z = 0 are ellipses.

(h) $y^2 + z^2 = 4$.

Hint: Restricted to the xy-plane, this would be a circle . Since x isn't given, translate along the x-axis. This surface is a circular cylinder with the x-axis its main axis.

(i) $x = z^2 + 4$

Hint: Since y is not present , translate curve $x = z^2 + 4$ (a parabola opening in the positive x-direction) along the y-axis. This surface is a parabolic cylinder.

EXAMPLE: Find the level surface f(x,y,z) = 3 where $f(x,y,z) = z - 1 - x^2 - y^2$.

Answer: The graph would be in 4-dimensions, but the level surface is 3-dimensional . Writing $z - 1 - x^2 - y^2 = 3$, we obtain $z = 4 + x^2 + y^2$, whose graph is a paraboloid opening upward along the positive z-axis, passing through (0,0,4).

EXAMPLE: Determine a function f of 2 variables and a function g of one variable such that F = g $\circ$ f, where $F(x,y) = 4 - x^2 - y^2$.

Answer: Use $f(x,y) = x^2 + y^2$ and g(t) = 4 - t.

NOTE: The use of level surfaces in graphing cannot be understated. For example, graph z = 1/xy.

Where xy is constant, or y = c/x , we get either set of curves, depending on the sign of c. The graph is unusual.

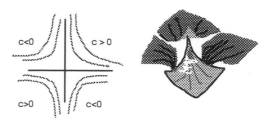

Explain how would you go about finding the graph of the function $f(x,y) = 1 - xy^2$?
Consider such things as level surfaces, symmetry, intercepts, etc. Write a list of the various
techniques we have used.

SECTION 13.2

*"For all knowledge and wonder (which is the seed of knowledge) is an impression of pleasure in
itself"...Francis Bacon*

LIMITS AND CONTINUITY

Many definitions are presented in your text, that you will need to learn and possibly,
memorize. The definition of limit in your text
$$\lim_{(x,y)\to(x_0,y_0)} f(x,y) = L$$
is behind most of what we do with functions in 3-space--this is calculus! In the case of a
surface, (x,y) can approach the limiting point in any direction. The concept of an (open)
interval on the line is generalized to the concept of an (open) **disk** in the plane, or an (open)
ball in 3-space or higher. We see the definition of boundary point and interior point.

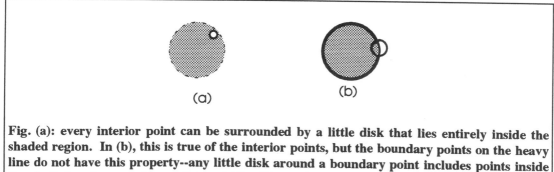

(a)   (b)

Fig. (a): every interior point can be surrounded by a little disk that lies entirely inside the
shaded region. In (b), this is true of the interior points, but the boundary points on the heavy
line do not have this property--any little disk around a boundary point includes points inside
the shaded region as well as points outside. The region in (a) does not have a boundary.

EXAMPLE: Evaluate the limit: $\displaystyle\lim_{(x,y)\to(0,0)} \frac{x^2 - xy}{x^2 - y^2}$.

Answer: MOTTO: Be cautious around denominators! Rewrite as

$$\lim_{(x,y)\to(0,0)} \frac{x^2 - xy}{x^2 - y^2} = \lim_{(x,y)\to(0,0)} \frac{x(x-y)}{(x-y)(x+y)} = \lim_{(x,y)\to(0,0)} \frac{x}{x+y}$$

As $(x,y) \to (0,0)$ along the line $y = x$, we have $\displaystyle\lim_{(x,y)\to(0,0)} \frac{x}{x+y} = \lim_{(x,y)\to(0,0)} \frac{x}{2x} = \frac{1}{2}$. As

$(x,y) \to (0,0)$ along the line $y = 0$, $\displaystyle\lim_{(x,y)\to(0,0)} \frac{x}{x+y} = \lim_{(x,y)\to(0,0)} \frac{x}{x} = 1$. There is no limit.

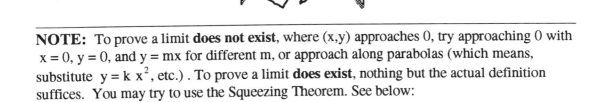

NOTE: To prove a limit **does not exist**, where (x,y) approaches 0, try approaching 0 with $x = 0$, $y = 0$, and $y = mx$ for different m, or approach along parabolas (which means, substitute $y = k\, x^2$, etc.) . To prove a limit **does exist**, nothing but the actual definition suffices. You may try to use the Squeezing Theorem. See below:

EXAMPLE: Explain why f is continuous, if

$$f(x,y) = \begin{cases} \dfrac{x^2 y^2}{x^2 + y^2} & \text{for } (x,y) \neq (0,0) \\ 0 & \text{for } (x,y) = (0,0) \end{cases}.$$

Answer: f is continuous at $(0,0)$ if $\displaystyle\lim_{(x,y)\to(0,0)} f(x,y) = f(0,0)$.

We use the Squeezing Theorem. Since $0 \le \dfrac{x^2 y^2}{x^2 + y^2} \le x^2 \left(\dfrac{y^2}{x^2 + y^2} \right) \le x^2$, we have

$\displaystyle\lim_{(x,y)\to(0,0)} f(x,y) = \lim_{(x,y)\to(0,0)} \frac{x^2 y^2}{x^2 + y^2} = 0$, which is f(0,0). So f is continuous.

(The graph of f looks like a handkerchief laid flat, then gathered up at the 4 corners.)

SECTION 13.3

" SNAFU--situation normal: all fouled up".. Anonymous

PARTIAL DERIVATIVES

Earlier, we found the derivative of a vector-valued function parametrized by t. But what gives the "derivative" of a function of several variables? Now the slope of the tangent can be found in many different ways, using different cross-sections of a surface.You will see several new concepts in the next sections: "partial derivatives", the "gradient", "directional derivative", the "differential", and the "tangent plane".

PARTIAL DERIVATIVES

When we consider a trace of f parallel to the x-axis, and take a derivative of f at a point (holding y fixed), we are taking the **partial derivative of f with respect to x** at those points. The partial derivative is denoted $\dfrac{\partial f}{\partial x}$, or f_x.

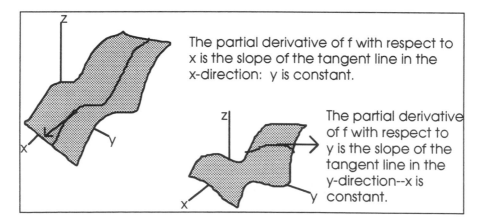

The partial derivative of f with respect to x is the slope of the tangent line in the x-direction: y is constant.

The partial derivative of f with respect to y is the slope of the tangent line in the y-direction--x is constant.

NOTE: Recall the definition of the derivative of a single-variable function, from Calculus I.

$$\frac{df}{dx}\bigg|_{x_0} = f'(x_0) = \lim_{h \to 0} \frac{f(x_0 + h) - f(x_0)}{h}.$$

Now, since by definition, we are holding y constant,

$$\frac{\partial f}{\partial x}(x_0, y_0) = \lim_{h \to 0} \frac{f(x_0 + h, y_0) - f(x_0, y_0)}{h}.$$

This partial derivative can also be denoted $\left.\dfrac{\partial f}{\partial x}\right|_{(x_0, y_0)}$ or $f_x(x_0, y_0)$.

The partial derivative of f with respect to y is found analogously and is denoted $\dfrac{\partial f}{\partial y}$ or f_y.

NOTE: Second Partials, "Mixed" Partials. We can take the "second partial derivative with respect to x", $\dfrac{\partial^2 f}{\partial x^2}$ (also denoted f_{xx}) and the second partial with respect to y $\dfrac{\partial^2 f}{\partial y^2}$ (or f_{yy}). We can also take "mixed" partials. For example $\dfrac{\partial^2 f}{\partial y \partial x}$, or f_{xy} represents first the partial derivative with respect to x, then with respect to y. In terms of notation, we think: $\dfrac{\partial^2 f}{\partial y \partial x} = \dfrac{\partial}{\partial x}\left(\dfrac{\partial f}{\partial y}\right) = (f_y)_x$. An **IMPORTANT RESULT** is that the "mixed" partials are equal when they are continuous. That is, $f_{xy} = f_{yx}$ if both terms are continuous.

THE CHAIN RULE

EXAMPLE (from Calculus I). Find f'(x) where $f(x) = \cos^5(x^3 + 1)$.
Answer: This function is the composite:

$$x \xrightarrow{\ (\ )^3 + 1\ } x^3 + 1 \xrightarrow{\ \cos(\ )\ } \cos(x^3 + 1) \xrightarrow{\ (\ )^5\ } \cos^5(x^3 + 1)$$

$\qquad\qquad\uparrow\qquad\qquad\qquad\uparrow\qquad\qquad\qquad\uparrow$

this deriv is 3 ( )　　　this deriv is - sin( )　　　this deriv is 5 ( )¯

So working backwards,

$$f'(x) = 5(\cos(x^3 + 1))^4 (-\sin(x^3 + 1))\quad (3x^2) = \quad -15x^2(\cos^4(x^3 + 1))(\sin(x^3 + 1)).$$

$\qquad\qquad\uparrow\qquad\qquad\qquad\uparrow\qquad\qquad\uparrow$

deriv of ( )⁵　　　deriv of cos ( )　　deriv of ( )³ + 1

EXAMPLE: Find the partial derivative with respect to x of the function
f(x,y) = $2x + e^{xy} + \sin(x^2 + y^2) + 4y^2 x$.

Answer:

$$\frac{\partial f}{\partial x} = 2 \quad + \quad e^{xy}y \qquad\qquad + \left(\cos(x^2+y^2)\right)(2x) \qquad\qquad + 4y^2 \qquad .$$

deriv of 2x + deriv of e^{xy} times deriv of   xy + deriv of   sin( ) times deriv of $x^2 + y^2$ + deriv of $4y^2x$

And $\dfrac{\partial f}{\partial y} = 0 + e^{xy}x + \left(\cos(x^2+y^2)\right)(2y) + 8yx$.

☞   In finding $\dfrac{\partial f}{\partial x}$ or f_x, think of y as constant. Replace "y" temporarily by "c"
if it helps you to remember this.

EXAMPLE: We will rework the example above. Find the partial derivative with respect to x of the function
f(x,y) $= 2x + e^{xy} + \sin(x^2+y^2) + 4y^2x$.

Answer: Temporarily replace all y's by the constant c-- assuming there are no c's in this equation!

f(x,y) $= 2x + e^{xc} + \sin(x^2+c^2) + 4c^2x$. Now
$\dfrac{\partial f}{\partial x} = 2 + e^{xc}c + \left(\cos(x^2+c^2)\right)(2x) + 4c^2$. Now change c's back to y's.
$\dfrac{\partial f}{\partial x} = 2 + e^{xy}y + \left(\cos(x^2+y^2)\right)(2x) + 4y^2$, and this is the answer in the example above.
(Of course, you won't want to do this forever!)

NOTE: For purposes of taking derivatives of exponential and logarithmic functions, use the equalities:

$$a^x = e^{\ln a^x} = e^{x\ln a}$$

$$\log_a x = \frac{\ln x}{\ln a}$$

EXAMPLE: Find the partial derivatives of
f(x,y)$= \left(\dfrac{y}{x}\right)^x + \dfrac{1}{\sqrt{x^2+3y^2}} + \sin\left(\dfrac{x}{y}\right)$.

Answer: We use the equalities above to write
$$\left(\frac{y}{x}\right)^x = e^{\ln(y/x)^x} = \exp\left(x\ln\frac{y}{x}\right) = \exp(x(\ln y - \ln x)) = e^{x\ln y - x\ln x}$$
So f(x,y) $= e^{x\ln y - x\ln x} + (x^2+3y^2)^{-1/2} + \sin(xy^{-1})$, and
$\dfrac{\partial f}{\partial x} = e^{x\ln y - x\ln x}(\ln y - x(1/x) - \ln x) - \dfrac{1}{2}(x^2+3y^2)^{-3/2}(2x) + \left(\cos(xy^{-1})\right)(y^{-1})$ and

$\dfrac{\partial f}{\partial y} = e^{x \ln y - x \ln x}(x / y) - \dfrac{1}{2}(x^2 + 3y^2)^{-3/2}(6y) + \left(\cos(xy^{-1})\right)(-xy^{-2})$. These expressions can be simplified.

✋ **CAUTION!** When we take partial derivatives, we are not necessarily differentiating implicitly as we did in Calculus I: x and y are independent variables and y does not depend on x.

EXAMPLE: Find the partial derivatives of $f(x,y,z) = e^{xy}(\ln(y + z))(\sin x)$.

Answer: We require 3 partial derivatives. And incidentally, this happens to be a triple product! Using "D" to denote derivative, the rule for taking the derivative of a triple product is

Df = (D lst factor)(2nd)(3rd) + (lst)(D 2nd factor)(3rd) + (lst)(2nd)(D 3rdfactor). Thus:

$\dfrac{\partial f}{\partial x} = e^{xy}y(\ln(y + z))(\sin x \ ) \ + \ 0 \ + \ e^{xy}(\ln(y + z))(\cos x)$

$\dfrac{\partial f}{\partial y} = e^{xy}x(\ln(y + z))(\sin x \ ) \ + \ e^{xy}\left(\dfrac{1}{y + z}\right)(\sin x \ ) \ + \ 0$

$\dfrac{\partial f}{\partial z} = 0 + e^{xy}\left(\dfrac{1}{y + z}\right)(\sin x \ ) \ + \ 0$

HOW TO DIFFERENTIATE AN INTEGRAL: REVIEW FROM CALCULUS I
If f(x) is continuous on an interval I, and c is in I, then $\dfrac{d}{dx}\displaystyle\int_c^x f(t)dt = f(x)$. In other words, to find $\dfrac{d}{dx}\displaystyle\int_c^x f(t)dt$, simply take the function $f(t)$ in the integrand and plug in x. (The derivative and integral in this form undo each other.)

EXAMPLE: Differentiating an integral from Calculus I. Find $\dfrac{d}{dx}\displaystyle\int_1^x \cos^2\sqrt{t + 3} \ \ dt$.

Answer: This is an integral we cannot evaluate, but we only intend to differentiate it.

$\dfrac{d}{dx}\displaystyle\int_1^x \cos^2\sqrt{t + 3} \ \ dt = \cos^2\sqrt{x + 3}$. Note: When the problem is in the form,

$\dfrac{d}{dx}\displaystyle\int_{cons\tan t}^x$, the answer is given by plugging "x" into the integrand.

EXAMPLE: Find $\dfrac{d}{dx}\displaystyle\int_1^{x^2} \cos^2\sqrt{t + 3} \ \ dt$.

Answer: This situation is different. Now we have x^2, not x, in the top limit. We use the Chain Rule.

(1) Plug in x^2 wherever you saw t in the integrand.

(2) Multiply this result by the derivative of x^2.

$$\frac{d}{dx}\int_1^{x^2}\cos^2\sqrt{t+3}\;dt = \left(\cos^2\sqrt{x^2+3}\right)(2x).$$

EXAMPLE: Partial derivatives of integrals. Find $\displaystyle\frac{\partial}{\partial x}\int_2^{x^2-3x+y}\sin t^2 dt$.

Answer: We plug the top limit into the integrand and multiply that result by the partial with respect to x of the top limit. $\displaystyle\frac{\partial}{\partial x}\int_2^{x^2-3x+y}\sin t^2 dt = \left(\sin(x^2-3x+y)^2\right)(2x-3)$.

EXAMPLE: Find $\displaystyle\frac{\partial}{\partial x}\int_{x+y}^{x^2-3x+y}\sin t^2 dt$.

Answer: $\displaystyle\int_{x+y}^{x^2-3x+y}\sin t^2 dt = \int_{x+y}^c\sin t^2 dt + \int_c^{x^2-3x+y}\sin t^2 dt = -\int_c^{x+y}\sin t^2 dt + \int_c^{x^2-3x+y}\sin t^2 dt.$ So

$$\frac{\partial}{\partial x}\int_{x+y}^{x^2-3x+y}\sin t^2 dt =$$

$$\frac{\partial}{\partial x}\left(-\int_c^{x+y}\sin t^2 dt + \int_c^{x^2-3x+y}\sin t^2 dt\right) = -\sin(x+y)^2(1) + \left(\sin(x^2-3x+y)^2\right)(2x-3)$$

EXAMPLE: Finding higher order partial derivatives and evaluating at a point.

If $f(x) = \sin^{-1}(xy)$, find $f_x\left(\frac{1}{2},\sqrt{3}\right)$ and $f_{xx}\left(\frac{1}{2},\sqrt{3}\right)$.

Answer:

$f_x = (1-(xy)^2)^{-1/2}(y)$

$f_{xx} = \left(-\frac{1}{2}(1-(xy)^2)^{-3/2}(-2xy)(y)\right)(y) = (1-(xy)^2)^{-3/2}(xy^3)$

and $f_{xx}(1/2,\sqrt{3}) = (1-(3/4))^{-3/2}(3\sqrt{3}/2) = 12\sqrt{3}$

EXAMPLE: Higher order partial derivatives. If $f(x) = e^{x+y^2}(\ln x)$, find f_{xy} and f_{yx} and determine if they are equal.

Answer:

$f_x = e^{x+y^2}(\ln x) + e^{x+y^2}\left(\frac{1}{x}\right)$ and $f_{xy} = e^{x+y^2}(2y)(\ln x) + e^{x+y^2}(2y)\left(\frac{1}{x}\right)$. Also ,

$f_y = e^{x+y^2}(2y)(\ln x)$ and $f_{yx} = e^{x+y^2}(2y)(\ln x) + e^{x+y^2}(2y)\left(\frac{1}{x}\right)$

The mixed partials are equal when they are continuous, which happens when $x > 0$.

SECTION 13.4

"...God is a mathematician of a very high order, and he used very advanced mathematics in constructing the universe"...Paul Dirac

Differentiability

---

f is differentiable at (x_0, y_0)

if there exists an open disk D centered at (x_0, y_0), and ε_1 and ε_2 such that
$$f(x,y) = f(x_0, y_0) + f_x(x_0, y_0)(x - x_0) + f_y(x_0, y_0)(y - y_0) +$$
$$\varepsilon_1(x,y)(x - x_0) + \varepsilon_2(x,y)(y - y_0)$$

for (x,y) in D, where ε_1 and $\varepsilon_2 \to 0$ as $(x,y) \to (x_0, y_0)$

What Does This Mean?

Geometrical Motivation: We will later see that the expression above *without the epsilon - terms* is the equation for the **tangent plane** at (x_0, y_0). This definition states that **f is approximated by its tangent plane** in a neighborhood of $f(x_0, y_0)$. (And this plane cannot be vertical.) The definition should call to mind the expression for the differential in first year calculus. The epsilon terms are indicators that f is not *exactly* given by its tangent plane, and that the tangent plane approximation gets better close to (x_0, y_0). The epsilon terms remind us of the higher order derivative terms in Taylor's Series.) Theorem 13.6 gives the conditions for when f is differentiable. Rarely, in practice, do we use the definition above to insure differentiability. Instead we use the "Guiding Condition" below.

The definition of differentiability says essentially, that f is well-approximated by a tangent plane

---

GUIDING CONDITION :

☞ **If f has continuous partial derivatives at a point, then f is differentiable there .**

EXAMPLE: Let f be given as $f(x) = e^{x+y^2} (\ln x)$. Where is f differentiable?

Answer: We computed the partial derivatives in an earlier example:

$$f_x = e^{x+y^2} (\ln x) + e^{x+y^2} \left(\frac{1}{x} \right) \quad \text{and} \quad f_y = e^{x+y^2} (2y)(\ln x). \text{ These partials}$$

are continuous everywhere but at points where $x \leq 0$. Thus f is differentiable for all (x,y) where $x > 0$.

THE CHAIN RULE IN SEVERAL VARIABLES

Chain Rule, First Version

If z is a function of x and y which are functions of t, we may find **dz/dt**. We use *derivative* notation, not partials because z can be expressed **entirely in terms of t**. The text illustrates it this way:

$$\frac{dz}{dt} = \frac{\partial z}{\partial x} \frac{dx}{dt} + \frac{\partial z}{\partial y} \frac{dy}{dt}$$

Geometrical (but non-rigorous) motivation for the Chain Rule: Write $z = f(x,y)$. Let Δz be the change in z as measured by the tangent plane (*which only approximates the actual change in z*).

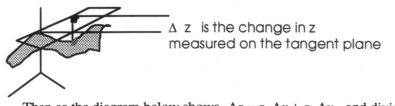

Δz is the change in z measured on the tangent plane

Then as the diagram below shows, $\Delta z = z_x \Delta x + z_y \Delta y$, and dividing by Δt,

$$\frac{\Delta z}{\Delta t} = z_x \frac{\Delta x}{\Delta t} + z_y \frac{\Delta y}{\Delta t} \quad \text{and in the limit}, \quad \frac{dz}{dt} = z_x \frac{dx}{dt} + z_y \frac{dy}{dt}$$

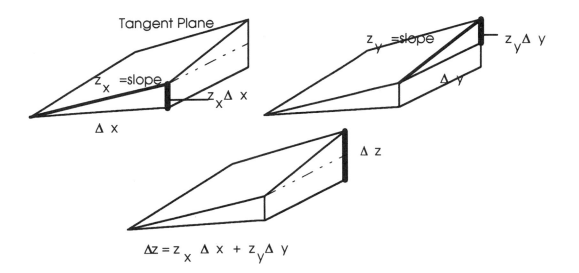

$$\Delta z = z_x \, \Delta x + z_y \Delta y$$

EXAMPLE: Compute dz/dt if $z = \sin x + \cos y$ and $x = \sqrt{t}$; $y = e^t$.

Answer: $\dfrac{dz}{dt} = \dfrac{\partial z}{\partial x}\dfrac{dx}{dt} + \dfrac{\partial z}{\partial y}\dfrac{dy}{dt}$, or $\dfrac{dz}{dt} = (\cos x)\dfrac{1}{2\sqrt{t}} + (-\sin y)e^t$ (and expressing this

entirely in terms of t) $\dfrac{dz}{dt} = (\cos \sqrt{t})\dfrac{1}{2\sqrt{t}} + (-\sin e^t)e^t$.

EXAMPLE: Compute dw/dt if $w = \dfrac{y}{x} - \dfrac{x}{z}$ and $x = t^2$, $y = \ln t$, $z = \sin t$.

Answer: Use a variation on the rule above: $\boxed{\dfrac{dw}{dt} = \dfrac{\partial w}{\partial x}\dfrac{dx}{dt} + \dfrac{\partial w}{\partial y}\dfrac{dy}{dt} + \dfrac{\partial w}{\partial z}\dfrac{dz}{dt}}$.

Rewrite: $w = yx^{-1} - xz^{-1}$, and

$\dfrac{dw}{dt} = (-yx^{-2} - z^{-1})(2t) + (x^{-1})\left(\dfrac{1}{t}\right) + (xz^{-2})(\cos t)$ or

$\dfrac{dw}{dt} = (-(\ln t)(t^2)^{-2} - (\sin t)^{-1})(2t) + ((t^2)^{-1})\left(\dfrac{1}{t}\right) + (t^2)(\sin t)^{-2}(\cos t)$

which should be simplified.

EXAMPLE: Implicit differentiation. Assume z is implicitly a function of x and y (z = z(x,y)) where $xz = x^2 \sin y + z\ln(xy)$. Compute $\dfrac{\partial z}{\partial x}$ and $\dfrac{\partial z}{\partial y}$.

Answer:

Take $\dfrac{\partial}{\partial x}$ of both sides, where we assume x and y are independent.

Since $xz = x^2 \sin y + z \ln(xy)$,

$$\frac{\partial}{\partial x}(xz) = \frac{\partial}{\partial x}(x^2 \sin y + z \ln(xy)) \text{ or}$$

$$x\frac{\partial z}{\partial x} + z = 2x \sin y + \ln(xy)\frac{\partial z}{\partial x} + z\left(\frac{1}{xy} \cdot y\right). \quad \text{Gather all } \frac{\partial z}{\partial x} \text{ terms to the left side, factor}$$

and solve .

$$(\frac{\partial z}{\partial x})(x - \ln(xy)) = 2x \sin y - z + \frac{z}{y}.$$

$$\frac{\partial z}{\partial x} = \frac{2x \sin y - z + z/y}{x - \ln(xy)}$$

THEN:

$$\frac{\partial}{\partial y}(xz) = \frac{\partial}{\partial y}(x^2 \sin y + z \ln(xy)) \text{ or}$$

$$x\frac{\partial z}{\partial y} + 0 = x^2 \cos y + \frac{\partial z}{\partial x}\ln(xy) + z\left(\frac{1}{xy}\right)(x). \quad \text{Gather all } \frac{\partial z}{\partial x} \text{ terms to the left side, factor}$$

and solve .

$$\left(\frac{\partial z}{\partial y}\right)(x - \ln(xy)) = x^2 \cos y + \frac{z}{y}$$

$$\frac{\partial z}{\partial x} = \frac{x^2 \cos y + z/y}{x - \ln(xy)}$$

NOTE: If you are unsure about implicit differentiation, put a box around each z term; this will remind you that z has "x" in it.

Since $x\boxed{z} = x^2 \sin y + \boxed{z}\ln(xy)$,

$$x\frac{\partial\boxed{z}}{\partial x} + \boxed{z} = 2x \sin y + \ln(xy)\frac{\partial\boxed{z}}{\partial x} + \boxed{z}\left(\frac{1}{xy}\right)(y).$$

Other steps are the same as above.

Chain Rule, Second Version

If f is composed of x **and** y, and these variables are both functions of u and v, we use:

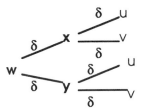

Use δ when the path has more than one branch

$$\frac{\partial f}{\partial u} = \frac{\partial f}{\partial x}\frac{\partial x}{\partial u} + \frac{\partial f}{\partial y}\frac{\partial y}{\partial u}$$

EXAMPLE:

Assume $f(x,y) = \cos x + e^y$ where $x = u + v^2$ and $y = \sin uv$. Compute $\frac{\partial f}{\partial u}$ and $\frac{\partial f}{\partial v}$.

Answer:

We have $\frac{\partial f}{\partial u} = \frac{\partial f}{\partial x}\frac{\partial x}{\partial u} + \frac{\partial f}{\partial y}\frac{\partial y}{\partial u}$ and $\frac{\partial f}{\partial u} = (-\sin x)(1) + (e^y)(\cos uv)(v)$.

Also, $\frac{\partial f}{\partial v} = \frac{\partial f}{\partial x}\frac{\partial x}{\partial v} + \frac{\partial f}{\partial y}\frac{\partial y}{\partial v}$ and $\frac{\partial f}{\partial v} = (-\sin x)(2v) + e^y(\cos uv)(u)$.

Expressing x and y in terms of u and v:

$\frac{\partial f}{\partial u} = (-\sin x)(1) + (e^y)(\cos uv)(v) = (-\sin(u+v^2))(1) + (e^{\sin uv})(\cos uv)(v)$ and

$\frac{\partial f}{\partial v} = (-\sin x)(2v) + e^y(\cos uv)(u) = (-\sin(u+v^2))(2v) + e^{\sin uv}(\cos uv)(u)$.

These expressions can be simplified.

SECTION 13. 5

"The knowledge at which geometry aims is the knowledge of the eternal"...Boris Pasternak

DIRECTIONAL DERIVATIVES

We have seen that the partials of f with respect to x and y are found by taking derivatives along slices of the surface of f, in the direction of the x- and y- axes respectively. It is possible to slice the surface in other directions, and find the derivative in that direction. We set the direction by means of a unit vector, $\mathbf{u} = u_1\mathbf{i} + u_2\mathbf{j}$. Then the directional derivative in the direction of $\mathbf{u}$ at a point is:

$$D_{\mathbf{u}}f(x_0, y_0) = f_x(x_0, y_0)u_1 + f_y(x_0, y_0)u_2$$

which is easily remembered as a dot product:

$$\boxed{D_{\mathbf{u}}f = (f_x\mathbf{i} + f_y\mathbf{j}) \cdot (u_1\mathbf{i} + u_2\mathbf{j})}$$

NOTE: The definition of the directional derivative generalizes to higher dimensions.

NOTE: We might have seen the directional derivative coming when the chain rule told us that

$$\frac{dz}{dt} = \frac{\partial z}{\partial x}\frac{dx}{dt} + \frac{\partial z}{\partial y}\frac{dy}{dt}.$$

This suggests that $\dfrac{dz}{dt} = (z_x\mathbf{i} + z_y\mathbf{j}) \cdot \left(\dfrac{dx}{dt}\mathbf{i} + \dfrac{dy}{dt}\mathbf{j}\right)$, which looks like the formula above.

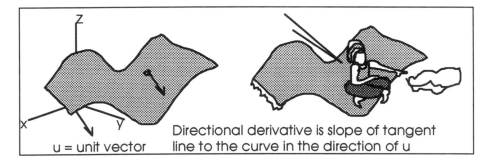

Directional derivative is slope of tangent line to the curve in the direction of u

u = unit vector

NOTE: The unit vector **u** lies on the xy-plane: it gives the direction in which the directional derivative is to be computed. The directional derivative, evaluated at a point, is a number corresponding to the **slope of the tangent line** to a cross-section of the surface in the direction of **u**. For example, when **u** = **i**, the directional derivative is just $D_u f = \dfrac{\partial f}{\partial x}$ and when **u** = **j**, $D_u f = \dfrac{\partial f}{\partial y}$.

EXAMPLE: Find the directional derivative of $f(x,y) = x^2 + y^2$ in the direction given by $\mathbf{u} = \dfrac{1}{\sqrt{2}}\mathbf{i} + \dfrac{1}{\sqrt{2}}\mathbf{j}$ at the point $(1,1)$. (This is a paraboloid and we are finding the derivative in the direction of $\mathbf{u}$ below.)

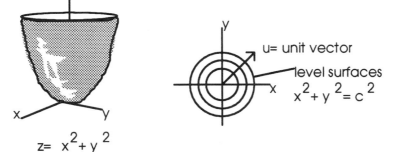

Answer: $D_{\mathbf{u}} f(x,y) = (f_x\mathbf{i} + f_y\mathbf{j}) \cdot (a_1\mathbf{i} + a_2\mathbf{j}) = (2x\,\mathbf{i} + 2y\,\mathbf{j}) \cdot (\dfrac{1}{\sqrt{2}}\mathbf{i} + \dfrac{1}{\sqrt{2}}\mathbf{j}) = \dfrac{2x}{\sqrt{2}} + \dfrac{2y}{\sqrt{2}}$

and at $(x,y) = (1,1)$, $D_{\mathbf{u}} f(1,1) = \dfrac{2(1)}{\sqrt{2}} + \dfrac{2(1)}{\sqrt{2}} = \dfrac{4}{\sqrt{2}}$.

EXAMPLE: Find the directional derivative of $f(x,y,z) = x^2 + y^2 + 2z$ in the direction given by $\mathbf{u} = \mathbf{i} - \mathbf{j} + \mathbf{k}$ at the point $P = (2, 1, 3)$.
Answer: $D_{\mathbf{u}} f(x,y,z) = (f_x\mathbf{i} + f_y\mathbf{j} + f_z\mathbf{k}) \cdot (a_1\mathbf{i} + a_2\mathbf{j} + a_3\mathbf{k}) =$
$(2x\mathbf{i} + 2y\mathbf{j} + 2\mathbf{k}) \cdot (\mathbf{i} - \mathbf{j} + \mathbf{k}) = 2x - 2y + 2$, and at $(x,y,z) = (2,1,3)$ this is
$D_{\mathbf{u}} f(x,y,z) = 2(2) - 2(1) + 2 = 4$

SECTION 13. 6

"Zounds! I never was so bethumped with words"...William Shakespeare

THE GRADIENT

We saw above that the directional derivative of f given by $z = f(x,y)$ was given by
$$D_{\mathbf{u}} = (f_x\mathbf{i} + f_y\mathbf{j}) \cdot (a_1\mathbf{i} + a_2\mathbf{j})$$

> **The gradient of f is the vector**
> $$\mathbf{grad}\, f = f_x\mathbf{i} + f_y\mathbf{j} = \frac{\partial f}{\partial x}\mathbf{i} + \frac{\partial f}{\partial y}\mathbf{j}$$
> The gradient is generally evaluated at a point (x_0, y_0).
> Then we say $\mathbf{grad}\, f = \nabla\mathbf{f}(x_0, y_0) = f_x(x_0, y_0)\mathbf{i} + f_y(x_0, y_0)\mathbf{j}$ (*)

(*) **NOTE:** The text uses the notation "**grad** f " for this vector. Another familiar notation for **grad** f is ∇**f**. **In this Guide we use both notations, together at first, and then interchangeably. It is probably good to become familiar with different notations for the same object.** (∇ is called the "del" operator. An operator acts on functions. The operator ∇ is defined as $\dfrac{\partial}{\partial x}($ $)$ **i** $+\dfrac{\partial}{\partial y}($ $)$ **j**, so that $\nabla f = \dfrac{\partial}{\partial x}($ f $)$ **i** $+\dfrac{\partial}{\partial y}($ f $)$ **j**.)

NOTE: From the work above, we can see that $D_{\mathbf{u}}$ $=$ **grad** f $\cdot$ **u.** The directional derivative in the direction of **u** (at a point) is the gradient there dotted with **u**.

EXAMPLE: Find the gradient of the function $f(x,y) = \dfrac{x-1}{xy-x^2}$.

Answer: Grad f $=\nabla$**f** $=\dfrac{(xy-x^2)(1)-(x-1)(y-2x)}{(xy-x^2)^2}$**i**$+\dfrac{(xy-x)(0)-(x-1)(x)}{(xy-x^2)^2}$**j.**

(Simplify!)

EXAMPLE: Find the gradient of $f(x,y) = x \sin xy$ at $(1, \pi)$.

Answer: Grad f $=\nabla$**f** $=(\,(\cos xy)(y) + \sin xy\,)$ **i** $+ (x\,(\cos xy)(x)\,)$ **j** . Now evaluated at $(1, \pi)$, **grad** f $(1, \pi)$ $= ((\cos\pi)(\pi) + \sin\pi)$ **i** $+ (\cos\pi)$ **j** $= -\pi$ **i** $-$ **j**

The gradient of f points in the direction of greatest increase of f

The text shows that if the function is given as $z = f(x,y)$, **in explicit form**, when the gradient is not 0, ∇**f** **points in the direction of steepest ascent of the function.** ‖∇f‖ **is the maximum rate of increase in the function in any direction.** However, there is another result regarding the gradient of level surfaces.

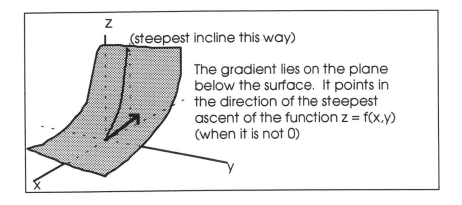

(steepest incline this way)

The gradient lies on the plane below the surface. It points in the direction of the steepest ascent of the function z = f(x,y) (when it is not 0)

EXAMPLE: Below, on the left, are some graphs and on the right, a bird's eye view trace in the xy-plane. With *our* eye we have tried to locate the gradient at a given point: recall that the gradient points in the direction of steepest ascent of z. We have attached the gradient to the given point, but it will actually emanate from the origin. (Why?) We are not able to find the *length* of the gradient by eye, however.

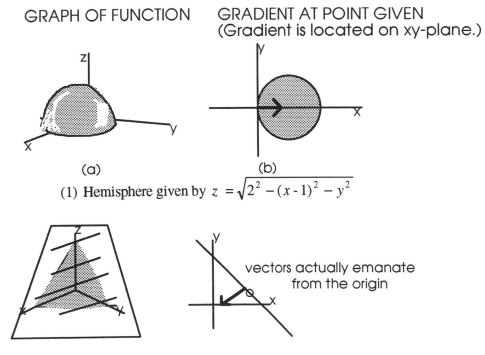

GRAPH OF FUNCTION   GRADIENT AT POINT GIVEN
(Gradient is located on xy-plane.)

(a)   (b)

(1) Hemisphere given by $z = \sqrt{2^2 - (x-1)^2 - y^2}$

vectors actually emanate from the origin

(2) Plane given by $z = -x - y + 1$

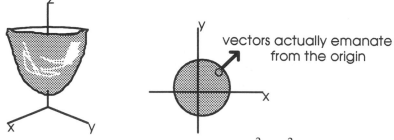

(3) Paraboloid given by $z = x^2 + y^2 + 2$

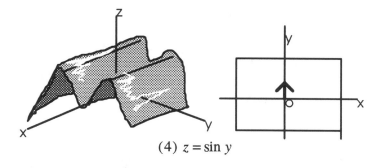

(4) $z = \sin y$

EXAMPLE: Compute the gradients of the functions at the points indicated, above.

(1) Hemisphere given by $z = \sqrt{2^2 - (x-1)^2 - y^2}$.

Answer: grad f $= \dfrac{-2(x-1)}{2\sqrt{2^2 - (x-1)^2 - y^2}}\mathbf{i} - \dfrac{2y}{2\sqrt{2^2 - (x-1)^2 - y^2}}\mathbf{j}$.

At $(0,0)$ this is $\dfrac{1}{\sqrt{3}}\mathbf{i}$. At the point $(1, 1)$, **grad f** $= -\dfrac{1}{\sqrt{3}}\mathbf{j}$.

(2) Plane given by $z = -x - y + 1$

Answer: grad f $= -\mathbf{i} - \mathbf{j}$. This is the gradient at any point (x_0, y_0).

(3) Paraboloid given by $z = x^2 + y^2 + 2$

Answer: grad f $= 2x\,\mathbf{i} + 2y\,\mathbf{j}$. **At (0,0) the gradient is 0**. We cannot say this gradient points in the direction of steepest ascent. Frankly, the gradient is confused--*all* directions point in the direction of steepest ascent!

EXAMPLE: (Hint). Let f(x,y) = sin y (graphed above). **Find the direction in which f increases most rapidly** at (0,0).

Answer: The gradient gives the direction of steepest ascent. **grad** f $= 0\,\mathbf{i} + \cos y\,\mathbf{j}$. This is largest when $y = 0$. Thus the directions of most rapid increase are **grad** f $= \pm\mathbf{j}$.

The gradient of a level surface is normal to the level surface.

When a function of two variables is given in explicit form, $z = f(x,y)$, the gradient, **grad** f $=$ $\frac{\partial f}{\partial x}\mathbf{i} + \frac{\partial f}{\partial x}\mathbf{j}$, evaluated at (x_0, y_0), points in the direction of steepest ascent of the function at that point. **The gradient is perpendicular to the level curves** (or surfaces) of the function, given by f(x,y) = constant. (Can you explain why this should be? What would be the case if there were a tangential component to the gradient?)

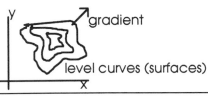

NOTE: We cannot graph a function of 3 variables, $w = F(x,y,z)$. However, we can graph the level surface given by $F(x,y,z) = 0$, because w, the 4th coordinate, is 0, which puts us into 3-space. **The gradient grad** F $= \frac{\partial F}{\partial x}\mathbf{i} + \frac{\partial F}{\partial x}\mathbf{j} + \frac{\partial F}{\partial x}\mathbf{k}$, **is a 3-dimensional vector,** which when evaluated at (x_0, y_0, z_0), **points in a direction normal to the level surface at that point.**

AN EXAMPLE THAT SHOWS THE DIFFERENCES

EXAMPLE: Find the gradient to the plane given in the form (a) $f(x,y) = z = - x - y + 1$ and in the form (b) $x + y + z - 1 = 0$.

Answer: Notice that in (a), the function is given explicitly, and in (b), implicitly.

In (a) we find **grad** f = -**i** - **j**. This vector, lying in the xy-plane, points in the direction of steepest ascent of z.

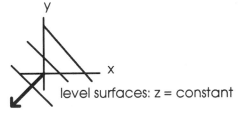

level surfaces: z = constant

gradient perpendicular to level surfaces

In (b) , we write $F(x,y,z) = x + y + z$ and find **grad** F =**i** + **j** + **k**. This vector is normal to the plane.

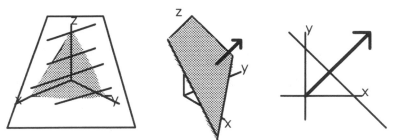

gradient is normal to plane

Note:

a.) In this example we did not need to evaluate the gradients at specific points.

b.) What affects the direction of the gradient, in part (b)? How might F have been written to reverse the direction of the gradient? Is the steepest ascent of this function in the direction shown in part (a)?

c.) We might talk about an upward pointing and a downward pointing normal to the plane. How might these be defined?

EXAMPLE: Compute the gradient of the function $z = x^2 + y^2 + 2$, whose graph is a paraboloid.

Answer: grad $f = f_x \mathbf{i} + f_y \mathbf{j} = 2x\,\mathbf{i} + 2y\mathbf{j}$. This vector (in the xy-plane) points in the direction of steepest ascent of the function at (x,y), unless **grad** $f = \mathbf{0}$. (See previous example.)

EXAMPLE: Compute the gradient of the level surface $F(x,y,z) = z - x^2 - y^2 - 2 = 0$ at the point (0, 1,3).

Answer: This function has a graph in 4-dimensions. The graphs of the 3-dimensional level surfaces are paraboloids. E.g., the graph of $z = 2 + x^2 + y^2$ is a paraboloid. The gradient to the level surface is **grad** $F = F_x \mathbf{i} + F_y \mathbf{j} + F_z \mathbf{k} = -2x\,\mathbf{i} - 2y\,\mathbf{j} + \mathbf{k}$, which evaluated at (0,1,3) is $-2\mathbf{j} + \mathbf{k}$. This vector is normal to the paraboloid at (0,1,3).

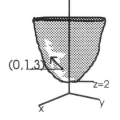

NOTE: NOW we can find a plane tangent to the graph of a function at a given point, although not all surfaces have tangent planes--beware of cusps, sharp places, ridges, etc. as in the surface of this crystal, shown under a microscope. below.

The tangent plane to z = f(x,y) at the point (x_0, y_0, z_0) that is written
implicitly as F(x,y,z) = 0 is given as:

grad F · (**vector on plane**) = 0, or

$$(F_x\mathbf{i} + F_y\mathbf{j} + F_z\mathbf{k})\Big|_{(x_0,y_0,z_0)} \cdot (x - x_0, y - y_0, z - z_0) = 0, \text{ or}$$

$$f_x(x_0, y_0, z_0)(x - x_0) + f_y(x_0, y_0, z_0)(y - y_0) + f_z(x_0, y_0, z_0)(z - z_0) = 0$$

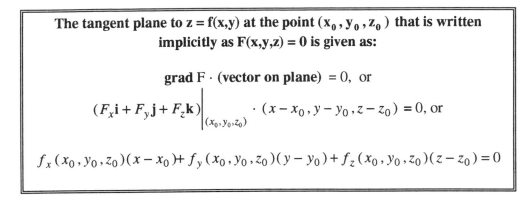

How to find the tangent plane to the graph of the function z = f(x,y).
- Write the function implicitly.
- Find the gradient.
- Find a vector on the plane--of the form $(x - x_0, y - y_0, z - z_0)$
- Set the dot product of this vector above with the gradient to 0.
- This is the equation of the tangent plane.

There is a theoretical analogy with Taylor's Series (or the differential) from first year calculus.
1. Can you remember the formula for the linear terms in the Taylor's Series for f?
$$f(x) = f(x_0) + f'(x_0)(x - x_0)$$
The tangent line is found from the linear terms. It is: $y = f(x_0) + f'(x_0)(x - x_0)$.

2. Then in two variables (tangent plane must not be vertical in this description),
$$f(x, y) = f(x_0, y_0) + f_x(x_0, y_0)(x - x_0) + f_y(x_0, y_0)(y - y_0)$$

and the tangent plane is

$$z = f(x_0, y_0) + f_x(x_0, y_0)(x - x_0) + f_y(x_0, y_0)(y - y_0)$$

EXAMPLE: **Find the tangent plane** to the graph of f(x,y) = xy - x at the point (3, 4, 9).
Answer:
- Write the equation z = xy - x implicitly, as F(x,y,z) = z - xy + x = 0.
- Find **grad** F. Here **grad** F = (-y +1)**i** -(x)**j** + **k**. Evaluate at (3,4,9): **grad** F = (-3,-3,1).
  This vector is normal to the surface.

- The formula
$$\mathbf{grad}\, F \cdot (\textbf{vector on plane}) = 0 \text{ becomes}$$
$$(-5\mathbf{i}-3\mathbf{j}+\mathbf{k}) \cdot (x-3)\mathbf{i} +(y-4)\mathbf{j} + (z-9)\,\mathbf{k} = 0, \text{ or } -3(x-3) - 3(y-4) + (z-9) = 0, \text{ or}$$
$$-3x - 3y + z +12 = 0.$$
- The tangent plane at $(3,4,9)$ is given by $z = 3x + 3y - 12$.

EXAMPLE: Find the plane tangent to the level surface $xyz = 2$ **at the point**
$(1/2, 3, 4/3)$.
Answer:
- The equation is written implicitly.
- Find $\mathbf{grad}\, F$. Here $\mathbf{grad}\, F = yz\mathbf{i} + xz\mathbf{j} + xy\mathbf{k}$ and we evaluate at $(1/2, 3, 4/3)$. So
  $\mathbf{grad}\, F = 4\mathbf{i} + 2/3\mathbf{j} + 3/2\mathbf{k}$.
- The formula
$$\mathbf{grad}\, F \cdot (\textbf{vector on plane}) = 0 \text{ becomes}$$
$$(4\mathbf{i} + 2/3\mathbf{j} + 3/2\mathbf{k}) \cdot ((x-1/2)\mathbf{i}+(y-3)\mathbf{j}+(z-4/3)\,\mathbf{k}) = 0, \text{ or } 4x+(2/3)y+(3/2)z = 6.$$

- This is the equation of the tangent plane.

Mathematical Check-Up

What skills from this chapter do you feel confident about? Reply in column 2 .
After a test or quiz, come back to this table and fill in the last column with a comment .

TYPE OF PROBLEM	do you feel in control?	after test comments
Navigating in 3-dimensions; graphing planes		
Graphing 3-dimensional surfaces		
Finding limits		
Finding directional derivatives		
Using the gradient (with an explicitly given function or with level surfaces)		
Finding partial derivatives of all orders, including mixed partial derivatives		
Using the chain rule, differentiating implicitly		
Finding the tangent plane to a surface		
Understanding the definitions (of limit, continuity, differentiability, etc.)		

Review Questions

These "loose", unformed questions are the types that test only the most basic skills, but might make an appearance on a test or quiz. Try to outline how you would solve the problem. In a study group, it would be good to make up some other examples of "good questions" , to solve together.

1. Find the limit and prove it is a limit, or explain why the limit does not exist :

$$\lim_{(x,y)\to(0,0)} \frac{xy}{x^2 + y^2}.$$

2. Sketch the surface....
3. Find the first order and second order partial derivatives of .... (don't forget the mixed partials.)
4. Find the unit normal vector to the graph of the surface given by.... Find the normal line to the surface..
5. Find the directional derivative of the function... in the direction...
6. Find the equation of the tangent plane to the function... at the point...
7. What condition guarantees that the mixed partials of a function of 2 variables are equal?
8. What is the maximum value possible for a directional derivative of f = ... at a point P= ..? In what direction does this maximum occur? What is the minimum value? In what direction does this minimum occur?
9. Find the derivatives dw/dt (where w is a function of t, x and y and x and y are functions of t) and $\dfrac{\partial w}{\partial u}$ (where w is a function of x and y, which are in turn functions of u and v.) Find $\dfrac{\partial^2 w}{\partial u^2}$.
10. What is meant by "a function of two variables is differentiable? Under what conditions is a function differentiable?
11. Implicitly differentiate...
12. Give the definition using "limit" of the partial derivative of f with respect to x.
13. Illustrate geometrically what the partial derivative at a point represents.
14. Find the gradient to a level surface of a given function at a point.
15. Find the tangent plane to a level surface of a function at a point.

Partial Answers: (1) The limit won't exist: come into (0,0) along the lines y = x and y = -x   (2) Do you have a strategy? Do you have some remembered guideline-examples in mind? (3) Practice needed here!   (4) One normal vector is $f_x\mathbf{i} + f_y\mathbf{j} - \mathbf{k}$ --this points downward because of -**k**. However, the unit normal is found by dividing this vector by its length. To find the equation of the normal line, go back to the section on finding the line given a point and a slope (direction) vector. (5) This is the grad f dot the unit direction vector. (6) See the equation of the tangent plane again. Memorize! The mixed partials must be continuous. (8). The maximum value occurs when we take the gradient. For the minimum value take the negative of the gradient. (9) Various forms of the chain rule needed.   (10) Review definition of differentiable--recall approximation by the tangent plane. Differentiability guaranteed when first partials are continuous. (11). Practice again.

(12) This is like the usual definition of derivative, but f is a function of 2 variables. Hold the y variable fixed...see text. (13) The slope of the tangent line in the plane parallel to the x-axis through the point (14) Don't forget that this gradient is perpendicular to the surface. (15) This goes to show these techniques generalize to higher dimensions. The level surface would be a 3-dimensional "surface". See the last worked example above.

Major Trouble Spot

A major trouble spot in this part of the course is that students don't recall how to differentiate functions of a single variable. Review, review!

SECTION 13.7

"A good listener is not only popular everywhere, but after a while, he gets to know something"....Wilson Misener

TANGENT PLANE APPROXIMATIONS AND DIFFERENTIALS

In the previous section we introduced the tangent plane. We briefly discussed an analogy with the tangent line, the differential, and Taylor's Series-- topics you saw in first year calculus. This section expands on those ideas.

THE CASE FOR A SINGLE VARIABLE (first year calculus):

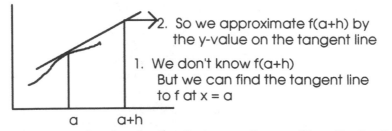

The linear approximation is what the tangent line would predict for f(x).

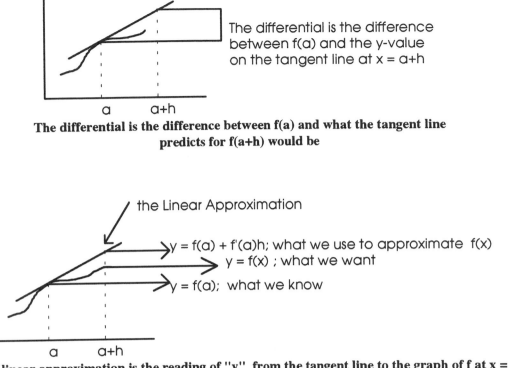

The differential is the difference
between f(a) and the y-value
on the tangent line at x = a+h

**The differential is the difference between f(a) and what the tangent line
predicts for f(a+h) would be**

the Linear Approximation

y = f(a) + f'(a)h; what we use to approximate f(x)

y = f(x) ; what we want

y = f(a); what we know

**The linear approximation is the reading of "y" from the tangent line to the graph of f at x =
a . The tangent line is given by y = f(a) + f'(a)h .**

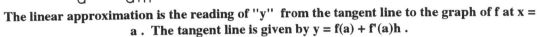

the Linear Approximation

the Differential

The difference between the Linear Approximation and the Differential

THE CASE FOR SEVERAL VARIABLES (Calculus NOW)

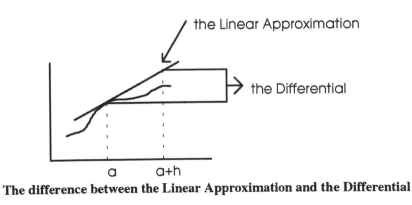

We do know f(x,y) here

The approximation

We don't know f(x,y) here, so we
approximate its value using the
tangent plane to the function
at a point where we do know f(x,y)

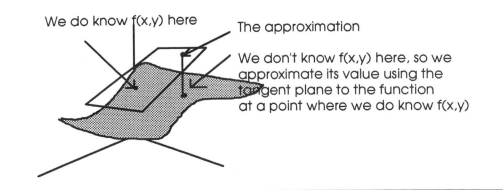

THE LINEAR APPROXIMATION:

$$f(x_0 + h, y_0 + k) \approx f(x_0, y_0) + f_x(x_0, y_0)h + f_y(x_0, y_0)k$$

which is the same as saying

$$f(x_0 + h, y_0 + k) \approx f(x_0, y_0) + \nabla f\Big|_{(x_0, y_0)} \cdot (h\mathbf{i} + j\mathbf{k})$$

In other words,
f(new point) is approximated by ....

       ... f(old point) + (gradient at old point) · (vector representing change)

THE DIFFERENTIAL:

In the expression $f(x_0 + h, y_0 + k) \approx f(x_0, y_0) + f_x(x_0, y_0)h + f_y(x_0, y_0)k$, the quantity by which we have changed $f(x_0, y_0)$ is the differential, namely,

the differential is $f_x(x_0, y_0)h + f_y(x_0, y_0)k = \nabla f\Big|_{(x_0, y_0)} (h\mathbf{i} + j\mathbf{k})$

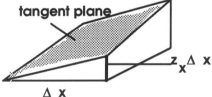

z_x is the slope of the tangent
plane in the x-direction

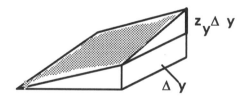

$z_y \Delta y$ is the slope of the tangent
plane in the y-direction

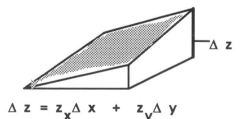

$$\Delta z = z_x \Delta x + z_y \Delta y$$

The TOTAL DIFFERENTIAL is

$$dz = z_x \, dx + z_y \, dy, \text{ or } df = f_x \, dx + f_y \, dy, \text{ or}$$

$$df = \frac{\partial f}{\partial x} dx + \frac{\partial f}{\partial y} dy$$

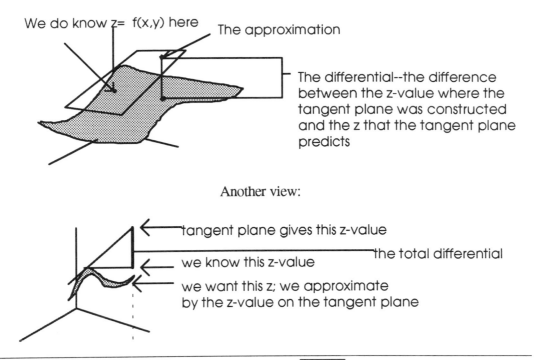

We do know z= f(x,y) here   The approximation

The differential--the difference between the z-value where the tangent plane was constructed and the z that the tangent plane predicts

Another view:

tangent plane gives this z-value

the total differential

we know this z-value

we want this z; we approximate by the z-value on the tangent plane

---

EXAMPLE: Approximate the value of $f(x,y) = \sqrt{x^2 - y^2}$ at $(2.99, 0.04)$.

Answer:

(i) Seeing the word "approximate" we use the linear approximation:
$$f(x_0 + h, y_0 + k) \approx f(x_0, y_0) + f_x(x_0, y_0)h + f_y(x_0, y_0)k$$

(ii) Choose (x_0, y_0) to be a point where we have information, say $(3, 0)$

(iii) Let h and k represent the differences between in x- and y- values between this point and our given point. So h = -.01 and k = .04.

(iv) Compute the partial derivatives:
$$f_x(x,y) = \frac{2x}{2\sqrt{x^2 - y^2}} = \frac{x}{\sqrt{x^2 - y^2}} \text{ and } f_x(3,0) = 1$$
$$f_y(x,y) = \frac{-2y}{2\sqrt{x^2 - y^2}} = \frac{-y}{\sqrt{x^2 - y^2}} \text{ and } f_y(3,0) = 0$$

(v) Plug all this into formula:
$$f(2.99, 0.04) \approx f(3,0) + 1(-.01) + 0(.04) = \sqrt{3^2 - 0^2} - .01 = 2.99.$$

EXAMPLE: Approximate the number $e^{.02} \ln(.99)$.

Answer: Let f(x,y) = $f(x,y) = e^x \ln y$ and let $(x_0, y_0) = (0,1)$, $so f(0,1) = e^0 \ln 1 = 0$.

Now h = 0.02 and k = -0.01,
$f_x = e^x \ln y$ and $f_x(0,1) = e^0 \ln 1 = 0$
$f_y = e^x(1/y)$ and $f_y(0,1) = e^0(1/1) = 1$

Then using the formula, $f(x_0 + h, y_0 + k) \approx f(x_0, y_0) + f_x(x_0, y_0)h + f_y(x_0, y_0)k$, we
have:
$e^{.02} \ln(.99) \approx e^0 \ln 1 + (0)(0.02) - (1)(0.01) = -0.01.$

EXAMPLE: Find df for $f(x, y) = \sin x^2 y$.

Answer: $df = \dfrac{\partial f}{\partial x} dx + \dfrac{\partial f}{\partial y} dy$ where $\dfrac{\partial f}{\partial x} = (\cos x^2 y)(2y)$ and $\dfrac{\partial f}{\partial y} = (\cos x^2 y)(x^2)$.

Thus, $df = (\cos x^2 y)(2xy)dx + (\cos x^2 y)(x^2)dy$.

Let x = ??

Form a Study Group! Your best learning is usually achieved
when you work with a group of others, if you meet on a regular basis. Students who work with
others have a better, more thorough understanding of the subject, and perform better on tests.

SECTION 13.8

*"If a man will begin with certainties, he shall end in doubts; but if he will be content to begin with
doubts, he shall end in certainties."...Francis Bacon*

EXTREME VALUES

This is a generalization of what we did in first year calculus when we found maxima and
minima. There, we found a continuous function on a closed interval assumed a maximum
and minimum. Definitions of a maximum (minimum) and relative maximum (minimum) for
functions of 2 or 3 variables are found in your text, in Definition 13.18. NOW:

The MAXIMUM MINIMUM THEOREM FOR TWO VARIABLES appears as
Theorem 13.21; read it carefully. The essential meaning is this: if f is continuous on a
closed bounded set in the plane, (closed means the set includes its boundary) then f has
maximum and minimum values taken on at points (x,y) in the set.

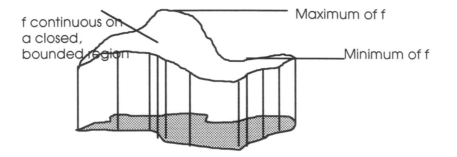

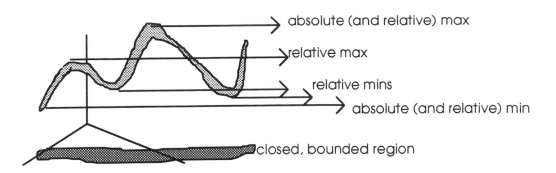

Note: there may be many <u>relative</u> maxima and minima as well as an <u>absolute</u> maximum and minimum. Even the absolute maximum and minimum may be taken on at more than one place.

METHOD: Finding a relative maximum or minimum . (We assume that our function is "nice" -- it has continuous partial derivatives and second partials that are continuous.)

1.) Find the points where the gradient is the **0** . I.e. find where both f_x and f_y are 0.

2.) The point(s) you have found is (are) possibly a relative max, a min, or....nothing at all. In first year calculus we checked the sign of the second derivative. Do you remember that result? Don't rely on this result any more! It was:

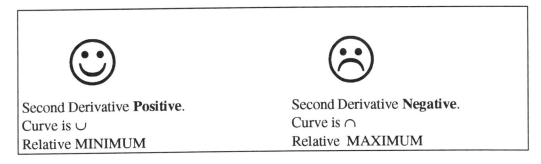

Second Derivative **Positive**.
Curve is ∪
Relative MINIMUM

Second Derivative **Negative**.
Curve is ∩
Relative MAXIMUM

Now the situation becomes far more complicated. Form the determinant:

$$D = \begin{vmatrix} f_{xx} & f_{xy} \\ f_{yx} & f_{yy} \end{vmatrix} = f_{xx}f_{yy} - f_{xyx}^{2}$$

where the mixed partials are equal because they are continuous.

Note: to find the determinant we multiply elements on the primary diagonal and subtact the product of elements on the secondary diagonal.

$$D = \begin{vmatrix} \blacksquare & \square \\ \square & \blacksquare \end{vmatrix} - \begin{vmatrix} \square & \blacksquare \\ \blacksquare & \square \end{vmatrix}$$

3.) Note the boxed term in D: $D = \begin{vmatrix} \boxed{f_{xx}} & f_{xy} \\ f_{yx} & f_{yy} \end{vmatrix}$

If $D > 0$ and $\boxed{f_{xx}} < 0$, we have a relative maximum

If $D > 0$ and $\boxed{f_{xx}} > 0$, we have a relative minimum

If $D < 0$, we have a saddle point, a "nothing".

If $D = 0$ we probably have a sheet-surface--this situation needs more work.

4.) In situations where the region in which we are working has a boundary, the Maximum - Minimum Theorem guarantees an absolute maximum and minimum (we require f to be continuous). To find the absolute max or min, we compare f values obtained using the method above with values of f on the boundary.

NOTE: The relative maximum or minimum is a **z-value**; it is taken on **AT** some (x,y).

SOME EASY EXAMPLES

If you think of these 3 **prototype** examples, you will probably remember the method above. Think of the 3 surfaces:

$(a)\quad z = x^2 + y^2 \qquad (b)\quad z = x^2 - y^2 \qquad (c)\quad z = -(x^2 + y^2)$

EXAMPLE: A local minimum. For the function $z = x^2 + y^2$, find the critical points and determine if they yield a local max, min or neither.

Answer:

Compute the first partial derivatives: $f_x = 2x$, $f_y = 2y$.

Compute the second partial derivatives: $f_{xx} = 2, f_{yy} = 2, f_{xy} = f_{yx} = 0$.

Find $D = \begin{vmatrix} \boxed{f_{xx}} & f_{xy} \\ f_{yx} & f_{yy} \end{vmatrix} = \begin{vmatrix} 2 & 0 \\ 0 & 2 \end{vmatrix} = 4 > 0$

Since the boxed position is positive as well, $f(0,0) = 0$ is a relative minimum.

EXAMPLE: A saddle point. For the function $z = x^2 - y^2$, find the critical points and determine if they yield a local max, min or neither.

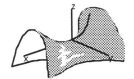

Answer: We find $f_x = 2x$ and $f_y = -2y$, and these partials are 0 at (0,0). Now since

$$D = \begin{vmatrix} f_{xx} & f_{xy} \\ f_{yx} & f_{yy} \end{vmatrix} = \begin{vmatrix} 2 & 0 \\ 0 & -2 \end{vmatrix} = -4 < 0,$$ at (0,0), we have a saddle point--the point (0,0) gives

neither a max nor a min. In fact (0,0) gives a max in one direction, a min in another.

EXAMPLE: A local maximum. For the function $z = -(x^2 + y^2)$, find the critical points and determine if they yield a local max, min or neither.

Answer: We find $f_x = -2x$ and $f_y = -2y$ are 0 at (0,0). Now

$$D = \begin{vmatrix} \boxed{f_{xx}} & f_{xy} \\ f_{yx} & f_{yy} \end{vmatrix} = \begin{vmatrix} -2 & 0 \\ 0 & -2 \end{vmatrix} = 4 > 0,$$ with the boxed position negative. So we have a

relative maximum. The maximum of f is 0, and is taken on at (0,0).

EXAMPLE: When D = 0. For the function $z = (y - x)^2$, find the critical points and determine if they yield a local max, min or neither.

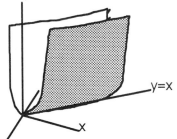

Answer: We have $f_x = -2(y - x)$ and $f_y = 2(y - x)$ and any point where y = x is a

critical point. Let's examine what these points yield: $D = \begin{vmatrix} f_{xx} & f_{xy} \\ f_{yx} & f_{yy} \end{vmatrix} = \begin{vmatrix} 2 & -2 \\ -2 & 2 \end{vmatrix} = 0$ for all of

these points. But from the picture, we see these points form the "spine" at the base of this parabolic cylinder -- they give local minima in all directions. But in the direction of the "spine", z is constant.

EXAMPLE: In general, D is not a constant: we must evaluate D at the critical point we are investigating. Let $f(x, y) = x^2 - e^{y^2}$ Find the critical points and determine if they yield a relative max, min, or saddle point.

Answer: $f_x = 2x$ and $f_y = e^{-y^2}(2y)$, so the critical point is (0,0). A chart of second

derivatives gives:

$f_{xx} = 2$	$f_{xy} = 0$
$f_{yx} = 0$	$f_{yy} = -e^{y^2}(2y)^2 + 2e^{-y^2}$

or, evaluted at $(0,0)$,

$f_{xx} = 2$	$f_{xy} = 0$
$f_{yx} = 0$	$f_{yy} = -2$

.

Thus $D = \begin{vmatrix} f_{xx} & f_{xy} \\ f_{yx} & f_{yy} \end{vmatrix} = \begin{vmatrix} 2 & 0 \\ 0 & -2 \end{vmatrix} = -4 < 0$. So f has a saddle point at (0,0).

EXAMPLE: Let $f(x,y) = x^2 + 6xy + y^2 - 3y + 5$. Find the critical points and determine if they yield a relative max, min, or saddle point.

Answer: $f_x = 2x + 6y$ and $f_y = 6x + 2y - 3$. We have two simultaneous equations:

2x + 6y = 0
6x + 2y = 3

and solving, the critical point is x = 9/16, y = -3/16 .

Now $D = \begin{vmatrix} f_{xx} & f_{xy} \\ f_{yx} & f_{yy} \end{vmatrix} = \begin{vmatrix} 2 & 6 \\ 6 & 2 \end{vmatrix} = 4 - 36 < 0$. So we have a saddle point at (0,0).

Note: When we have a quadric surface with a saddle point, we know it is some type of hyperboloid.

EXAMPLE: Find the extreme values of $f(x,y) = y^2 - x^2$ on the disk $x^2 + y^2 \le 3^2$.

Answer: The critical values we have seen above occur at (0,0) which give a saddle point. By the Maximum-Minimum Theorem, however, f must have a maximum and a minimum in this region, and since they are not occurring inside the disk, they must occur on the boundary. When $x^2 + y^2 = 3^2$,
$$f(x,y) = y^2 - x^2 = (3^2 - x^2) - x^2 = 9 - x^2$$
where x is in [-3,3]. This is a single-variable problem now. The minimum value of f occurs when x = ±3, with y = 0; there f(±3,0)= 0. The maximum f occurs when x = 0, with y = ±3. There, f (0,±3) = 9.

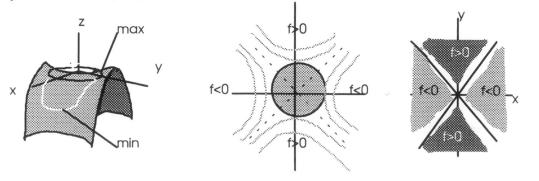

 **WORD PROBLEMS!!**

EXAMPLE: Find three non-negative numbers whose sum is 24 and whose product is as large as possible, and find their product.

Answer: Let the numbers be x,y,z. Then $x + y + z = 24$. Let $P = xyz$, or replacing z, $P = xy(24-x-y)$, or $P = 24xy - x^2y - xy^2$. Now we want (x,y) such that $P_x = P_y = 0$. We require:

$P_x = 24y - 2xy - y^2 = y(24 - 2x - y) = 0$ so $y = 0$ or $24 - 2x - y = 0$

$P_y = 24x - x^2 - 2xy = x(24 - x - 2y) = 0$ so $x = 0$ or $24 - x - 2y = 0$.

Note: We have the following candidates: (0,0), (0, 24), (24,0). But all of these yield a product $P = 0$. The last candidate is found by solving simultaneously:

$y + 2x = 24$
$2y + x = 24$

from which we obtain $(x,y) = (8,8)$. We check D, using

$P_{xx} = -2y$	$P_{xy} = 24 - 2x - 2y$
$P_{yx} = 24 - 2x - 2y$	$P_{yy} = -2x$

At (8,8), $D = \begin{vmatrix} -16 & -8 \\ -8 & -16 \end{vmatrix} > 0$ with $f_{xx} < 0$, giving a relative max. Will (8,8) yield the absolute maximum value of P? What is the "boundary" of the region in question? We have $x \geq 0$, $y \geq 0$, and the restriction that $x + y + z = 24$, or that $z = 24 - x - y \geq 0$, or that $x + y \leq 24$. We can't draw $P(x,y,z) = xyz$--its graph is in 4-dimensions! But we draw the x-y region in the plane:

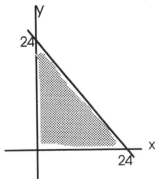

Since $P = 0$ on the boundary of this region, it will not provide the maximum . Thus, P is maximal when $x = 8$, $y = 8$, and $z = 8$. The maximum product is 512.

EXAMPLE: An open-topped box has a volume of 4 cu. m. Find the dimensions of such a box having the smallest possible surface area.

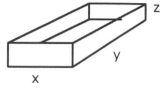

Answer: Letting V = volume, V= xyz = 4. Now surface area, S, is given as
S = 2xz + 2yz + xy. From the volume equation, z = 4/xy. Substituting, S = 8/y + 8/x + xy.
Now $S_x = -8/x^2 + y = 0$ and $S_y = -8/y^2 + x = 0$ from which we have that $yx^2 = xy^2 = 8$
and x = y. Thus x = y = 2 (from either equation), and z = 1. The graph of z = 1/xy is given
in Section 13.1, and the graph of z = 4/xy is similar. There is no boundary for the set of
(x,y) values in this problem, but we see from the fact that S = 8/y + 8/x + xy that as x or y
approach 0, we get increasing surface area. Thus (2,2) must be the point at which the
minimum surface area occurs. This minimum surface area is S = 8/2 + 8/2 + (2)(2) = 12.

SECTION 13.9

*"Always fall in with what you're asked to accept. Take what is given, and make it over your way.
My aim in life has always been to hold my own with whatever's going. Not against: with
it."...Robert Frost*

LAGRANGE MULTIPLIERS

"I do not know"...Lagrange(1736-1813)

In this section, we maximize and minimize a surface subject to constraints. For functions of
two variables, the constraint is usually that (x,y) lie on a curve.

For example, we might want the maximum and minimum of $z = y^2 - x^2$ over the circle
$x^2 + y^2 = 9$ -- meaning that (x,y) must lie on the circle.

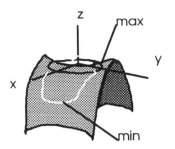

THE METHOD: Suppose we are to maximize or minimize f, a function of x and y, subject to a boundary constraint given implicitly by g(x,y) = 0. Then we set

$$\text{grad } f = \lambda \text{ grad } g$$

where λ is some parameter we do not know. This equation puts conditions on λ in terms of x and y, and/or constraints on x and y. We investigate all possibilities by plugging in feasible solutions into the equation for f.

NOTE: The rationale for this equation is as follows: at a maximum or minimum, **grad f**, which points in the direction of steepest ascent of the function, should be perpendicular to the level surfaces. A vector perpendicular to the level surface is given by **grad g**. This is why we seek points for which **grad f** is some multiple of **grad g**. If **grad f** were not perpendicular to the level surface, but had a tangential component along the level surface, then f would be larger (or smaller) in some other direction.

EXAMPLE: Let $f(x,y) = y + x^2$. Find the extreme values of f subject to the given constraint: $x^2 + y^2 \le 4$. Note: g(x,y) = $x^2 + y^2 - 4$.

Answer:
i) Check the interior. (Here is where relative max/mins show up.) The interior of the disk consists of points (x,y) such that $x^2 + y^2 < 4$. Since grad f = (2x) **i** + **j** , which is never the **0**-vector, there is no relative maximum or minimum. Therefore the absolute max and min must fall on the boundary of the disk, namely, the circle $x^2 + y^2 = 4$

(ii) Check the boundary. Form the quantity
$$\text{grad } f = \lambda \text{ grad } g$$
or here,   2x **i** + **j** = λ (2x **i** + 2y **j**). Then
$$2x = \lambda(2x) \text{ and}$$
$$1 = \lambda(2y)$$

(iii) Find what these equations say about λ, x, and y to obtain critical pints. From the first
equation, λ=1, or x = 0 , and from the second equation, λ = 1/(2y).

➔ **If x = 0,** from the graph of the circle, y = 2. This gives: $\boxed{f(0,2) = 2}$.

➔ **If λ = 1,** then from the second equation, y = 1/2. And from the equation of the circle,
     $x = \pm\sqrt{4 - (1/2)^2} = \pm\sqrt{15}/2$. This gives $\boxed{f(\pm\sqrt{15}/2,\ 1/2) = 17/4}$.

➔ **If λ = 1/(2y),** then from the first equation, we must have y = 2 or x = 0, situations already
     investigated.

(iv) Compare the value of f at all points found on the interior and boundary; the largest value of f is the maximum and the smallest, the minimum. So the minimum is $\boxed{f(0,2)=2}$ and the maximum, $\boxed{f(\pm\sqrt{15}/2,\ 1/2)=17/4}$.

EXAMPLE: Find the extreme values of $f(x,y)=x^3+x^2+\dfrac{y^2}{3}$ given the constraint $x^2+y^2\le 36$.

Answer:

(i) First locate critical points in the interior of the disk by setting first order partials to 0.

$f_x=3x^2+2x=0$ when $x=0$, or $x=-2/3$. For these values of x, y = 0.
$f_y=2y/3=0$ when y = 0. For this value of y, x = 0.

(ii) Use critical points (0,0) and (-2/3, 0) in $D=\begin{vmatrix} f_{xx} & f_{xy} \\ f_{yx} & f_{yy} \end{vmatrix}=\begin{vmatrix} 6x+2 & 0 \\ 0 & 2/3 \end{vmatrix}$.

➔ At **(0,0)**, $D=(6(0)+2)(2/3)-0=4/3>0$, and $f_{xx}=2>0$. **So we have a relative minimum: f(0,0) = 0.**

➔ At **(-2/3,0)**, $D=(6(-2/3)+2)(2/3)-0=-4/3$, which gives a saddle point.

(iii) Now look at the boundary: write $g(x,y)=x^2+y^2-36$. Now we set

$$\text{grad } f = \lambda \text{ grad } g$$

from which we find $3x^2+2x=\lambda(2x)$ and $2y/3=\lambda(2y)$. From the second equation, y = 0 or $\lambda=1/3$.

(iv) Get the information from the above equation. We will put it in a table.

➔ **If y = 0** then in the circle equation, x = ±6. We will investigate the behavior of f at (±6,0).

➔ **If $\lambda=1/3$** then in the first equation, $3x^2+2x=2x/3$ or $3x^2+4x/3=0$, and x = 0 or x = -4/9. If x = 0, then y = ±6, so we will investigate the behavior of f at (0,±6). If x = -4/9, from the equation of the circle, $y=\pm(10/9)\sqrt{29}$, so we will investigate (-4/9, $\pm(10/9)\sqrt{29}$).

(v) Compare. Putting all critical points together with the interior point (0,0) we have:

(x,y)	f(x,y)	Classification
(0,0)	0	Relative Minimum
(6,0)	252	Absolute Maximum
(-6,0)	-180	Absolute Minimum
(0,6)	12	*
(0,-6)	12	*
(-4/9, $\pm(10/9)\sqrt{29}$)	107.9	*

(*) These are not critical points.

(vi) Conclude. The absolute maximum of f is 252, occurring at (6,0) and the absolute minimum, -180 occurring at (-6,0).

Can you:

1. Find the change along the plane tangent to a surface at a given point, given a small change in x and a small change in y? (This is finding the differential.)
2. Find a linear approximation to a function near a given point?
3. Use linear approximations in a word problem--for example, the volume of a cylinder   is increasing as a function of r as well as h? Estimate the change in the volume if the radius is increased by ... and the height is increased by ...
4. Find the relative maxima and minima of a given function?
5. State a theorem that guarantees when a function has an absolute maximum and minimum?
6. Use the method of Lagrange multipliers to find where the absolute maximum and minimum for a function on a constrained region?

Taking Stock

What skills from this chapter do you feel confident about? What do you intend to do to improve areas of potential weakness?

Type of Problem	Do you feel in control?	Plan of action
Differentiating skills from 1st year calculus?		
Finding partial derivatives, second partials?		
Definitions (e.g. differentiability, the limit definition of the partial derivative, "saddle point", etc.) and theorems and conditions (when is a function differentiable, when is an absolute max/min guaranteed, etc.?)		
Graphing functions		
Finding the differential, the linear approximation to a function ...		
Finding relative max/mins; using "D" to determine them		
Using Lagrange parameters/multipliers		
Finding gradient, directional derivative		
Finding the tangent plane, the normal vector, etc.		

Summarizing : We have seen the following concepts, related to the
first year calculus notion of derivative:

      ❶ partial derivatives

      ❷ gradient

      ❸ directional derivative

      ❹ the total differential (df)

      ❺ tangent plane

      ❻ differential

      ❼ linear approximation to the function

      ❽ determinant of second partials

For every concept, discuss: **WHAT** it is, **WHY** it's important, and **WHEN** we use it.

⌘ Give an Example of ⌘

⌘ A function of x and y which has a saddle point and no absolute maximum and minimum.

⌘ A function of x and y which has an absolute maximum; one with an absolute minimum.

⌘ A function of x and y which has no minimum.

⌘ A function of x and y which takes on its maximum and minimum on the boundary of a
     closed, bounded region, but not on the interior of this region.

⌘ A situation where you might approximate a function of x and y by values from the
     tangent plane at a point.

⌘ A word problem in which it is necessary to find the maximum or minimum of a function
     of x and y.

A NOTE ON MEMORIZATION

Generally, we are told that memorization is an inferior method of learning, a "rote" type of
learning, and we may feel there's no point in memorizing, because we're going to forget the
material anyway. However, memorization is extremely important in mathematics. We don't
memorize as an END in itself, but as a MEANS for learning. We can't do mathematics unless we
have a source of ideas, definitions, theorems, and some recall of previously solved problems to fall
back on. After all, we memorize a great deal more in a history, language, physics, or engineering
courses. Be sure you don't succumb to the idea that anything important can be looked up.
Highlighted areas of the text are not for back-up reference; they should be learned and memorized.
If you ROUTINELY memorize important information as you go along, rather than to cram at the
last minute, memorization will feel more naturally like part of the learning process, and what you
memorize will stay with you.

CHAPTER 14

"[Mathematics'] characters are triangles, circles, and other geometrical figures, without which it is humanly impossible to understand a single word of it; without these, one is wandering about in a dark labyrinth."...Gallileo Gallilei

Multiple integrals are used extensively in the real world, where problems have many variables, generally not just two or three. With practice, you will eventually see some methods for setting up multiple integrals . But don't lose sight of the theory. Keep asking questions. Some of the results in this chapter have proofs that are at a level beyond us at this stage ; other remarkable results can be obtained swiftly . Don't be surprised if, at the beginning, you can't seem to visualize the graphs of functions in 3-space. There is no special ability or talent here that experience cannot help you gain.   Practice a few good examples carefully; rehearse in your mind the graphs of surfaces you saw in Section 13.1-- and the visualization will come.

SECTION 14.1

"Scholarship is less than sense. Therefore seek intelligence"...the Panchatantra

MULTIPLE INTEGRALS

In first year calculus we found the area under a curve. Here we find the volume under a surface.

HIGHLIGHTS OF THE THEORY

Read in the text the definition of the multiple integral of f(x,y) over the region R.

(i) For the integral to exist, R must be a bounded region in the xy-plane and f must be continuous on R. Denoted $\iint\limits_R f(x,y)dA$, the integral is defined to be the unique quantity between the lower and upper sums of all partitions of R.

(ii) When f is non-negative, the integral $\displaystyle\iint_R f(x,y)\,dA$ can be interpreted as the volume

under f, over the region R.

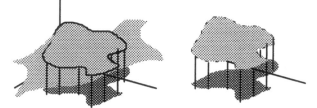

(iii) An alternative definition to that in (i) is the Riemann Sum definition, similar to that which we saw in Calculus I.

$$\iint_R f(x,y)\,dA = \lim_{\|P\|\to 0}\sum_{k=1}^{n} f(x_k, y_k)\,\Delta A_k.$$

(Understand what all the notation means. See your text!)

To find the volume under the surface, we take a limit of the sum of the volumes of many small rectangular solids.

EXAMPLE: In the diagrams below, Let R be the triangular region and R' the rectangular region . Let f(x,y) = x + y over the region R. (The graph of f is a plane.) Let P be a partition of R' consisting of squares with sides 1 unit long. Find the upper and lower sums, $U_f(P)$ and $L_f(P)$ for f over the region R.

Answer:

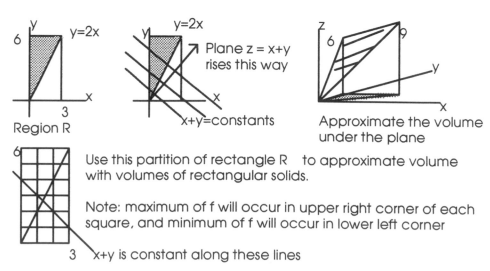

Region R

Plane z = x+y rises this way

x+y=constants

Approximate the volume under the plane

Use this partition of rectangle R   to approximate volume with volumes of rectangular solids.

Note: maximum of f will occur in upper right corner of each square, and minimum of f will occur in lower left corner

x+y is constant along these lines

Answer:

$U_f(P)$: The Upper Sum of this partition will be computed using base areas equal to 1 ($\Delta A_k = 1$) and the height of each rectangular solid given by $M_i = f$(upper right hand corner point). Using the grid below as a guide,

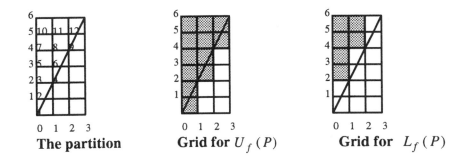

The partition   **Grid for** $U_f(P)$ **Grid for** $L_f(P)$

$U_f(P) = M_1\Delta A_1 + M_2\Delta A_2 + M_3\Delta A_3 + ... + M_n\Delta A_n$

$U_f(P) = $ f(1,1)1 + f(1,2) 1 + f(1,3) 1 + f(1,4)1 + f(1,5)1 + f(1,6)1 + f(2,4)1 + f(2,3) 1 +

     f(2,5)1 + f(2,6)1 + f(3,5)1 + f(3,6)1

    $=$ (1+1) + (1+2) + (1+3) + (1+4) + (1+5) + (1+6) + (2+4) + (2+5) + (2+6) +

     (2+3)1 + (3+5) + (3+6) = 70.

$L_f(P)$: In finding the Lower Sum, only rectangular solids completely in the region R are used. The Lower Sum is computed using $\Delta A_k = 1$ and as height, $m_i = f$(lower left hand corner):

$L_f(P) = m_3\Delta A_3 + m_4\Delta A_4 + m_5\Delta A_5 + ... + m_n\Delta A_n$

$L_f(P) = $ f(0,2)1 + f(0,3)1 + f(0,4)1 + f(0,5)1 + f(1,4)1 + f(1,5)1

    $=$ 2 + 3 + 4 + 5 + 6 = 46.

NOTE: When the function and the region are suitable, the double integral can be found by evaluating the "iterated" integral. This means we can evaluate the double integral by evaluating nested single integrals, one at a time, generally working from the inside to the outside integral.

EXAMPLE: Evaluate the iterated integral: $\displaystyle\int_1^3 \int_0^2 yx^2\,dx\,dy$.

Answer:

$$\int_1^3 \left(\int_0^2 yx^2\,dx\right)dy = \int_1^3 y\cdot\left(\frac{x^3}{3}\Big|_0^2\right)dy = \int_1^3 y(\frac{8}{3}-0)\,dy = \frac{8}{3}\int_1^3 y\,dy = \frac{8}{3}\left(\frac{x^2}{2}\right)\Big|_1^3 = \frac{8}{3}\left(\frac{9}{2}-\frac{1}{2}\right) = \frac{32}{3}.$$

EXAMPLE: Evaluate $\displaystyle\int_{4}^{5}\int_{0}^{x} e^{x+y}\,dy\,dx$.

Answer:

$$\int_{4}^{5}(\int_{0}^{x} e^{x} e^{y}\,dy)\,dx = \int_{4}^{5} e^{x}\left(e^{y}\Big|_{0}^{x}\right)\,dx = \int_{4}^{5} e^{x}(e^{x}-1)\,dx = \int_{4}^{5} e^{2x}-e^{x}\,dx = \left(\frac{e^{2x}}{2}-e^{x}\right)\Big|_{4}^{5} =$$

$$e^{10}/2 - e^{5} - e^{8}/2 + e^{4}.$$

VERTICALLY AND HORIZONTALLY SIMPLE REGIONS
We find the integral in different ways over these different regions R:

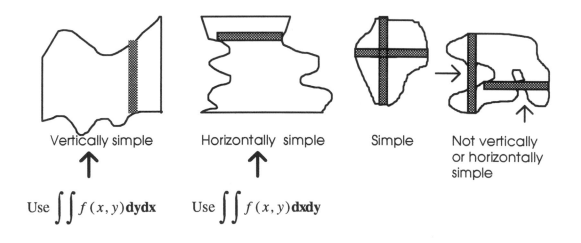

Vertically simple   Horizontally simple   Simple   Not vertically or horizontally simple

Use $\displaystyle\iint f(x,y)\,\mathbf{dy\,dx}$ Use $\displaystyle\iint f(x,y)\,\mathbf{dx\,dy}$

EXAMPLE: Region is Simple. Let A be the region bounded by the lines y = 2x, x = 1, x = 4. Express the double integral $\displaystyle\iint_{R} (x+y)\,dA$ as an iterated integral .

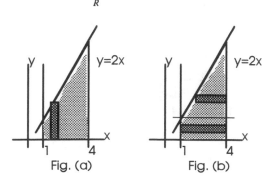

Fig. (a)   Fig. (b)

Answer: This region is simple. We can use either method (a) or (b) below corresponding to the diagrams above: (a) is easier.

$(a) \displaystyle\iint_{R} (x+y)dA = \int_{1}^{4}\int_{0}^{2x}(x+y)\,dydx$ where we integrate with respect to y first.

From Fig.(a), y ranges from its lower limit 0 to its upper limit y = 2x; then x ranges from 1 to 4.

$(b) \displaystyle\iint_{R} (x+y)dA = \int_{0}^{2}\int_{1}^{4}(x+y)\,dxdy + \int_{2}^{8}\int_{y/2}^{4}(x+y)\,dxdy$

Referring to Fig. (b), we divide the integral into a sum. In the first integral, we integrate over a rectangle, where x ranges from 1 to 4, and y from 0 to 2. In the second integral, we integrate over a triangle, where x goes from y / 2 (the line y = 2x) to 4, and y from 2 to 8.

EXAMPLE: Express $\displaystyle\iint_{R} (x^2 + y)dA$ as an iterated integral over the region R which

is bounded by the graphs of $y = 1 + x^2$ and $y = 9 - x^2$.

Answer: Graph the region.

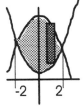

-2   2

We will integrate with respect to y first. Intersections of the region come at

$1 + x^2 = 9 - x^2$ or $2x^2 = 8$ or $x = \pm 2$. The integral is $\displaystyle\int_{-2}^{2}\int_{1+x^2}^{9-x^2} x^2 + y \, dydx$.

EXAMPLE: Express $\displaystyle\iint_{R} (x^2 + y)dA$ as an iterated integral over the region R,

bounded by y = x, y = 1 and y = 5-x.

Answer: Find the intersection points of these lines.

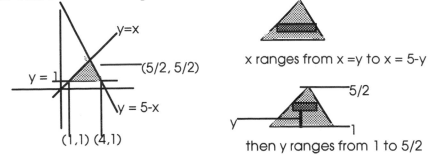

y=x

y = 1   (5/2, 5/2)

y = 5-x

(1,1) (4,1)

x ranges from x =y to x = 5-y

5/2

1

then y ranges from 1 to 5/2

Then integrating first with respect to x, we have $\iint\limits_{R} (x^2 + y)dA =$

$$\int\limits_{1}^{5/2} \int\limits_{y}^{5-y} (x^2 + y)\, dxdy \ .$$

EXAMPLE: Express as an integral the volume of the solid region bounded by the coordinate planes and the planes x + y = 1 and x + 4y + 2z = 6.

Answer: Note that z is not restricted by the plane x + y = 1, which in the usual orientation of axes, is vertical. The plane $z = 3 - \dfrac{1}{2}x - 2y$ provides either the "roof" or "floor" of this region: we need to know where z is positive ("roof") and where z is negative ("floor"). Since z will be positive if $\dfrac{1}{2}x + 2y < 3$, or x + 4y < 6 (which is left of the line x + 4y = 6) we see that z is positive over the entire shaded region. This region is vertically simple.

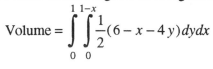

$$\text{Volume} = \int\limits_{0}^{1} \int\limits_{0}^{1-x} \frac{1}{2}(6 - x - 4y)\,dydx$$

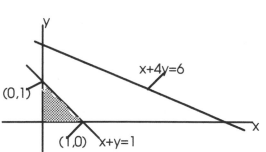

EXAMPLE: Express the volume of the region bounded above by the sphere $x^2 + y^2 + z^2 + 1$ and below by the xy-plane.

Answer: We find the volume of this hemisphere by first finding the volume under the sphere over the first quadrant of the xy-plane, then multiplying the result by 4.

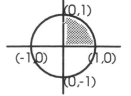

$$V = 4\int_0^1 \int_0^{\sqrt{1-x^2}} \sqrt{1-x^2-y^2}\,dy\,dx =$$

$$4\int_0^1 \int_0^{\sqrt{1-x^2}} \sqrt{(1-x^2)-y^2}\,dy\,dx = 4\int_0^1 \int_0^a \sqrt{a^2-y^2}\,dy\,dx \text{ where we let } a^2 = 1-x^2.$$

(In the first quadrant, this term is always non - negative.)

$$= 4\int_0^1 a \int_0^{\pi/2} \sqrt{a^2 - a^2 \sin^2\theta}\,(\cos\theta)\,d\theta\,dx \text{ where we let } y = a\sin\theta \text{ and } dy = a\cos\theta\,d\theta.$$

$$4\int_0^1 a^2 \int_0^{\pi/2} \cos^2\theta\,d\theta\,dx = 4\int_0^1 a^2 \int_0^{\pi/2} \frac{1+\cos 2\theta}{2}\,d\theta\,dx = 4\int_0^1 a^2 \left(\frac{\theta}{2}+\frac{\sin 2\theta}{4}\right)\Bigg|_0^{\pi/2} dx = 4\cdot\frac{\pi}{4}\int_0^1 a^2\,dx =$$

$$\pi\int_0^1 1-x^2\,dx = \pi(x-\frac{x^3}{3})\Bigg|_0^1 = \frac{2\pi}{3}$$

This is the volume of a hemisphere of radius 1. (The volume of a sphere is $V = \frac{4}{3}\pi r^3$.)

EXAMPLE: The iterated integral $\displaystyle\int_{-4}^{4} \int_{-\sqrt{4^2-x^2}}^{\sqrt{4^2-x^2}} 5\,dy\,dx$ represents the volume of a solid region,

D. Sketch the region D.

Answer: The height of the solid region can be taken as $z = 5$. The region that is the base of the solid is a disk of radius 4. Thus the solid is a cylinder with height equal to 5.

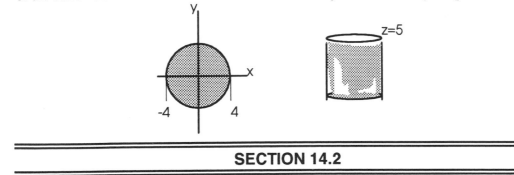

SECTION 14.2

"Let no one unversed in geometry enter here"...over the gates of Plato's Academy

DOUBLE INTEGRALS IN POLAR COORDINATES

When we deal with circular areas, we change coordinates from Cartesian to **polar**; with this change comes a change in the expression dA under the integral: we use

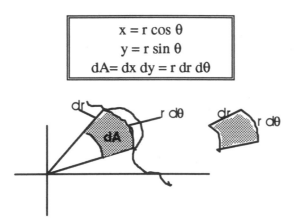

$$x = r \cos \theta$$
$$y = r \sin \theta$$
$$dA = dx \, dy = r \, dr \, d\theta$$

EXAMPLE: Express the integral below in terms of polar coordinates, and evaluate.

$\displaystyle\int_R (x^2 + y^2)^{1/2} \, dA$, where R is the region bounded by the limacon $r = 2 + \sin \theta$.

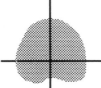

Answer: With the substitution above, we recognize that $x^2 + y^2 = r^2$, so the integral

becomes $\displaystyle\int\int_{?\ ?} (r^2)^{1/2} \, r dr d\theta = \int\int_{?\ ?} r^2 dr d\theta$ **where we have replaced dxdy by r drdθ.**

$$\boxed{\textbf{DON'T FORGET the "r" in the expression r drd}\theta\textbf{!}}$$

To find the limits of integration, we note that over the region described by the limacon, r ranges from 0 to $2 + \sin \theta$ and θ from 0 to 2π. Thus, we have

$$\int_0^{2\pi} \int_0^{2+\sin\theta} r^2 dr d\theta = \int_0^{2\pi} \frac{r^3}{3}\Big|_0^{2+\sin\theta} d\theta = \frac{1}{3}\int_0^{2\pi} (2 + \sin\theta)^3 \, d\theta = \frac{1}{3}\int_0^{2\pi} (8 + 12\sin\theta + 6\sin^2\theta + \sin^3\theta) \, d\theta =$$

$$\frac{1}{3}\int_0^{2\pi} [8 + 12\sin\theta + 3(1 - \cos 2\theta) + \sin\theta(1 - \cos^2\theta)] d\theta =$$

$$\frac{1}{3}(8\theta - 12\cos\theta + 3\theta - \frac{3}{2}\sin 2\theta - \cos\theta + \frac{\cos^3\theta}{3})\Big|_0^{2\pi} = \frac{11\theta}{3}\Big|_0^{2\pi} = \frac{22\pi}{3}$$

(We have shortened our work by recognizing that all sine and cosine terms cancel when integrated between 0 and 2π.)

EXAMPLE: Find an integral that expresses the solid region inside the sphere $x^2 + y^2 + z^2 = 9$, outside the cylinder $x^2 + y^2 = 1$, and above the xy-plane.

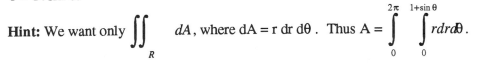

Hint: The shaded "washer" at right, above, represents the area over which we integrate z $= \sqrt{9 - x^2 - y^2}$. Notice that r ranges from an inner radius of 1 (the radius of the cylinder) to an outer radius of 3 (the radius of the sphere). Thus

$$V = \iint_{?\ ?} \sqrt{9 - x^2 - y^2}\, dxdy = \int_0^{2\pi} \int_1^3 \sqrt{9 - r^2}\, rdrd\theta$$. This can be integrated with the

substitution u = r^2.

EXAMPLE: Find the area in polar coordinates of the region interior to the cardioid r = 1+sin θ.

Hint: We want only $\iint_R dA$, where dA = r dr dθ . Thus A = $\int_0^{2\pi} \int_0^{1+\sin\theta} rdrd\theta$.

EXAMPLE: Change the integral $\int_{-3}^{3} \int_0^{\sqrt{9-x^2}} e^{\sqrt{x^2+y^2}}\, dxdy$ to one in terms of polar coordinates.

Answer: The region of integration is the upper half of a disk of radius 3:

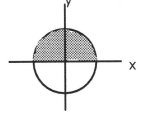

Thus "r" will range from 0 to 3 and "θ" from 0 to π. The integral is $\int_0^{\pi} \int_0^3 e^r rdrd\theta$.

SECTION 14.3

"If you can meet with Triumph and Disaster/ And treat those two impostors just the same"
..Rudyard Kipling

SURFACE AREA

The derivation of the formula for surface area is fascinating and beautiful. Read it carefully in your text.

$$S = \iint_R \sqrt{f_x^2 + f_y^2 + 1}\, dA$$

EXAMPLE: Find the integral that expresses the surface area of the portion of the paraboloid $z = 4 - x^2 - y^2$ that is above the plane $z = 1$.

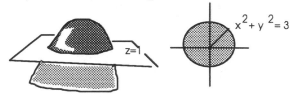

Answer: The paraboloid intersects the plane when $1 = 4 - x^2 - y^2$, or $x^2 + y^2 = 3$, so the region over which we integrate is the disk given by: $x^2 + y^2 \leq 3$. Now in Cartesian coordinates $S = \iint_R \sqrt{(-2x)^2 + (-2y)^2 + 1}\, dxdy$. This integral is difficult to evaluate, and

so we rewrite the integral instead in polar coordinates: $\displaystyle\int_0^{2\pi}\int_0^{\sqrt{3}} \sqrt{4r^2 + 1}\ rdrd\theta$.

(To evaluate this integral, let $u = 4r^2 + 1$.)

EXAMPLE: Find the integral that expresses the surface area of the portion of the sphere $x^2 + y^2 + z^2 = 9$ that is inside the cylinder $x^2 + y^2 = 1$.

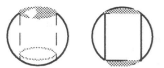

Answer: First we will set up the integral in Cartesian coordinates and convert to polar coordinates. We use z which describe the top hemisphere; then multiply our result by 2.

$$z = f(x,y) = \sqrt{9 - x^2 - y^2} \text{ so } f_x = \frac{-2x}{2\sqrt{9 - x^2 - y^2}} = \frac{-x}{\sqrt{9 - x^2 - y^2}}; \quad f_y = \frac{-y}{\sqrt{9 - x^2 - y^2}}$$

Thus $S = 2\iint\limits_{R} \sqrt{f_x^2 + f_y^2 + 1} \, dA =$

$$2\int\limits_{0}^{2\pi}\int\limits_{0}^{1} \frac{3}{\sqrt{9 - r^2}} \, r\, dr d\theta, \text{ which we evaluate with the substitution } u = 9 - r^2. \text{ Thus}$$

$$2\iint\limits_{?\ ?} \sqrt{\left(\frac{-x}{\sqrt{9 - x^2 - y^2}}\right)^2 + \left(\frac{-y}{\sqrt{9 - x^2 - y^2}}\right)^2 + 1} \, dxdy = 2\iint\limits_{?\ ?} \sqrt{\frac{r^2}{9 - r^2} + 1} \, r\, dr d\theta =$$

$$S = 2\left(-\frac{1}{2}\int\limits_{0}^{2\pi}\int\limits_{9}^{8} 3u^{-1/2} \, dud\theta\right) = -3\int\limits_{0}^{2\pi} 2\sqrt{u}\Big|_{9}^{8} \, d\theta = -6(\sqrt{8} - 3)(2\pi) = 12\pi(3 - 2\sqrt{2}).$$

SECTION 14.4

"The negligent who say: 'Some day, some other day--the thing is petty, small, Demands no thought at all,' Are, heedless, headed straight For that repentant state That ever comes too late"...the Panchatantra, c. 200 B.C.

TRIPLE INTEGRALS

Triple integrals generalize what we did with double integrals. In the examples below we will use the triple integral for

(i) **Evaluating volumes of 3-dimensional objects.** In this case,

$$V = \iiint\limits_{?} dV = \iiint\limits_{?\ ?\ ?} dxdydz \ .$$

But watch out! Sometimes it will be easier to use $\iiint\limits_{?\ ?\ ?} dzdydx$, or variations, depending

on the solid.

(ii) **Finding the volume "under" a 4-dimensional surface** and over a 3-dimensional

region. In this case the integral is V = $\iiint\limits_{?\ ?\ ?} f(x,y,z) \, dzdydx$ (we may rearrange the

order of dz, dy,dx). The function w = f(x,y,z) gives a mapping whose graph would be in 4-space. How do we visualize these spaces? We don't!

NOTE: When finding the limits of integration of multiple integrals such as

$$\int_?\int_?\int_? f(x,y,z)\ dzdydx$$, ask: what does z have to do with x and y? Then -- what does y

have to do with x? The limits of the **final** integration should be numbers. Often we "stack"
the order of integration so that the inner integrals have more variables and the outer
integrals, fewer.

I can't visualize these pictures in 3-dimensional space!

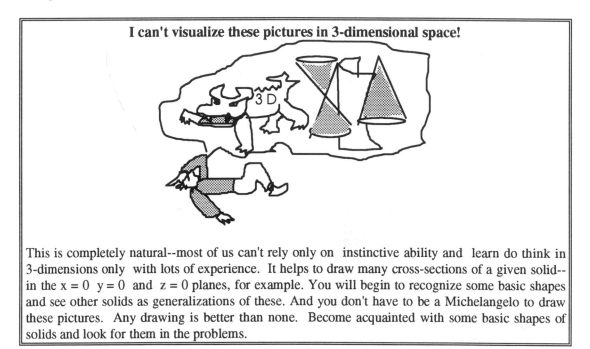

This is completely natural--most of us can't rely only on instinctive ability and learn do think in
3-dimensions only with lots of experience. It helps to draw many cross-sections of a given solid--
in the x = 0 y = 0 and z = 0 planes, for example. You will begin to recognize some basic shapes
and see other solids as generalizations of these. And you don't have to be a Michelangelo to draw
these pictures. Any drawing is better than none. Become acquainted with some basic shapes of
solids and look for them in the problems.

NOTE: This is a hint regarding the graphing of straight lines (*it's never too late!*). The
vector a **i** + b **j** is normal to the straight line ax + by = c. In the same way, ai + bj + ck is
normal to the plane ax + by + cz = d.

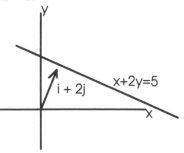

EXAMPLE: Evaluate the iterated integral: $\displaystyle\int_0^1 \int_1^{2z} \int_0^y (x-yz)\,dx\,dy\,dz$.

Answer:

$$\int_0^1 \int_1^{2z} \int_0^y (x-yz)\,dx\,dy\,dz = \int_0^1 \int_1^{2z} \left(\int_0^y x-yz \,\boxed{dx} \right) dy\,dz = \int_0^1 \int_1^{2z} \left(\frac{x^2}{2} - xyz \right) \Bigg|_0^y \, dy\,dz =$$

$$\int_0^1 \int_1^{2z} \left(\frac{y^2}{2} - y^2 z \right) \boxed{dy} \, dz = \int_0^1 \left(\frac{y^3}{6} - \frac{y^3}{3} z \right) \Bigg|_1^{2z} dz = \int_0^1 \left(\frac{8z^3}{6} - \frac{8z^4}{3} - \frac{1}{6} + \frac{z}{3} \right) \boxed{dz} =$$

$$\int_0^1 \left(\frac{4z^3}{3} - \frac{8z^4}{3} - \frac{1}{6} + \frac{z}{3} \right) dz = \frac{z^4}{3} - \frac{8z^5}{15} - \frac{z}{6} + \frac{z^2}{6} \Bigg|_0^1 = -\frac{1}{5} .$$

EXAMPLE: Find limits for the integral $\displaystyle\iiint_D e^z dV$ where D is the solid region bounded by the planes $x + y + z = 4$, $y = x$, $x = 1$, $x = 2$, $z = 0$, $y = 0$.

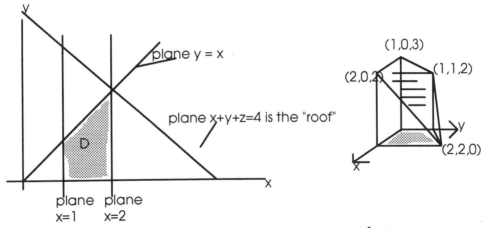

Answer: D is the solid region over which we integrate $f(x,y,z) = e^z$. The actual graph of w $= f(x,y,z)$ lies in 4-dimensions.

1.) Finding the limits of integration for $\displaystyle\iiint_D e^z \textbf{dxdydz}$ with the order of integration dx dy dz, would require expressing this integral as a sum of 2 integrals, over the 2 pieces of D, below, just to do the integration, dxdy. Even still, x and y would have to be expressed in terms of z, for limits of z as **numbers.**

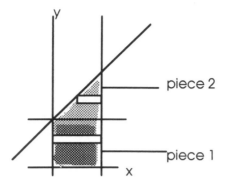

As a general rule, it is a good idea to stack the order of integration in a way such that the outer limits rely on fewer variables.

2.) Now if we integrate first with respect to y and z, our integral is $\displaystyle\int_{?}^{} \int_{0}^{4-x-y} \int_{0}^{x} e^z \,\mathbf{dy\,dz\,dx}$.

The outer limits (the limits for x) must be constant. But after integrating with respect to y and z, we would be left with an integral of the form:

$$\int_{\text{constant}}^{\text{constant}} (\text{Function of x and y}) \, dx$$, which again would not be hard to find.

3.) As a result of these considerations, the integral is best expressed when the order of integration is dz dy dx. And our integral becomes $\displaystyle\int_{1}^{2} \int_{0}^{x} \int_{0}^{4-x-y} e^z \,\mathbf{dz\,dy\,dx}$.

EXAMPLE: Find the integral which expresses the volume of the solid region bounded by above by the paraboloid $z = 4(x^2 + y^2)$, below by the plane z = -2, and on the sides by the parabolic sheet $y = x^2$ and the plane y = x.

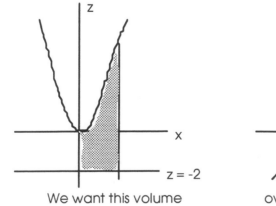

We want this volume

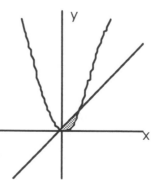

over this region in the z=0 -plane

Answer: We see z ranges from -2 to variable z-values on the paraboloid: $z = 4(x^2 + y^2)$. For x and y limits, draw a diagram in the xy-plane. Now we see that y goes from the lower curve,
$y = x^2$ to the upper curve: y = x. And x, which must have numerical (number) limits,
ranges from x = 0 to x = 1. Thus, the volume is $\displaystyle\int_0^1 \int_{x^2}^x \int_{-2}^{4(x^2+y^2)} dz\,dy\,dx$.

EXAMPLE: Find the integral which expresses the volume of the solid region bounded above by the elliptic paraboloid $z = 2x^2 + 3y^2$, on the sides by the parabolic sheet $y^2 = 1 - x$ and the plane x = 0 and below by the plane z = 0. (Hint: integrate with respect to x before y.)

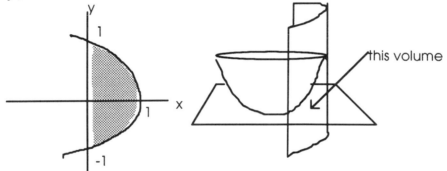

Answer: The region over which we integrate is horizontally simple, so we integrate with respect to x first (thus the hint): Volume $= \displaystyle\int_{-1}^1 \int_0^{1-y^2} \int_0^{2x^2+3y^2} dz\,dx\,dy$.

SECTION 14.5

"Nature, to be commanded, must be obeyed"...Francis Bacon

TRIPLE INTEGRALS IN CYLINDRICAL COORDINATES

Many integrals are easier to evaluate with cylindrical coordinates . These coordinates combine the usual polar coordinates for x and y, with the addition of "z". Any point in 3-space can be expressed with these coordinates.

The variables we use are r, θ, and z with dxdydz given below.

For x use **r cos θ**
For y use **r sinθ**
z is just **z**
For dx dy dz use **r dr dθ dz**

Volume is dV = rdr dθ dz

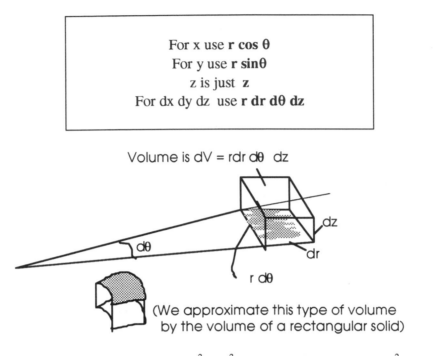

(We approximate this type of volume
by the volume of a rectangular solid)

NOTE: When you see the combination $x^2 + y^2$, recognize it instantly as r^2.

USEFUL REFERENCES : Study the graphs in Chapter 13 as well as the many graphs in your text. Relate these graphs to their equations so you will recognize them. The graphs that are easily written with these coordinates are circular cylinders, cones, paraboloids, and spheres.

EXERCISE: Write the equations of these graphs in cylindrical coordinates. Assume cross-sections are circular.

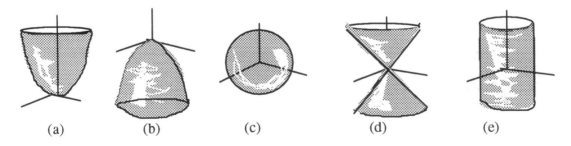

(a)   (b)   (c)   (d)   (e)

Answers:

a) Paraboloid $\quad z = x^2 + y^2$ or $z = r^2$

b) Paraboloid $\quad z = -(x^2 + y^2)$ or $z = -r^2$

c) Sphere $\qquad x^2 + y^2 + z^2 = a^2$ or $r^2 + z^2 = a^2$

d) Cone $\qquad z^2 = a^2(x^2 + y^2)$ or $z^2 = a^2 r^2$

e) Cylinder $\qquad x^2 + y^2 = a^2$ or $r^2 = a^2$

NOTE: In the past, we have written a circle as $x^2 + y^2 = r^2$. Now, however, we want to distinguish a fixed "r" from a variable "r", so we write a circle as $x^2 + y^2 = a^2$

NOTE: These curves from the past are useful to know because they are expressed easily in polar coordinates and appear in the exercises.

When a > 0:

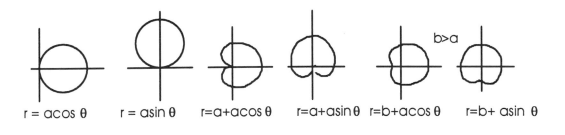

r = a cos θ $\qquad$ r = a sin θ $\qquad$ r = a+a cos θ $\qquad$ r = a+a sin θ $\qquad$ r = b+a cos θ $\qquad$ r = b+ a sin θ

EXAMPLE: Write the equation $x^2 + y^2 + 3z^2 = 9$ in cylindrical coordinates: .

Answer: Using $x^2 + y^2 = r^2$, this equation is $r^2 + 3z^2 = 9$. Without the factor "3" this would be the equation of a sphere; *with* the "3" we have an ellipsoid which is circular with respect to x and y, and elliptical with respect to x and z and also, with respect to y and z.

EXAMPLE: Write $5x^2 + 5y^2 - z^2 = 0$ in cylindrical coordinates and describe the graph of this set of points.

Answer: This is a cone of two nappes: $z^2 = 5r^2$. The cross-sections in the y = 0 and x = 0 planes are the lines $z = \pm\sqrt{5}\ x$ or $z = \pm\sqrt{5}\ y$.

EXAMPLE: Evaluate the iterated integral: $\displaystyle\int_1^2\int_0^\pi\int_0^1 r\cos\theta\,drd\theta dz$.

Answer: As usual we find the inside integral first:

$$\int_1^2\int_0^\pi\int_0^1 r\cos\theta\,drd\theta dz=\int_1^2\int_0^\pi\left(\frac{r^2}{2}\bigg|_0^1\right)(\cos\theta)d\theta\,dz=\frac{1}{2}\int_1^2\int_0^\pi\cos\theta d\theta dz=\ \frac{1}{2}\int_1^2(-\sin\theta)\bigg|_0^\pi\,dz=0.$$

Note that the "$\cos\theta$" factor in the integrand of the original integral has integration limits of π and 0. We could have integrated with respect to θ first, and saved ourselves work .

EXAMPLE: Express the triple integral $\displaystyle\iiint\limits_D xz\,dV$ as an iterated integral in cylindrical

coordinates, where D is the solid region in the first octant bounded by the sphere $x^2+y^2+z^2=9$, and the circular cylinder $r=\cos\theta$.

Answer:

(The graph of the cylinder $r=\cos\theta$ is is recognizable from the equation of the circle in polar coordinates given above: note z does not enter the picture.)

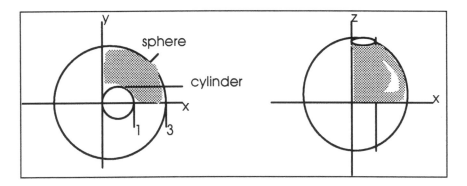

We translate the "xz" term under the integral into cylindrical coordinates: $xz=(r\cos\theta)z$. Note that z goes from 0 to values of z on the sphere. Also observe that r goes from values on the cylinder

$r=\cos\theta$ to values of r on the sphere, and (our integration is taking place in the first octant)

θ goes from 0 to $\pi/2$. Thus, our integral becomes $\displaystyle\int_0^{\pi/2}\int_{r\cos\theta}^3\int_0^{\sqrt{9-r^2}}(r\cos\theta)z\,dz(rdrd\theta)$.

EXAMPLE: Find the integral that represents the solid region inside the cone $z = r$ and between the planes $z = 2$ and $z = 4$.

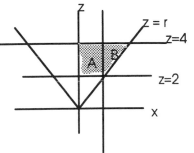

Answer: The order of integration you choose will alter the looks of the integral(s) you arrive at.

(i) If we integrate using dr dz dθ, we can represent the volume as Volume = $\displaystyle\int_0^{2\pi}\int_2^4\int_0^z r\,dr\,dz\,d\theta$.

(ii) If we imagine the shaded areas A and B twirled around the z axis to form solid regions, then

Volume = (Volume of solid region created by A) + (Volume of solid region created by B) =

$$\int_0^{2\pi}\int_0^2\int_2^4 r\,dz\,dr\,d\theta + \int_0^{2\pi}\int_2^4\int_r^4 r\,dz\,dr\,d\theta$$

EXAMPLE: Find the integral that represents the solid region bounded above by the sphere $x^2 + y^2 + z^2 = 3$ and below by the paraboloid $z = x^2 + y^2$.

Answer: First we find where these 2 surfaces intersect. Substituting z from the paraboloid equation into that of the sphere , $z + z^2 = 2$ and solving, z = 1 or -2, where z = 1 is the only feasible solution. Thus, these curves intersect in the circle $x^2 + y^2 = 1$.

Sphere and paraboloid intersect in a circle.

For limits, z will range from values of "r" on the paraboloid ($z = r^2$) to values of z on the

sphere ($z = \sqrt{3 - r^2}$). Thus Volume = $\displaystyle\int_0^{2\pi}\int_0^1\int_{r^2}^{\sqrt{3-r^2}} r\,dz\,dr\,d\theta$.

SECTION 14.6

"At the round earth's imagined corners"...John Donne

SPHERICAL COORDINATES

When an airplane is in motion, we can speak of "yaw" and "pitch". Here "yaw" corresponds to "azimuthal" rotation in the xy-plane by angle θ, and "pitch" to rotation measured downward from the z-axis by the angle ϕ (phi), where ϕ lies in $[0, \pi]$. Reserving "r" for polar coordinates, we use "ρ" (rho) for the distance of a point from the origin.

These coordinates work efficiently with spherical surfaces.

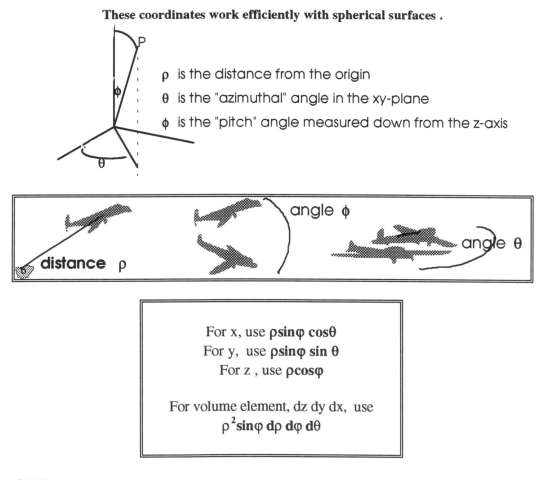

ρ is the distance from the origin

θ is the "azimuthal" angle in the xy-plane

ϕ is the "pitch" angle measured down from the z-axis

For x, use $\rho\sin\varphi\,\cos\theta$
For y, use $\rho\sin\varphi\,\sin\theta$
For z, use $\rho\cos\varphi$

For volume element, dz dy dx, use
$\rho^2\sin\varphi\,d\rho\,d\varphi\,d\theta$

NOTE: In spherical coordinates we express a point with this order of variables (ρ, φ, θ).

EXAMPLE: Express in rectangular (Cartesian) coordinates the point given in spherical coordinates as $(2, \pi, \pi/2)$.

Answer:

$x = \rho\sin\varphi\,\cos\theta = 2 \sin\pi \cos\pi/2 = 0$

$y = \rho\sin\varphi\,\sin\theta = 2 \sin\pi \sin\pi/2 = 0$

$z = \rho\cos\varphi = 2\cos\pi = -2$.
The given point is expressed as (0,0,-2) in rectangular coordinates.

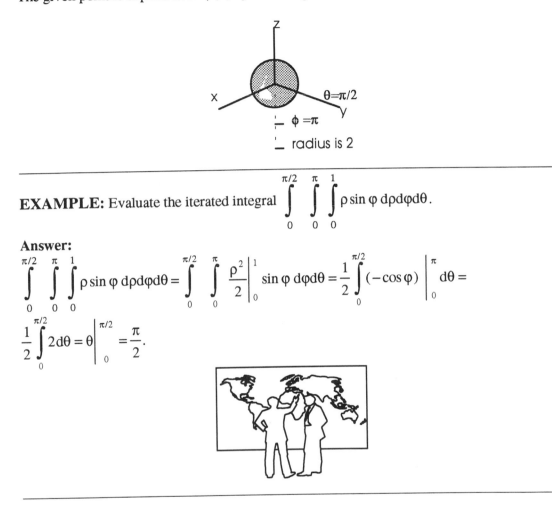

EXAMPLE: Evaluate the iterated integral $\displaystyle\int_0^{\pi/2}\int_0^{\pi}\int_0^1 \rho\sin\varphi\, d\rho d\varphi d\theta$.

Answer:

$$\int_0^{\pi/2}\int_0^{\pi}\int_0^1 \rho\sin\varphi\, d\rho d\varphi d\theta = \int_0^{\pi/2}\int_0^{\pi} \frac{\rho^2}{2}\bigg|_0^1 \sin\varphi\, d\varphi d\theta = \frac{1}{2}\int_0^{\pi/2}(-\cos\varphi)\bigg|_0^{\pi} d\theta =$$

$$\frac{1}{2}\int_0^{\pi/2} 2 d\theta = \theta\bigg|_0^{\pi/2} = \frac{\pi}{2}.$$

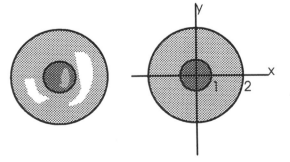

EXAMPLE: Express the integral $\displaystyle\iiint_D (x^2 + y^2)\,dV$ as an iterated integral in spherical

coordinates, where D is the solid region between the two spheres $x^2 + y^2 + z^2 = 1$ and
$x^2 + y^2 + z^2 = 4$.

Answer: The angle φ will range from 0 to π and θ will range from 0 to 2π. The variable radius ρ will range from 1 on the inner sphere, to 2, on the outer. Meanwhile, the quantity in the integrand, $x^2 + y^2$, must be expressed as

$(\rho\sin\varphi\cos\theta)^2 + (\rho\sin\varphi\sin\theta)^2 = \rho^2\sin^2\varphi$. Thus our integral becomes

$$\int_0^{2\pi}\int_0^{\pi}\int_1^2 (\rho^2\sin^2\varphi)(\rho^2\sin\varphi)d\rho d\varphi d\theta = \int_0^{2\pi}\int_0^{\pi}\int_1^2 (\rho^4\sin^3\varphi)d\rho d\varphi d\theta.$$ To integrate

$\int\sin^3\varphi d\varphi$, rewrite as $\int\sin\varphi(1-\cos^2\varphi)d\varphi$ and let $u = \cos\varphi$.

EXAMPLE: Express the integral $\iiint_D \sqrt{z}\,dV$ as an iterated integral in spherical

coordinates, where D is the region in the first octant bounded by the sphere $x^2 + y^2 + z^2 = 9$ and the planes $z = 0$, $x = \sqrt{3}y$ and $x = y$. (Note: the line $x = \sqrt{3}y$ is the line $y = \dfrac{1}{\sqrt{3}}x$.)

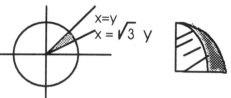

Answer: The variable that will need careful attention is θ: on the line $x = y$, we have $\theta = \pi/4$, and on the line $y = \dfrac{1}{\sqrt{3}}x$, we have $\tan\theta = \dfrac{1}{\sqrt{3}}$ and $\theta = \pi/6$. Thus, expressing $\sqrt{z}$ as $\sqrt{\rho\cos\varphi}$, we have

$$\int_{\pi/6}^{\pi/4}\int_0^{\pi/2}\int_0^3 \sqrt{\rho\cos\varphi}\ \rho^2\sin\varphi d\rho d\varphi d\theta = \int_{\pi/6}^{\pi/4}\int_0^{\pi/2}\int_0^3 \rho^{5/2}(\cos\varphi)^{1/2}\sin\varphi d\rho d\varphi d\theta.$$

EXAMPLE: Express the integral which gives the volume of the solid region bounded above by the sphere $x^2 + y^2 + z^2 = 2$ and below by the upper nappe of the cone $z^2 = x^2 + y^2$.

Answer: We did a similar example using cylindrical coordinates--notice the ease with which we can do this now, with spherical coordinates.

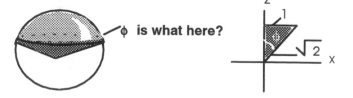

We want to know where the sphere intersects the cone, or where
$z^2 = 2 - x^2 - y^2 = x^2 + y^2$. The answer is: where $x^2 + y^2 = 1$; namely, on a circle of radius 1. Now the variable radius, ρ, measured from the origin, ranges from 0 to $\sqrt{2}$ while θ goes around the circle, taking on values from 0 to 2π. Meanwhile φ goes from 0 (north pole) to the circle at the intersection. From the diagram above, $\varphi = \pi / 4$ at the

intersection. Thus, Volume$= \displaystyle\int_0^{2\pi} \int_0^{\pi/4} \int_0^{\sqrt{2}} \rho^2 \sin\varphi \ d\rho d\varphi d\theta$.

EXAMPLE: Express the integral which gives the volume, V, of the solid region between the spheres $x^2 + y^2 + z^2 = 1$ and $x^2 + y^2 + z^2 = 4$ and **above** the upper nappe of the cone $3z^2 = x^2 + y^2$.

Answer: As with the previous problem, we find where the spheres intersect the cone. To find the intersection of the cylinder with the sphere of radius 1 we solve:
$x^2 + y^2 + z^2 = 1$ and $3z^2 = x^2 + y^2$, or
$\rho^2 = 1$ and $3\rho^2 \cos^2\varphi = (\rho\sin\phi\cos\phi)^2 + (\rho\sin\phi\sin\phi)^2 = \rho^2 \sin^2\varphi$, or
$3\rho^2 \cos^2\varphi = \rho^2 \sin^2\varphi$, or
$3\cos^2\varphi = \sin^2\varphi$, since $\rho^2 = 1$ on the sphere.
This occurs when $\tan^2\varphi = 3$, or $\tan\varphi = \sqrt{3}$. So $\varphi = \pi / 3$.
The intersection of $x^2 + y^2 + z^2 = 4$ and $3z^2 = x^2 + y^2$ is found similarly.

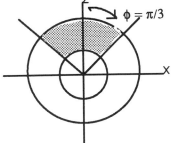

The integral is then $V = \displaystyle\int_0^{2\pi} \int_0^{\pi/3} \int_1^{2} \rho^2 \sin\varphi \ d\rho d\varphi d\theta$. If we wanted the volume below the

upper nappe, we would send φ from $\pi/3$ to π.

REVIEW

What are cylindrical coordinates? spherical coordinates? Give the formulas that relate x, y, and z to these other coordinates, and give the expression for dV for these other coordinate systems. Give examples of integrals posed in Cartesian coordinates that would be better solved with other coordinates.

SECTION 14.7

"..Half a proper gardener's work is done upon his knees"...Rudyard Kipling

MOMENTS AND CENTER OF GRAVITY

The formulas for moments and the coordinates which give the center of gravity are in your text. Learn them well enough to consider them memorized. Learning *one* set of formulas, for $\bar{x}$ and M_{xy}, for example, should be enough to stimulate your recall of all the formulas.

DEFINITIONS FROM THIS SECTION

♦ **Moment** about the xy plane: $M_{xy} = \iiint\limits_{D} \delta(x,y,z)\,dzdydx$, where $\delta(x,y,z)$ gives the

density at the point (x,y,z) etc.

♦ **Center of gravity:** $(\bar{x}, \bar{y}, \bar{z})$, where $\bar{x} = \dfrac{M_{yz}}{m}$, etc. (Called "**centroid**" if mass

density is constant. Center of gravity and centroid may not be the same.)

♦ **Mass** $m = \iiint\limits_{D} \delta\,dV$, where $\delta(x,y,z)$ is the given density of mass per unit volume .

EXAMPLE: Determine the center of gravity of the plane region inside the circles $x^2 + (y+1)^2 = 1$ and $(x+1)^2 + y^2 = 1$.

Answer: This region is symmetric with respect to the lines y = x and y = -x-1. Thus the center of gravity falls on the axes of symmetry, and $(\bar{x}, \bar{y}) = (-1/2, -1/2)$. (The center of gravity is the same as the centroid, since density is not specified.)

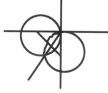

EXAMPLE: Find the centroid of the solid region in the first octant bounded by the planes x = 0 and y = 0 and the parabolids

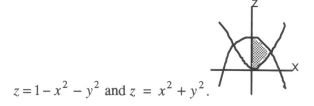

$z = 1 - x^2 - y^2$ and $z = x^2 + y^2$.

Answer: The intersection of these paraboloids is the circle, $x^2 + y^2 = \dfrac{1}{2}$. With δ taken to be 1,

$$M_{xy} = \iiint_D z\,dV = \int_0^{\pi/2} \int_0^{1/\sqrt{2}} \int_{r^2}^{1-r^2} zr\ dz\ dr\ d\theta = \frac{\pi}{32}\quad \text{(after some work!)}$$

$$M_{xz} = \iiint_D y\,dV = \int_0^{\pi/2} \int_0^{1/\sqrt{2}} \int_{r^2}^{1-r^2} (r\sin\theta)r\ dz\ dr\ d\theta = \frac{\sqrt{2}}{30}$$

$$M_{yz} = \iiint_D x\,dV = \int_0^{\pi/2} \int_0^{1/\sqrt{2}} \int_{r^2}^{1-r^2} (r\cos\theta)r\ dz\ dr\ d\theta = \frac{\sqrt{2}}{30}\quad \text{(That this is the same as the}$$

previous moment could be seen by symmetry.) Now the density is assumed to be 1, so

$$m = \iiint_D dV = \int_0^{\pi/2} \int_0^{1/\sqrt{2}} \int_{r^2}^{1-r^2} r\,dz\,dr\,d\theta = \frac{\pi}{16}\ .\ \text{Thus}$$

$$\overline{x} = \frac{M_{yz}}{m} = \frac{8\sqrt{2}}{15\pi}\ ,\quad \overline{y} = \frac{M_{xz}}{m} = \frac{8\sqrt{2}}{15\pi}, \text{and}\quad \overline{z} = \frac{M_{xy}}{m} = \frac{1}{2}\ \text{are the coordinates of the centroid.}$$

EXAMPLE: Find the centroid for the pyramid with vertices $(2,0,0)$, $(0,2,0)$, $(-2,0,0)$, $(0,-2,0)$, and $(0,0,2)$

Answer: It is easy to see that $\overline{x} = 0$ and $\overline{y} = 0$, but we must find $\overline{z}$. It will suffice to find $\overline{z}$ in the first octant . To do this, we need the equation for the plane that creates the shaded face, below, and it is easy to see that the plane , given by $ax + by + cz = d$, is $x + y + z = 2$. We use this plane in the first octant to find M_{xy}.

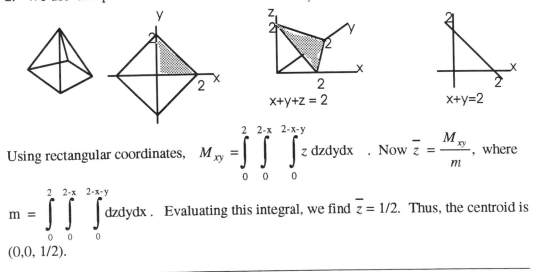

Using rectangular coordinates, $M_{xy} = \int_0^2 \int_0^{2-x} \int_0^{2-x-y} z\ dz\,dy\,dx$. Now $\overline{z} = \dfrac{M_{xy}}{m}$, where

$$m = \int_0^2 \int_0^{2-x} \int_0^{2-x-y} dz\,dy\,dx\ .\ \text{Evaluating this integral, we find } \overline{z} = 1/2.\ \text{Thus, the centroid is}$$
$(0,0,\ 1/2)$.

EXAMPLE: Find M_{xy} for the solid region bounded above by the sphere $x^2 + y^2 + z^2 = 4$ and below by the upper nappe of the cone $z^2 = x^2 + y^2$.

Answer: (We have seen this "ice cream cone" object many times.) These two surfaces intersect in the circle, $x^2 + y^2 = 2$, of radius $\sqrt{2}$. Note, in cylindrical coordinates, the

equation of the top half of the sphere is $z = \sqrt{4 - r^2}$. Thus $M_{xy} = \int\limits_{0}^{2\pi} \int\limits_{0}^{\sqrt{2}} \int\limits_{r}^{\sqrt{4-r^2}} z r\, dz\, dr\, d\theta$.

EXAMPLE: Find M_{xy} for a solid that occupies the region inside the sphere $x^2 + y^2 + z^2 = 4$ and inside the cylinder $x^2 + y^2 = 3$ and has mass density $\delta(x, y, z) = z^2$:

$M_{xy} = \int\limits_{0}^{2\pi} \int\limits_{0}^{\sqrt{3}} \int\limits_{-\sqrt{4-r^2}}^{\sqrt{4-r^2}} z(z^2) r\, dz\, dr\, d\theta$, which can be seen without evaluation to be 0. (The

z^3-term inside the integrand is odd and has limits of integration which differ only in sign.) The terms $\bar{x}$ and $\bar{y}$ will be 0, by symmetry and because the density is not affected by x and y. The centroid of this region is (0,0,0).

SECTION 14.8

"Truth has no special time of its own. Its hour is now--always"...Albert Schweitzer

CHANGE OF VARIABLE

The coordinate systems that we know so far are the following.

rectangular, volume element dzdydx
cylindrical, volume element r dzdrdθ
spherical, volume element $\rho^2 \sin \varphi d\rho d\varphi d\theta$

In the real world, however, these are not necessarily the ones we want--we might like to invent our own. For example, a statistician, who is working with a scatter plot like that in the first diagram on the following page, might wish to have a coordinate system like that in the second diagram.

If a scatter plot fills out the type of ellipse at left, we might prefer to change the coordinate frame to that at right.

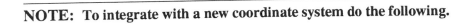

NOTE: To integrate with a new coordinate system do the following.

(1) Express old coordinates in terms of new.

(2) Compute the determinant **J** of the matrix whose terms are $\left(\dfrac{\partial(old)}{\partial(new)}\right)$. The quantity J

is called the **Jacobian**.

(3) If we call the new volume element dV' and the old volume element dV, we have:

> **(old volume element) = | J | (new volume element) or**
> $$dV = | J | dV'$$

New variables for old!

(4) Change all variables under the integral sign to new variables; use the new volume
element from (3), and integrate with respect to the new limits.

NOTE: This is a generalization of the "u-substitution" we saw in first year calculus. Here
is an example:

To integrate $\int x^2 \sin x^3 dx$ we let $u = x^3$. Then $x = u^{1/3}$. We took the "1×1" determinant

$|J| = \left|\dfrac{\partial u}{\partial x}\right| = \left|\dfrac{1}{3}u^{-2/3}\right|$. Then we had the integral in terms of u , with

$$\textbf{dx} = \boxed{\dfrac{1}{3}\textbf{u}^{-2/3}\textbf{du}}\text{ , or}$$

old dx - volume element   = I J I $\times$ new du - volume element

Thus, $\int u^{2/3}(\sin u)\boxed{\dfrac{1}{3}\textbf{u}^{-2/3}\textbf{du}} = \dfrac{1}{3}\int \sin u \; du$ which we can solve more readily .

NOTE: To compute a 3 x 3 determinant:

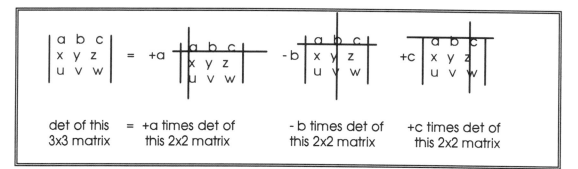

det of this   = +a times det of   - b times det of   +c times det of
3x3 matrix   this 2x2 matrix   this 2x2 matrix   this 2x2 matrix

EXAMPLE: Derive the formula for integrals using the cylindrical coordinate system.
Answer:
(1) Express old coordinates in terms of new:

        Old coordinates (rectangular) x, y, z
        New coordinates (cylindrical) r, θ , z

        Write $x = r \cos \theta$, $y = r \sin \theta$, $z = z$.

(2) Find the matrix of partials of **old coordinates with respect to new**: take the
    determinant:

$$J = \frac{\partial(x,y,z)}{\partial(r,\theta,z)} = \begin{vmatrix} \dfrac{\partial x}{\partial r} & \dfrac{\partial x}{\partial \theta} & \dfrac{\partial x}{\partial z} \\[2mm] \dfrac{\partial y}{\partial r} & \dfrac{\partial y}{\partial \theta} & \dfrac{\partial y}{\partial z} \\[2mm] \dfrac{\partial z}{\partial r} & \dfrac{\partial z}{\partial \theta} & \dfrac{\partial z}{\partial z} \end{vmatrix}$$

Notice the notation on the left side of the equality.

$$J = \begin{vmatrix} \dfrac{\partial x}{\partial r} & \dfrac{\partial x}{\partial \theta} & \dfrac{\partial x}{\partial z} \\[2mm] \dfrac{\partial y}{\partial r} & \dfrac{\partial y}{\partial \theta} & \dfrac{\partial y}{\partial z} \\[2mm] \dfrac{\partial z}{\partial r} & \dfrac{\partial z}{\partial \theta} & \dfrac{\partial z}{\partial z} \end{vmatrix} = \begin{vmatrix} \cos\theta & -r\sin\theta & 0 \\ \sin\theta & r\cos\theta & 0 \\ 0 & 0 & 1 \end{vmatrix} = (\cos\theta)(r\cos\theta) + (r\sin\theta)(\sin\theta) + 0 = r$$

(3) Take absolute value $|J| = |r| = r$. Thus a volume element **dxdydz** becomes **"r" drdθdz** in cylindrical coordinates. In fact, $|J|$ is the "r"-factor.

EXAMPLE: Compute the Jacobian of the transformation $x = 2u - 3v$, $y = u + v$.
Answer: We are given *"old" expressed in terms of "new"*. We want

$$J = \begin{vmatrix} \dfrac{\partial x}{\partial u} & \dfrac{\partial x}{\partial v} \\[2mm] \dfrac{\partial y}{\partial u} & \dfrac{\partial y}{\partial v} \end{vmatrix} = \begin{vmatrix} 2 & -3 \\ 1 & 1 \end{vmatrix} = 2(1) - (-3)(1) = 5$$

This means dxdy = 5 du dv when we write x and y in terms of u and v.

NOTE: What if we accidentally wrote this matrix as $\begin{vmatrix} \dfrac{\partial x}{\partial u} & \dfrac{\partial y}{\partial u} \\[2mm] \dfrac{\partial x}{\partial v} & \dfrac{\partial y}{\partial v} \end{vmatrix}$, writing rows as columns? The answer is the same!

EXAMPLE: Evaluate the integral $\displaystyle\iint\limits_R x^2 dA$ where R is the region bounded by the ellipse $9x^2 + 25y^2 = 1$. Use an appropriate transformation.
Answer:

Since the ellipse is given by $\dfrac{x^2}{1/9} + \dfrac{y^2}{1/25} = 1$ or $\left(\dfrac{x}{1/3}\right)^2 + \left(\dfrac{y}{1/5}\right)^2 = 1$, we think it would

be nice to use $u = \dfrac{x}{1/3} = 3x$ and $v = \dfrac{y}{1/5} = 5y$; then we would be integrating over the region bounded by a *circle*, $u^2 + v^2 = 1$.

(1) We express **old in terms of new**. So **x = u/3, y = v/5**.

(2) Now form

$$J = \begin{vmatrix} \dfrac{\partial x}{\partial u} & \dfrac{\partial x}{\partial v} \\[2mm] \dfrac{\partial y}{\partial u} & \dfrac{\partial y}{\partial v} \end{vmatrix} = \begin{vmatrix} \dfrac{1}{5} & 0 \\[2mm] 0 & \dfrac{1}{3} \end{vmatrix} = \dfrac{1}{15}, \text{ and } |J| = \dfrac{1}{15}.$$

(3) We now know the "dxdy" term will be replaced by (1/15) dudv.

(4) Now we change the integral into terms of u and v, so

$$\iint_R x^2 dA = \iint_{??\ ?} (u/3)^2 \frac{1}{15} dudv.$$ We will be integrating over a disk in the u-v plane.

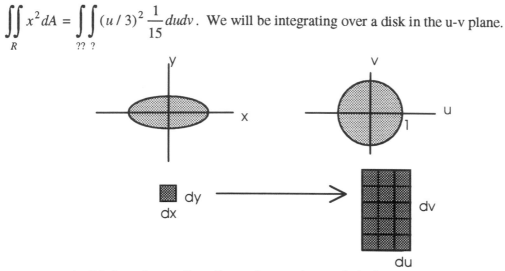

In this transformation, the volume element dxdy corresponds
to (1/15) dudv

So now in order to perform an easy integration, we change to polar coordinates! We use yet another substitution:

$$u = r \cos \theta, \quad v = r \sin \theta \quad \text{and} \quad dudv = rdr\, d\theta.$$

And obtain:

$$\iint_R x^2 dA = \iint_{??\ ?} (u/3)^2 \frac{1}{15} dudv =$$

$$\int_0^{2\pi}\int_0^1 \frac{(r\cos\theta)^2}{9}\cdot\frac{1}{15} rdrd\theta = \frac{1}{135}\int_0^{2\pi}\int_0^1 r^3 \cos^2\theta\, drd\theta = \frac{1}{135}\int_0^{2\pi}\left(\frac{r^4}{4}\Big|_0^1\right)\cos^2\theta d\theta =$$

$$\frac{1}{135}\cdot\frac{1}{4}\int_0^{2\pi}\cos^2\theta d\theta = \frac{1}{540}\left(\frac{2\theta+\sin 2\theta}{4}\right)\Big|_0^{2\pi} = \frac{\pi}{540}.$$

EXAMPLE: **This is a good, easy prototypical example to learn .** Evaluate the integral

$$\iint_R \left(\frac{x-y}{x+y}\right)^4 dA$$ by a suitable change of variables, where R is the region bounded by the

lines $x - y = 1$, $x - y = 2$, $x + y = 1$ and $x + y = 3$.

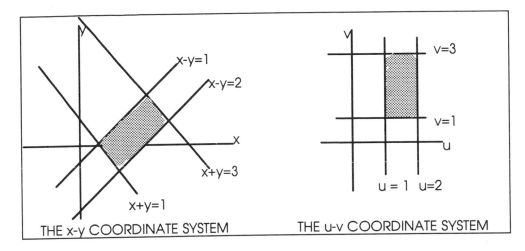

THE x-y COORDINATE SYSTEM   THE u-v COORDINATE SYSTEM

Answer: It would be awkward to integrate over this region in rectangular coordinates. We let u = x - y, v = x + y. (In this choice, the boundaries of the region of integration guide us; it helps that what we integrate bears resemblance to these terms.) With this substitution the boundaries of the parallelogram above become u = 1, u = 2, v = 1, v = 3-- a rectangle!

(1) Express **old in terms of new**: we must solve for x and y:

u = x + y
v = x - y
So x = (u+v)/2 and y = (u-v)/2.

(2) Find the **absolute value of the Jacobian**:

$$J = \begin{vmatrix} \dfrac{\partial x}{\partial u} & \dfrac{\partial x}{\partial v} \\ \dfrac{\partial y}{\partial u} & \dfrac{\partial y}{\partial v} \end{vmatrix} = \begin{vmatrix} 1/2 & 1/2 \\ 1/2 & -1/2 \end{vmatrix} = -(1/2) \text{ and } |J| = 1/2.$$

(3) Express the integral in terms of the new variables:

$$\iint\limits_{R} \left(\frac{x-y}{x+y}\right)^4 dA = \int_1^3 \int_1^2 \left(\frac{v}{u}\right)^4 \frac{1}{2} du\, dv = \frac{1}{2}\int_1^3 \int_1^2 u^{-4} v^4 du\, dv, \text{ which is easily evaluated.}$$

EXAMPLE: This is a good, medium-level prototypical example to learn.

Evaluate the integral $\iint\limits_{R} e^{(2x-y)/(x+y)} dA$ where A is the region bounded by the lines x = 2y, y = 2x, x + y = 1, and x + y = 2. (The first two lines, rewritten, are y = x/2 and y = 2x.)
Answer:
Choice of variable: It seems obvious to let one variable, say v, be x + y. There are many substitutions we can use for u. We could use u = x, for example. However, motivated to make the integrand look simple, we let u = 2x - y and v = x + y.

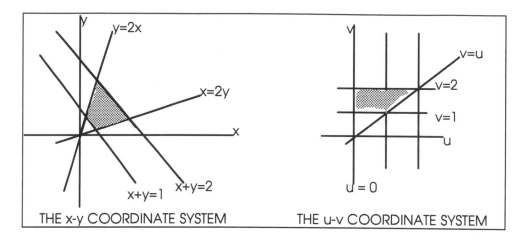

THE x-y COORDINATE SYSTEM   THE u-v COORDINATE SYSTEM

Finding limits--sometimes you must solve for x and y

       when $x + y = 1$, $v = 1$

       when $x + y = 2$, $v = 2$

       when $y = 2x$, $u = 0$, and

       when $y = x/2$, $u = ???$

To find this last limit, solve for x and y in terms of u and v .

$$u = 2x - y$$
$$v = x + y$$

$$\text{so } x = \frac{u + v}{3} \text{ and } y = \frac{2v - u}{3}$$

Thus **when $y = x/2$,** $\dfrac{2v - u}{3} = \dfrac{u + v}{6}$ and **$u = v$** !

(2) Find $J = \begin{vmatrix} \dfrac{\partial x}{\partial u} & \dfrac{\partial x}{\partial v} \\ \dfrac{\partial y}{\partial u} & \dfrac{\partial y}{\partial v} \end{vmatrix} = \begin{vmatrix} 1/3 & 1/3 \\ -1/3 & 2/3 \end{vmatrix} = \left(\dfrac{1}{3}\right)\left(\dfrac{2}{3}\right) - \left(\dfrac{1}{3}\right)\left(-\dfrac{1}{3}\right) = \dfrac{1}{3}$ and $|J| = 1/3$.

(3) Form the integral: $\displaystyle\iint_R e^{(2x-y)/(x+y)} dA = \int_1^2 \int_0^v e^{u/v}\left(\frac{1}{3}\right) du\, dv$.

To integrate this, recall that on the first integration, v is a constant with respect to u:

$$\int_1^2 \int_0^v e^{u/v}\left(\frac{1}{3}\right) du\, dv = \frac{1}{3}\int_1^2 ve^{u/v}\bigg|_{u=0}^{u=v} dv = \frac{1}{3}\int_1^2 v(e-1)dv = \frac{(e-1)}{3}\cdot\frac{v^2}{2}\bigg|_1^2 = \frac{(e-1)}{2}$$

I Hate Tests!

Try getting together with a group of friends to predict what will be on the next test. For example, each student could suggest 3 problems that would make good test questions. Try to find some variation in the way questions can be asked-- here are some examples

1. **Evaluate** the integral ☐

2. **Sketch the region of integration** for the integral ☐

3. Write the integral ☐ with **the order of integration reversed**. Then evaluate both integrals.

4. Write down the integral that expresses the volume under ☐ and above ☐ using

   (1) Cartesian coordinates (2) Cylindrical coordinates, and (3) Spherical coordinates

5. **Find the centroid** of the region bounded by ☐. If the density is ☐, find the center of gravity of the region.

6. **Find the moment,** M_{xy} , for the region bounded by ☐

7. **Find the Jacobian** for the transformation given by ☐

8. **Sketch the new region of integration** of the integral ☐ when the change of coordinates, ☐ , is made.

9. **Evaluate the integral** ☐ by performing a change of coordinates.

WHAT COORDINATE SYSTEM SHOULD I USE?

When you are asked to find the volume of solids bounded by surfaces, the choice of coordinates may seem overwhelming at first. However, if you review the worked examples in your text and the problems from the homework, you will develop an ability to determine in advance which coordinates will work most efficiently. Does the volume appear to come from the origin in "conical wedges?" (Problems with spheres and cones). These problems are frequently best solved with spherical coordinates. Do little pieces of volume appear be rectangular solids? (Problems with planes, squared off regions, etc.) Then use Cartesian coordinates. With any circular shapes, that are not obviously spherical, try cylindrical coordinates.

Create a Review Sheet

The best reviews are the ones you create yourself. List here or on another sheet of paper or 2, the most important concepts you have seen in this chapter, things to memorize, things that confuse you, errors you are likely to make, sample examples that might be useful to memorize. Compare your review sheet with that of a friend.

CHAPTER 15

"Hold fast the time! Guard it! Watch over it, every hour, every minute! Unregarded it slips away, like a lizard, smooth, slippery, faithless..."...Thomas Mann

In Chapter 11, we saw vector functions of a single variable, often denoted t. In this chapter, we define vector functions of many variables. A "vector field" is obtained by assigning vectors to points (x,y,z). If you have had courses dealing with electromagnetism, fluid flow, heat distribution, or magnetism for example, you may be familiar with vector fields. One of the high points of all of calculus is the remarkable Stokes' Theorem, a generalization of the Fundamental Theorem of Caculus that you saw in Calculus I.

SECTION 15.1

" I burned my life that I might find/ A passion wholly of the mind"...Louise Bogan

VECTOR FIELDS

NOTE: Quick review. R^2 is notation for $R \times R$ or $X \times Y$, the plane, or 2-space, in which points are the ordered pairs (x,y). ("Ordered" means the order in which we list x and y is important.) Similarly, R^3 is notation for 3-space, in which points are the ordered triples (x,y,z).

NOTE: A vector field in 3-space, F(x,y,z), is a rule assigning to the point (x,y,z) in 3-space, a unique vector. This vector may be thought of as being "attached" to the point (x,y,z), as long as we continue to remember that vectors actually emanate from the origin. In R^2, if we assign to every point (x,y), the gradient of a differentiable function, for example, we create a two-dimensional vector field. Given a vector field, the **divergence** and **curl** of the vector field are interesting concepts.

NOTE: The vector field may be given as **F = Mi + Nj + Pk** where M, N, P are functions of x, y, and z . As an example, **F** can be thought of as a vector giving the velocity of a fluid at the point (x,y,z).

DIVERGENCE

Divergence is a *scalar quantity* which measures the *rate of flow of a mass (with respect to time) per unit volume, from the point (x,y,z).*

If **F** = M**i** + N**j** + P**k** is a vector field, where M, N, P are functions of x, y, and z, then

$$\text{div } \mathbf{F}(x,y,z) = \frac{\partial M}{\partial x} + \frac{\partial N}{\partial y} + \frac{\partial P}{\partial z}.$$

Or div $\mathbf{F}(x,y,z) = M_x + N_y + P_z.$

NOTE: Divergence gives the rate of change in the **i**-component of the velocity field **F** with respect to x, plus the rate of change in the **j**- component of **F** with respect to y, plus the rate of change in the **k**-component with respect to z.)

"Flux" is the *rate of flow of a mass, such as a mass of fluid, entering or leaving a closed region.* Divergence is then the flux density. It measures

$$\frac{\text{Flux across the boundary of the region}}{\text{Volume of the region}}.$$

The mass flow may be escaping from a **source**, and flowing positively toward a **sink**. Find the source by asking where the divergence is positive. Find the sink by asking where the divergence is negative.

CURL

Curl is a vector which measures *the tendency of the fluid to rotate, or curl about an axis.* The direction of the curl vector gives the direction of this rotation, and the length of the curl vector measures the speed of the rotation.

If **F** = M**i** + N**j** + P**k** is a vector field, where M, N, P are functions of x, y, and z,

$$\text{curl } \mathbf{F}(x,y,z) = \left(\frac{\partial P}{\partial y} - \frac{\partial N}{\partial z} \right)\mathbf{i} - \left(\frac{\partial P}{\partial x} - \frac{\partial M}{\partial z} \right)\mathbf{j} + \left(\frac{\partial N}{\partial x} - \frac{\partial M}{\partial y} \right)\mathbf{k} \text{ . Or}$$

$$\text{curl } \mathbf{F}(x,y,z) = \left(P_y - N_z \right)\mathbf{i} - \left(P_x - M_z \right)\mathbf{j} + \left(N_x - M_y \right)\mathbf{k}$$

(While fluid might best be thought of as rotating in a plane, curl combines the effect of the rotation of the fluid about all planes into a general tendency of the fluid to rotate about one main axis.)

NOTE: Here is a useful mnemonic (for the memory) device .

Menmosyne was the Greek goddess of memory.

An operator is a function on functions. The "del" operator ∇ operates on vector fields $\mathbf{F}$ (x,y,z) and is given by:

$$\nabla = \frac{\partial}{\partial x}(\;\;)\,\mathbf{i}\; + \frac{\partial}{\partial y}(\;\;)\,\mathbf{j}\; + \frac{\partial}{\partial z}(\;\;)\,\mathbf{k}$$

NOTE: We can rewrite both the curl and divergence with the symbol ∇:

$$\text{div } \mathbf{F} = \nabla \cdot \mathbf{F}$$

i.e div $\mathbf{F} = \left(\frac{\partial}{\partial x}(\;\;)\,\mathbf{i}\; + \frac{\partial}{\partial y}(\;\;)\,\mathbf{j}\; + \frac{\partial}{\partial z}(\;\;)\,\mathbf{k}\right)\cdot(M\,\mathbf{i} + N\,\mathbf{j} + P\,\mathbf{k}) = \frac{\partial M}{\partial x} + \frac{\partial N}{\partial y} + \frac{\partial P}{\partial z}.$

$$\text{curl } \mathbf{F} = \nabla \times \mathbf{F}$$

i.e. $\mathbf{curl}\ \mathbf{F}\; = \begin{vmatrix} \mathbf{i} & \mathbf{j} & \mathbf{k} \\ \frac{\partial}{\partial x} & \frac{\partial}{\partial y} & \frac{\partial}{\partial z} \\ M & N & P \end{vmatrix} = \left(\frac{\partial P}{\partial y} - \frac{\partial N}{\partial z}\right)\mathbf{i} - \left(\frac{\partial P}{\partial x} - \frac{\partial M}{\partial z}\right)\mathbf{j} + \left(\frac{\partial N}{\partial x} - \frac{\partial M}{\partial y}\right)\mathbf{k}$

NOTE: curl (**grad f**) = 0 and div (**curl F**) = 0. (When we say **curl F = 0**, we use **0** to mean the **0**-vector.)

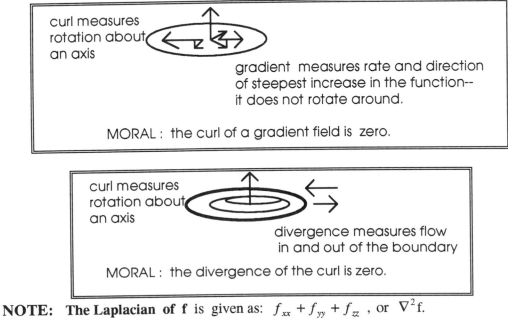

curl measures rotation about an axis

gradient measures rate and direction of steepest increase in the function-- it does not rotate around.

MORAL : the curl of a gradient field is zero.

curl measures rotation about an axis

divergence measures flow in and out of the boundary

MORAL : the divergence of the curl is zero.

NOTE: **The Laplacian of f** is given as: $f_{xx} + f_{yy} + f_{zz}$, or $\nabla^2 f$.

Laplace's Equation is: $\nabla^2 f = 0$, which means $f_{xx} + f_{yy} + f_{zz} = 0$. A function which satisfies Laplace's equation is said to be **harmonic.**

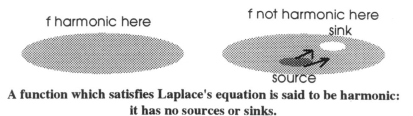

A function which satisfies Laplace's equation is said to be harmonic:
it has no sources or sinks.

NOTE: If the function f gives the heat on a disk at any point, f is harmonic if there are no hot or cold spots--the distribution of heat is uniform.

EXAMPLE: Compute the curl and divergence of the vector field given by
$F(x,y,z) = x^2\mathbf{i} + y\mathbf{j} + 3\mathbf{k}$.

Answer:
(1) div $\mathbf{F}(x,y,z) = \nabla \cdot \mathbf{F}(x,y,z) =$

$(\dfrac{\partial}{\partial x}\mathbf{i} + \dfrac{\partial}{\partial y}\mathbf{j} + \dfrac{\partial}{\partial z}\mathbf{k})(x^2\mathbf{i} + y\mathbf{j} + 3\mathbf{k}) = \dfrac{\partial x^2}{\partial x} + \dfrac{\partial y}{\partial y} + \dfrac{\partial 3}{\partial z} = 2x + 1$

(2) **curl** $\mathbf{F}(x,y,z) = \nabla \times \mathbf{F}(x,y,z) = \begin{vmatrix} \mathbf{i} & \mathbf{j} & \mathbf{k} \\ \dfrac{\partial}{\partial x} & \dfrac{\partial}{\partial y} & \dfrac{\partial}{\partial z} \\ M & N & P \end{vmatrix} =$

$\begin{vmatrix} \mathbf{i} & \mathbf{j} & \mathbf{k} \\ \dfrac{\partial}{\partial x} & \dfrac{\partial}{\partial y} & \dfrac{\partial}{\partial z} \\ x^2 & y & 3 \end{vmatrix} = \mathbf{i}(\dfrac{\partial 3}{\partial y} - \dfrac{\partial y}{\partial z}) - \mathbf{j}(\dfrac{\partial 3}{\partial x} - \dfrac{\partial x^2}{\partial z}) + \mathbf{k}(\dfrac{\partial y}{\partial x} - \dfrac{\partial x^2}{\partial y}) = 0$

Why Do We Have To Do the Long Problems?

Even though these problems may not occur on tests, these problems are useful because they help you to synthesize the material, and put it together in ways that will help you understand and remember it. Rarely in the real world will you be asked to "compute this gradient" or "find that integral". Instead your boss will be looking for someone who knows how to translate a complicated real world situation with many features, into a simpler mathematical one and recall the theory associated to provide a solution.

NOTE: Theorem 15.4 gives conditions under which a vector field is a gradient field. We pose this problem:

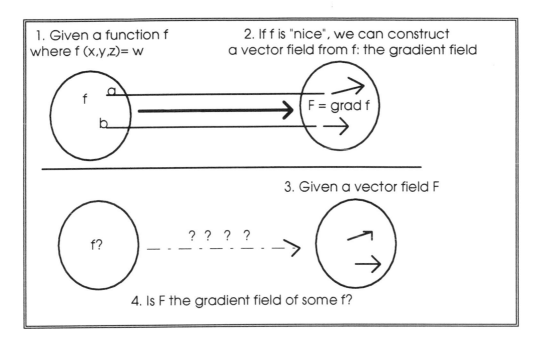

1. Given a function f where f (x,y,z)= w

2. If f is "nice", we can construct a vector field from f: the gradient field

f

a

b

F = grad f

3. Given a vector field F

f?

? ? ? ?

4. Is F the gradient field of some f?

The answer is : **YES!** *provided* two conditions are met.

(1) curl F = 0 (From above, **F = grad f** $\Rightarrow$ **curl F = 0**. So we need this condition at least: it is necessary.)

(2) F is defined in all of 3-space, R^3. (This condition can be relaxed).

EXAMPLE: Recovering a function from its gradient. Find the function f for which $\mathbf{F}(x,y,z) = (\sin xy)\mathbf{i} + (\cos xy)\mathbf{j}$ is the gradient.

Answer: **F** is defined on all of 3-space, although there is no z-variable in F. This does not stop us. We ask: is **curl F = 0**?

I.e., is $\mathbf{curl\ F} = \begin{vmatrix} \mathbf{i} & \mathbf{j} & \mathbf{k} \\ \dfrac{\partial}{\partial x} & \dfrac{\partial}{\partial y} & \dfrac{\partial}{\partial z} \\ \sin xy & \cos xy & 0 \end{vmatrix} = \mathbf{i}(0) - \mathbf{j}(0) + \mathbf{k}\left(\dfrac{\partial \cos xy}{\partial x} - \dfrac{\partial \sin xy}{\partial y}\right) = 0$?

But $\dfrac{\partial \cos xy}{\partial x} - \dfrac{\partial \sin xy}{\partial y} = $ y(-sinxy) -x(cosxy) $\neq 0$, so this vector field is *not* the gradient vector field of any function.

EXAMPLE: Recovering a function from its gradient in 2 dimensions. Find the function f for which $\mathbf{F}(x,y,z) = e^y\mathbf{i} + (xe^y + y)\mathbf{j}$ is the gradient vector field.

Answer: **F** is once again defined on all of 3-space, although no z-variable is present in the definition of F. We check first: is **curl F = 0**? But

$$\text{curl } \mathbf{F} = \begin{vmatrix} \mathbf{i} & \mathbf{j} & \mathbf{k} \\ \dfrac{\partial}{\partial x} & \dfrac{\partial}{\partial y} & \dfrac{\partial}{\partial z} \\ e^y & xe^y + y & 0 \end{vmatrix} = \mathbf{i}(0) - \mathbf{j}(0) + \mathbf{k}\left(\dfrac{\partial(xe^y + y)}{\partial x} - \dfrac{\partial e^y}{\partial y}\right) = 0. \text{ Thus we have 2}$$

main equations:

(1) $\dfrac{\partial f}{\partial x} = e^y$ so that $f(x,y,z) = xe^y + g(y)$ (Note: we treat y as a constant!)

(2) $\dfrac{\partial f}{\partial y} = xe^y + y$

Now compare information about partials with respect to y using (1) and (2)

(3) Since $f(x,y,z) = xe^y + g(y)$

$\dfrac{\partial f}{\partial y} = xe^y + g'(y) = xe^y + y.$ So

$g'(y) = y$ and $g(y) = \dfrac{y^2}{2} + C.$ Thus

$f(x,y,z) = xe^y + \dfrac{y^2}{2} + C$

Last step: check that $\nabla f = \mathbf{F}$! This checks!

EXAMPLE: Recovering a function from its gradient in 3 dimensions. Find the function f for which $\mathbf{F}(x,y,z) = (2xz + 1)\mathbf{i} + (2y(z+1))\mathbf{j} + (x^2 + y^2 + 3z^2)\mathbf{k} = \text{grad } f$.

Answer: (NOTE: the preliminary work should be done as in part (2) above: show **curl F = 0** and show **F** is defined on all of 3-space.)

We must have 3 main equations :

(1) $\dfrac{\partial f}{\partial x} = 2xz + 1$ so that $f(x,y,z) = x^2 z + x + g(y,z)$ and

(2) $\dfrac{\partial f}{\partial y} = 2y(z+1)$ so that $f(x,y,z) = y^2(z+1) + h(x,z)$ and

(3) $\dfrac{\partial f}{\partial z} = x^2 + y^2 + 3z^2$, STOP

Now compare information about partials with respect to y using (1) and (2)

(4) From (1) and (2): $\dfrac{\partial f}{\partial y} = \dfrac{\partial g}{\partial y} = 2y(z+1)$ and $g(y,z) = y^2(z+1) + h(z)$

(5) Thus $f(x,y,z) = x^2 z + x + g(y,z) = x^2 z + x + y^2(z+1) + h(z).$

Now compare information about partials with respect to Z using latest result and (3) : Solve

(6) From (3) and (5) : $\dfrac{\partial f}{\partial z} = x^2 + y^2 + 3z^2 = x^2 + y^2 + h'(z)$

so $3z^2 = h'(z)$ and $h(z) = z^3 + C$. Thus

(7) $f(x,y,z) = x^2 z + x + y^2(z+1) + z^3 + C$

Last step: check that $\nabla f = F$! This checks.

EXAMPLE: Prove
div $(f(x,y,z)\mathbf{F}(x,y,z)) = f(x,y,z)$div $\mathbf{F}(x,y,z) + ($**grad** $f(x,y,z))\cdot \mathbf{F}(x,y,z)$.

Answer: div $(f(x,y,z)\mathbf{F}(x,y,z)) =$
div $(f\,M\,\mathbf{i} + f\,N\,\mathbf{j} + f\,P\,\mathbf{k}) = (f\,M)_x + (f\,N)_y + (f\,P)_z = $ (by the product rule)

$(f\,M_x + f_x M) + (f\,N_y + f_y N) + (f\,P_z + f_z P) =$

$\Big((f\,M_x) + (f\,N_y) + (f\,P_z)\Big) + \Big((f_x M) + (f_y N) + (f_z P)\Big) =$

$(f)($div $\mathbf{F}) + ($**grad** $f)\cdot(\mathbf{F})$

SECTION 15.2

"By being so long in the lowest form [at Harrow] I gained an immense advantage over the cleverer boys"...Winston Churchill

LINE INTEGRALS

NOTE: (Read Definitions 15.5 and 15.6 carefully in your text!)

PREVIOUSLY we saw integrals over a region, that looked like this:

$\displaystyle\iint\limits_{R} f(.....)\,dA$ or $\displaystyle\iiint\limits_{D} f(.....)\,dV$ where A represented the area of the 2-dimensional

region R, and V represented the volume of the 3-dimensional solid region D.

....

NOW we will be forming integrals over a curve, C, that look like : $\displaystyle\int\limits_{C} f(.....)\,ds$ where ds

represents arc length, and C is a curve.

What is Allowed of the Curve C?

> **C is a piecewise smooth (differentiable) curve with finite length ("rectifiable").**

C can be the type of curve below, for example. Note the curve C may be "closed" as in the case of the first 2 pictures, may have some rare cusps and sharp points, but must be of finite length.

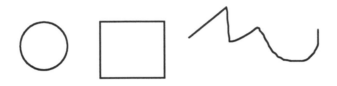

What is the Line Integral of a Function?

We want to understand what is meant by $\int_C f(.....)ds$. The **line integral of a continuous function** on a piecewise smooth curve C with finite length is defined by

$$\int_C f(x,y,z)ds = \lim_{\|P\|\to 0} \sum_{k=1}^{n} f(x_k,y_k,z_k)\Delta s_k$$

$$= \text{ limit of a sum of } : \; f \text{ (some point on the kth partition) } \times \text{ length of the line segment on that partition}$$

It is an integral that is a limit of sums of these types of areas:

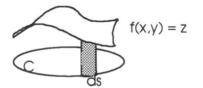

f(x,y) = z

How Do We Compute It?

We use: $\int_C f(x,y,z)ds = \int_C f(x,y,z)\dfrac{ds}{dt}dt$. The curve C generally comes parametrized in

terms of t, as the vector function $\mathbf{r}(t) = x(t)\,\mathbf{i} + y(t)\mathbf{j} + z(t)\,\mathbf{k}$. Then if t lies in [a,b], we use:

$$\int\limits_{a}^{b} f(x(t), y(t), z(t)) \left\| \frac{d\mathbf{r}}{dt} \right\| dt$$

What is the Line Integral of a Vector Field?

The **integral of a vector field F over the curve C** is different. Instead of talking about a real-valued function f(x,y,z), we now are dealing with a vector field **F**(x,y,z). The curve C is described by a *vector* : **r** (x,y,z) = x **i** + y **j** + z **k**. Then the line integral of **F** over C is defined as:

$$\int\limits_{C} \mathbf{F} \cdot d\mathbf{r} = \int\limits_{C} \mathbf{F}(x, y, z) \cdot \mathbf{T}(x, y, z) ds$$

where **T** is the unit tangent vector to the curve C. Note the integral is taking place **with respect to arc length**. We will rarely use this format , because this integral is difficult to evaluate. It does help us to clarify the meaning of a line integral, however.

The line integral is not easy to visualize. We are interested only in the contribution of the vector field F, in the direction of the curve.

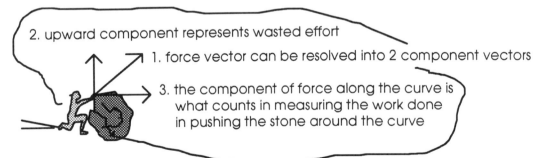

2. upward component represents wasted effort

1. force vector can be resolved into 2 component vectors

3. the component of force along the curve is what counts in measuring the work done in pushing the stone around the curve

For example, when we compute "work", we are only interrested in the component which contributes to pushing the stone AROUND the curve.

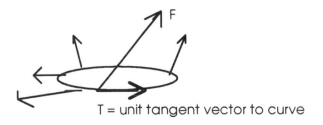

F

T = unit tangent vector to curve

NOTE: When we find line integrals, the curve must be "orientable". For example, we do not permit this sort of thing. It would be rare to run into this sort of curve, anyway.

two-way street

How Do We Compute The Line Integral?

The curve C is usually given as a *vector function* of t, by means of a "radius" or "ray" function: $\mathbf{r}(t)= x(t)\,\mathbf{i} + y(t)\mathbf{j} + z(t)\,\mathbf{k}$. We express all variables in the integral in terms of t. While it is easy to understand the meaning of the line integral from the formula we saw

earlier, namely, $\displaystyle\int_C \mathbf{F} \cdot \mathbf{dr} = \int_C \mathbf{F}(x,y,z) \cdot \mathbf{T}(x,y,z)\,ds$, this formula is not easy to use.

Instead we use:

$$\int_{t=a}^{b} \mathbf{F}(x(t), y(t), z(t)) \cdot \frac{d\mathbf{r}}{dt}\, dt$$

This is the integral of a dot product of two vectors. Note the change from the previous definition: we used "ds" earlier and we use "dt" now.

NOTE: The derivation of this formula, from the previous one, is extremely interesting. Please come to understand it, by reading your text carefully.

What if C's Parametrization Changes?

If C is a curve such as the one below, we will be using different parametrizations on the different "pieces".

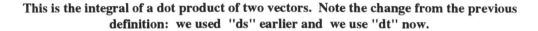

The curve C is comprised of 3 pieces on which the parametrization will be different.

Then we use results (that hopefully, conform to common sense, but can be validated mathematically) that

$$\int_C = \int_{C_1} + \int_{C_2} + \int_{C_3}$$

We also use $\displaystyle\int_{C} = -\int_{-C}$. (In other words, if we trace over a curve *backwards*, we should

have *the negative of the integral traced forwards.* Can you see why, from the definition?)

Another Way to Write the Line Integral of a Vector Field

If we write $\mathbf{F}(x,y,z) = M(x,y,z)\,\mathbf{i} + N(x,y,z)\,\mathbf{j} + P(x,y,z)\,\mathbf{k}$, then

$$\int_{C} \mathbf{F}(x,y,z)\cdot\frac{d\mathbf{r}}{dt}dt \text{ can also be found as}$$

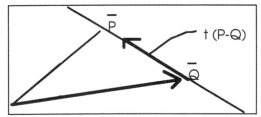

$$\int_{t=a}^{b} M(x,y,z)\frac{dx}{dt}dt + \int_{t=a}^{b} N(x,y,z)\frac{dy}{dt}dt + \int_{t=a}^{b} P(x,y,z)\frac{dz}{dt}dt$$

NOTE: Sometimes we integrate over paths that are straight line segments. In this case, use the equation for a line segment found earlier in Section 12.4:

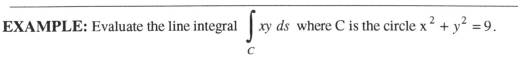

Let $\mathbf{X}$ be the vector (x,y,z). Then the line segment between vectors $\mathbf{P}$ and $\mathbf{Q}$ is given by $\mathbf{X} = \mathbf{Q} + t(\mathbf{P}\text{-}\mathbf{Q})$, which can also be written as $\mathbf{X} = t\,\mathbf{P} + (1\text{-}t)\,\mathbf{Q}$. This segment is directed from $\mathbf{Q}$ to $\mathbf{P}$, as we can see by finding where we are when $t = 0$ and $t = 1$.

Examples

EXAMPLE: Evaluate the line integral $\displaystyle\int_{C} xy\,ds$ where C is the circle $x^2 + y^2 = 9$.

Answer:
(1) Obtain a parametrization in terms of t, of the circle: we use $x = 3\cos t$, $y = 3\sin t$.
(2) Rewrite the integral as

$$\int_C xy\ ds = \int_0^{2\pi} (3\cos t)\ (3\sin t)\ \left\|\frac{d\mathbf{r}}{dt}\right\|\ dt\ \text{ where}$$

$$\mathbf{r} = (3\cos t)\,\mathbf{i} + (3\sin t)\,\mathbf{j}$$

$$\frac{d\mathbf{r}}{dt} = (-3\sin t)\mathbf{i} + (3\cos t)\mathbf{j},\ \text{ and}$$

$$\left\|\frac{d\mathbf{r}}{dt}\right\| = \sqrt{(-3\sin t)^2 + (3\cos t)^2} = 3$$

(3) Evaluate $\displaystyle\int_C xy\ ds = \int_0^{2\pi}(3\cos t)\ (3\sin t)\ \left\|\frac{d\mathbf{r}}{dt}\right\| dt\ = 27\int_0^{2\pi}(\cos t\sin t)\ dt$. This is easily

evaluated, with $u = \sin t$ or the identity $\sin 2t = 2\sin t\cos t$. In fact, this integral is 0.

EXAMPLE: Express the line integral $\displaystyle\int_C xy^2\ ds$ in terms of t,

where C is the curve given by $\mathbf{r}(t) = t^2\,\mathbf{i} - t\,\mathbf{j} + t\,\mathbf{k}$, for $0 \le t \le 1$.

Answer:

(1) On the curve, we have $x = t^2$, $y = -t$, and $z = t$.

(2) Now $\| \mathbf{r}'(t) \| = \| (2t)\mathbf{i} - \mathbf{j} + \mathbf{k}\| = \sqrt{4t^2 + 2}$

(3) The integral becomes $\displaystyle\int_0^1 (t^2)(-t)^2 \sqrt{4t^2 + 2}\ dt$.

EXAMPLE: Evaluate $\displaystyle\int_C xy\,ds$ where C is the path :

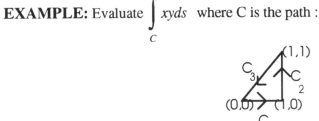

Answer:

(1) We break the curve into 3 pieces and use $\displaystyle\int_C = \int_{C_1} + \int_{C_2} + \int_{C_3}$. Note that on

C_1, only x varies. On C_2, points have x-coordinate 1, and y varies. On C_3, $x = y$.

(2) We can use the **NOTE** below. But here we will use parametrizations of the pieces in
terms of t (slower but more consistent with the other types of integrals we do):

on C_1: $\mathbf{r}(t) = t\mathbf{i}$ and $\| \mathbf{r}'(t)\| = 1$
on C_2: $\mathbf{r}(t) = \mathbf{i} + t\mathbf{j}$ and $\| \mathbf{r}'(t)\| = 1$
on C_3: $\mathbf{r}(t) = (1-t)\mathbf{i} + (1-t)\mathbf{j}$ and $\|\mathbf{r}'(t)\| = \sqrt{(-1)^2 + (-1)^2} = \sqrt{2}$.
(Refer to the NOTE directly following this example.)

(3) We are to integrate the function $f(x,y) = xy$. What are x and y on these curves?

We write $\mathbf{r}(t) = x(t)\mathbf{i} + y(t)\mathbf{j} + z(t)\mathbf{k}$. Then

on C_1: $xy = t \cdot 0 = 0$ since $y = 0$ on the line,

on C_2: $xy = 1 \cdot t = t$,

on C_3: we are on the line $y = x = 1 - t$, so $xy = (1-t)^2$

(4) The integral becomes

$$\int_C xy\,ds = \int_0^1 0\,dt + \int_0^1 t\,dt + \int_0^1 (1-t)^2 \sqrt{2}\,dt = \frac{3+5\sqrt{2}}{6}.$$

NOTE: We might write:

$$\int_C xy\,ds = \int_0^1 0\,dt + \int_0^1 t\,dt + \boxed{\int_1^0 t^2 \sqrt{2}\,dt} = \frac{3+5\sqrt{2}}{6}, \text{ where we parametrize x and y as t, on}$$

the segment C_3, running time "backwards" from 1 to 0. This is a faster method than the method above.

✋ **CAUTION!** There is a difference between $\displaystyle\int_C xy\ \mathbf{ds}$ and $\displaystyle\int_C xy\ \mathbf{dx} + xy\ \mathbf{dy}$. We

evaluate the latter integral in the next example.

EXAMPLE: Find $\displaystyle\int_C xy\,dx + xy\,dy$ where C is the triangular path above.

Answer: We use:
on C_1: $\mathbf{r}(t) = t\mathbf{i}$ so $x = t$, and $y = 0$, and $dx/dt = 1$,

on C_2: $\mathbf{r}(t) = \mathbf{i} + t\mathbf{j}$ so $x = 1$ and $y = t$, and $dy/dt = 1$,

on C_3: $\mathbf{r}(t) = (1-t)\mathbf{i} + (1-t)\mathbf{j}$ so $dx/dt = dy/dt = -1$.

Now: $\displaystyle\int_C xydx + xydy = \int_0^1 x\cdot 0\frac{dx}{dt}\ dt + \int_0^1 1\cdot y\frac{dy}{dt}\ dt + \int_0^1 (-x)^2\frac{dx}{dt}\ dt = \frac{1}{6}.$

☞ It might have occurred to us immediately , without going through the "t" substitution that this integral is:

$\displaystyle\int_C xydx + xydy = \int_0^1 x\cdot 0\frac{dx}{dt}\ dt + \int_0^1 1\cdot y\frac{dy}{dt}\ dt + \int_0^1 (-x)^2\frac{dx}{dt}\ dt = \frac{1}{6},$ because this integral is sufficiently simple.

EXAMPLE: Evaluate the line integral of the vector field $\mathbf{F}(x,y,z) = z\mathbf{i} - y\mathbf{j} + xy\mathbf{k}$ where C is the curve given by $\mathbf{r}(t) = 3\mathbf{i} - (\cos t)\mathbf{j} + 5t\mathbf{k}$ for $0 \le t \le 1$.

Answer: On this curve, x = 3, y = -cos t, and z = 5t, and $\mathbf{r}'(t) = (\sin t)\mathbf{j} + 5\mathbf{k}$. We now form the integral of the dot product:

$\displaystyle\int_{t=a}^{b} \mathbf{F}(x(t),y(t),z(t))\cdot\frac{d\mathbf{r}}{dt}dt = \int_0^1 (5t\mathbf{i} - (-\cos t)\ \mathbf{j} + (3)(-\cos t)\mathbf{k})\cdot((\sin t)\mathbf{j} + 5\mathbf{k})\ dt =$

$\displaystyle\int_0^1 \cos t \sin t - 15\cos t\ dt$, which can be evaluated as two integrals.

EXAMPLE: Evaluate:

$\displaystyle\int_C ydx + zdy + xdz$ where C is composed of the line segments from $(1,1,1)$ to $(3,3,3)$ and from $(3,3,3)$ to $(4,3,3.)$ (This is not a closed path.)

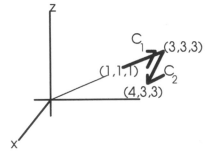

Answer:
(1) We have
        on C_1: $\mathbf{r}(t) = t\mathbf{i} + t\mathbf{j} + t\mathbf{k}$ with $1 \le t \le 3$
        on C_2: $\mathbf{r}(t) = t\mathbf{i} + 3\mathbf{j} + 3\mathbf{k}$ with $3 \le t \le 4$.

We could also have used the following.

Since C_1 is the line segment between $(1,1,1)$ and $(3,3,3)$, we could have used
$$\mathbf{r}(t) = t(3\mathbf{i} + 3\mathbf{j} + 3\mathbf{k}) + (1-t)(\mathbf{i} + \mathbf{j} + \mathbf{k}) \text{ with } 0 \le t \le 1.$$

Since C_2 is the line segment between $(3,3,3)$ and $(4,3,3)$, we could have used
$$\mathbf{r}(t) = t(4\mathbf{i} + 3\mathbf{j} + 3\mathbf{k}) + (1-t)(3\mathbf{i} + 3\mathbf{j} + 3\mathbf{k}) \text{ with } 0 \le t \le 1.$$

(2) Thus
$$\text{on } C_1: \ dx/dt = dy/dt = dz/dt = 1$$
$$\text{on } C_2: \ dx/dt = 1, \ dy/dt = dz/dt = 0$$

(3) Now we have

$$\int_C ydx + zdy + xdz = \int_{C_1} ydx + zdy + xdz + \int_{C_2} ydx + zdy + xdz =$$

$$\int_{\frac{1}{3}}^{3} t\frac{dx}{dt}dt + t\frac{dy}{dt}dt + t\frac{dz}{dt}dt + \int_{\frac{3}{4}}^{4} 3\frac{dx}{dt}dt + 3\frac{dy}{dt}dt + t\frac{dz}{dt}dt =$$

$$\int_{1}^{3} t(1)dt + t(1)dt + t(1)dt + \int_{3}^{4} 3(1)dt + 3(0)dt + t(0)dt =$$

$$\int_{1}^{3} 3t\,dt + \int_{3}^{4} 3\,dt = \int_{1}^{3} 3t\,dt + \int_{3}^{4} 3\,dt = \frac{3t^2}{2}\bigg|_{1}^{3} + 3t\bigg|_{3}^{4} = 13+3 = 16.$$

SECTION 15.3

"A man of words and not of deeds/Is like a garden full of weeds"...Nursery Rhyme

THE FUNDAMENTAL THEOREM OF LINE INTEGRALS

> Please read the theory carefully in this section, in your text. We
> cannot reproduce it here, only interpret it casually. This does not
> take the place of the understanding your text can provide.

If a function is "just right"-- its integral along a curve will depend only on the end points.

> ### The FUNDAMENTAL THEOREM OF CALCULUS
> stated that:
> If f is continuous on [a,b] then there is an antiderivative F of f and
> $$\int_{a}^{b} f(x)dx = F(b) - F(a) \text{ where } F'(x) = f(x)$$

**This theorem essentially tells us that the integral, under the "right" circumstances depends
only on the value of the antiderivative of f at the endpoints of the interval. Perhaps this is not**

enormously surprising on a straight line segment; but now our functions are defined on a variety of curves between 2 pointsd. And we are saying : if the function is "just right", the integral from a to b will be the same on any of the paths below! This is useful, because if the integral along all the paths below are equal, we would prefer to integrate on the straight line segment because of its simpler parametrization.

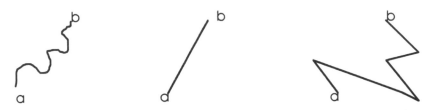

Now we find what types of functions are "just right" . Basically, a function is "just right" if it is a gradient.

The Fundamental Theorem of Line Integrals

Let C be an oriented curve with initial point (x_0, y_0, z_0) and terminal point (x_1, y_1, z_1).
Let f be a function of 3 variables that is differentiable on C and assume grad f is continuous on C, Then

$$\int_C \mathbf{grad} f \cdot d\mathbf{r} = f(x_1, y_1, z_1) - f(x_0, y_0, z_0)$$

NOTE: Observe how much like the Fundamental Theorem of Calculus this result is. The F.T.C. guaranteed under similar conditions an antiderivative. And of course, it's easy to find an antiderivative of something that is already known to be a derivative, like the gradient!

With the conditions of the boxed result above met, we say:
(1) The integral is **independent of the path**.
(2) The vector field **F** is **conservative**.
(3) **F = grad** f for some f
(3) The integral around a closed, oriented curve C in the domain of f is 0.

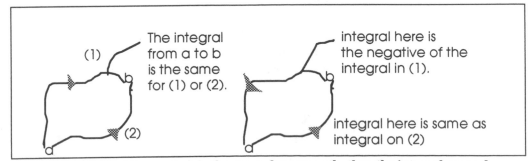

   If the integral from a to b is the same, for any path, then the integral around a closed curve is 0.

EXAMPLE: Show the line integral $\displaystyle\int_C (2xy^2 + 1)dx + 2x^2ydy$ is independent of the path if C is any piecewise smooth curve in the xy plane from (0,2) to (2,4). Evaluate the integral.
Answer:

♦ We need to show that $\mathbf{F}(x,y,z) = (2xy^2 + 1)\mathbf{i} + 2x^2y\mathbf{j}$ is a gradient; it will be, if

$$\mathbf{curl\ F} = \mathbf{0}, \text{ or } \mathbf{curl\ F} = \begin{vmatrix} \mathbf{i} & \mathbf{j} & \mathbf{k} \\ \dfrac{\partial}{\partial x} & \dfrac{\partial}{\partial y} & \dfrac{\partial}{\partial z} \\ 2xy^2+1 & 2x^2y & 0 \end{vmatrix} = \mathbf{0}, \text{ or if } \frac{\partial}{\partial x}(2x^2y) = \frac{\partial}{\partial y}(2xy^2+1),$$

which is true, both sides equalling 4xy.

♦ We also need to show the gradient is a continuous vector function. This is the case, because the components are continuous.

We will find the integral in two ways.

(1) **Going immediately for the integral:** Since the integral is independent of the path, we may use for the path the straight line segment y = x + 2. Let x = t, so y = t + 2, where t goes from 0 to 2.

Then $\displaystyle\int_C (2xy^2 + 1)dx + 2x^2ydy = \int_0^2 (2t(t+2)^2 + 1)\frac{dx}{dt}dt + 2t^2(t+2)\frac{dy}{dt}dt =$

$\displaystyle\int_0^2 (2t(t+2)^2 + 1)dt + 2t^2(t+2)dt = \int_0^2 (4t^3 + 12t^2 + 8t + 1)dt = 66.$

(2) **Finding the antiderivative:** Here we seek f such that

$$(2xy^2 + 1)\mathbf{i} + (2x^2y)\mathbf{j} = \mathbf{grad}f = \frac{\partial f}{\partial x}\mathbf{i} + \frac{\partial f}{\partial y}\mathbf{j}.$$

(We saw these problems in the last chapter.)

$$\frac{\partial f}{\partial x} = (2xy^2 + 1) \text{ we have } f(x, y) = x^2 y^2 + x + g(y)$$

$$\frac{\partial f}{\partial y} = (2x^2 y) \text{ so differentiating the above with respect to y}, \ 2x^2 y + g'(y) = 2x^2 y.$$

Then $g'(y) = 0$ and $g(y) = C$ and one antiderivative is found using $C = 0$.

$$f(x, y) = x^2 y^2 + x.$$

With the antiderivative we can integrate:

$$\int_C (2xy^2 + 1)dx + 2x^2 ydy = x^2 y^2 + x \Big|_{(0,2)}^{(2,4)} = (2^2 4^2 + 2) - (0^2 2^2 + 0) = 66.$$

EXAMPLE: Evaluate the line integral $\displaystyle\int_C (2xy^2 + 1)dx + 2x^2 ydy$ where C is a circle of

raidus 1, centered at the origin.

Answer: Since we have shown this integral is independent of the path, the integral around any closed curve, including C, is 0.

EXAMPLE: Evaluate the line integral

$$\int_C (y)dx + (x + z)dy + ydz \text{ where C is parametrized by}$$

$$\mathbf{r}(t) = \frac{t^2 + 1}{t^2 - 1}\mathbf{i} + \cos \pi t\mathbf{j} + 2t \sin \pi t\mathbf{k} \text{ as t goes from 0 to } 1/2.$$

Answer:

♦ We need to show that $\mathbf{F}(x,y,z) = y\mathbf{i} + (x + z)\mathbf{j} + y\mathbf{k}$ is some function's gradient; it will be if

$$\mathbf{curl}\ \mathbf{F} = \begin{vmatrix} \mathbf{i} & \mathbf{j} & \mathbf{k} \\ \dfrac{\partial}{\partial x} & \dfrac{\partial}{\partial y} & \dfrac{\partial}{\partial z} \\ y & (x+z) & y \end{vmatrix} = \mathbf{0} = 0\,\mathbf{i} + 0\mathbf{j} + 0\,\mathbf{k}.$$

But this is true, since we can verify that

$$\frac{\partial}{\partial y}(y) - \frac{\partial}{\partial x}(x + z) = 0,$$

$$\frac{\partial}{\partial y}(y) - \frac{\partial}{\partial z}(x + z) = 0, \text{ and}$$

$$\frac{\partial}{\partial x}(y) - \frac{\partial}{\partial z}(y) = 0.$$

♦ We also need to show the gradient is a continuous vector function. This is the case, because the component functions are continuous.

Now that we have the result that the integral is independent of the path we can use either method from the previous example to evaluate the integral. In the first method, we would integrate on the straight line from $\mathbf{r}(0) = (-1,1,0)$ to $\mathbf{r}(1/2) = (-5/3,0,1)$. The more challenging method is to find the antiderivative. Using the method in Section 15.1:

$(a)\dfrac{\partial f}{\partial x} = y$ and we have $f(x,y,z) = xy + g(y,z)$

$(b)\dfrac{\partial f}{\partial y} = (x + z)$ and $f(x,y,z) = xy + zy + h(x,z)$

$(c)\dfrac{\partial f}{\partial z} = y$

Differentiating f in (a) with respect to y and setting that result equal to $\dfrac{\partial f}{\partial y}$ in (b):

$\dfrac{\partial f}{\partial y} = x + z = x + g_y(y,z)$ so $g_y(y,z) = z$ and $g(y,z) = yz + h(z)$

So from (b), $f(x,y,z) = xy + yz + h(z)$.

Now using (c), $\dfrac{\partial f}{\partial z} = y = y + h'(z)$ and $h'(z) = 0$ so $h(z) = C$.

So one antiderivative for f is $f(x,y,z) = xy + yz$. We use this antiderivative.

$$\int_C (y)dx + (x + z)dy + ydz = xy + yz \Big|_{(-1,1,0)}^{(-5/3,0,1)} = 0 + 1 = 1 \ .$$

SECTION 15.4

"Happy as the grass was green"..Dylan Thomas

Green's Theorem

From the text: Let R be a simple region in the xy-plane (we are in 2-space now!) with a piecewise smooth boundary C oriented counterclockwise. Let M and N be functions of two variables having continuous partial derivatives on R. Then

$$\int_C Mdx + Ndy = \iint_R \left(\frac{\partial N}{\partial x} - \frac{\partial M}{\partial y} \right) dA.$$

CAUTION! We are in **2-space** now!

Some Observations

(1) A simple region is one that is vertically and horizontally simple.

(2) The quantity in Green's Theorem, $\left(\dfrac{\partial N}{\partial x} - \dfrac{\partial M}{\partial y} \right)$ is the very quantity we wanted to be 0,

in the previous section, in order of the integral to be independent of the path. In this theorem, C is a closed curve. So the theorem says that if this quantity is 0, the integral around the closed curve will be 0. But it says more.

(3) The theorem changes an area integral into a line integral and vice-versa. This is remarkable!

(4) If the region does have holes (it is called annular if it resembles a washer), we can often find the integral by integrating in pieces. (The examples you see will generally not involve simple regions.) Read in your text how such a region is broken up into simple pieces and "sewn up" again.

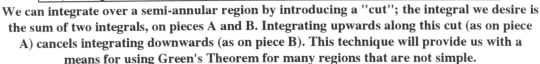

We can integrate over a semi-annular region by introducing a "cut"; the integral we desire is the sum of two integrals, on pieces A and B. Integrating upwards along this cut (as on piece A) cancels integrating downwards (as on piece B). This technique will provide us with a means for using Green's Theorem for many regions that are not simple.

Green's Theorem can be restated as :

$$\int_C \mathbf{F} \cdot d\mathbf{r} = \iint_R \ (\operatorname{curl} \mathbf{F}) \cdot \mathbf{k}\, dA$$

$$\int_C \mathbf{F} \cdot \mathbf{n}\, ds = \iint_R \ \operatorname{div} \mathbf{F}(x, y)\, dA \ \ .$$

The first equation says: the integral of the tangential component of the vector field along a curve equals the integral of rate of the field's "tendency" to rotate over the   interior. The second equation says: the integral of the normal component of the vector field around the boundary of the region is the integral of the divergence (which measures the rate of in-and-out flow of fluid) over the area of the region. What's important in both of these equations is that what's going on _inside_ the region is, in some sense, determined by what happens _on the boundary._

NOTE: To find the area of a simple region, we can use Green's Theorem . The area is the integral

$$A = \int_C x \, dy = -\int_C y \, dx = \frac{1}{2}\int_C x \, dy - y \, dx .$$ Remember, C is oriented counterclockwise.

EXAMPLE: Find the area enclosed by a circle of radius r, using Green's Theorem.

Answer: We use $A = \frac{1}{2}\int_C x \, dy - y \, dx$, and parametrize the circle by x = r cos t, y = r sin t,

for t in $[0, 2\pi]$. Then

$$A = \frac{1}{2}\int_C x \, dy - y \, dx = A = \frac{1}{2}\int_0^{2\pi} (r \cos t)(r \cos t) \, dt - (r \sin t)(-r \sin t) \, dt =$$

$$A = \frac{1}{2}\int_0^{2\pi} r^2 \, dt = \frac{r^2 t}{2}\Big|_0^{2\pi} = \pi r^2 .$$

EXAMPLE: Using a known area to find an integral. Find the integral

$\int_C y \, dx + 3 \, dy$ where C is the quarter of the circle $x^2 + y^2 = 9$ in the lst quadrant and the

intervals [0,3] and [3,0] on the x and y-axes.

Answer: Assume that C is oriented counterclockwise. Note that C and R meet the specifications of Green's Theorem.

Since by Green's Theorem , we have $\int_C M \, dx + N \, dy = \iint_R \left(\frac{\partial N}{\partial x} - \frac{\partial M}{\partial y} \right) dA ,$

$$\int_C y \, dx + 3 \, dy = \iint_R \left(\frac{\partial (3)}{\partial x} - \frac{\partial y}{\partial y} \right) dA = -\iint_R dA = -(\text{area of the region}) = -\frac{\pi 3^2}{4} = -9\pi / 4 .$$

NOTE: If the curve had the opposite orientation (clockwise), our answer would have been positive.

EXAMPLE: Evaluating a double integral using Green's Theorem. Find the integral
$\int_C (y\cos x)dx + (x\sin y)dy$ where C is the curve given below .

Answer: (Always check that the conditions of Green's Theorem are met: components of the field have continuous partial derivatives, region is simple; curve is smooth.)

The curve is oriented in the clockwise direction. Then

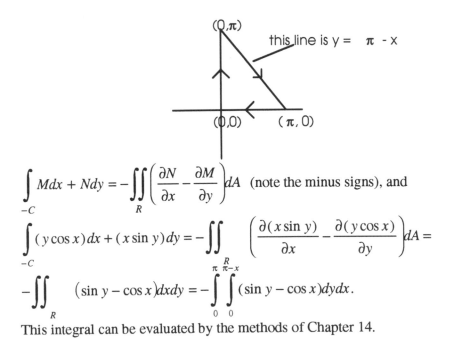

$\int_{-C} Mdx + Ndy = -\iint_R \left(\frac{\partial N}{\partial x} - \frac{\partial M}{\partial y} \right) dA$ (note the minus signs), and

$\int_{-C} (y\cos x)dx + (x\sin y)dy = -\iint_R \left(\frac{\partial(x\sin y)}{\partial x} - \frac{\partial(y\cos x)}{\partial y} \right) dA =$

$-\iint_R (\sin y - \cos x)dxdy = -\int_0^\pi \int_0^{\pi-x} (\sin y - \cos x)dydx.$

This integral can be evaluated by the methods of Chapter 14.

EXAMPLE: Find the integral $\int_C (y^3 + y)dx + 3y^2 xdy$ where C is the circle $x^2 + y^2 = 9$, oriented counterclockwise.

Answer: (The conditions of Green's Theorem are met: components of the field have continuous partial derivatives, region is simple; curve is smooth.)

Then $\int_C Mdx + Ndy = \iint_R \left(\frac{\partial N}{\partial x} - \frac{\partial M}{\partial y} \right) dA$ and

$\int_C (y^3 + y)dx + 3y^2 xdy = \iint_R \frac{\partial(3y^2 x)}{\partial x} - \frac{\partial(y^3 + y)}{\partial y} dA = \iint_R 3y^2 - (3y^2 + 1)dA =$

$-\iint_R dA = -\pi(3^2) = -9\pi.$

EXAMPLE: Use Green's Theorem to evaluate $\int_C \mathbf{F} \cdot d\mathbf{r}$ where $F(x,y) = y\mathbf{i} + 3x\mathbf{j}$ and C is the square with vertices (0,0), (1,0), (0,1), and (1,1).

Answer: The conditions of Green's Theorem are met. Now, although the notation $\int_C \mathbf{F} \cdot d\mathbf{r}$ might imply we should express quantities in terms of t, we can nevertheless write this integral as:

$$\int_C y\,dx + 3x\,dy = \iint_R \left(\frac{\partial(3x)}{\partial x} - \frac{\partial(y)}{\partial y} \right) dA = \iint_R (3-1)\,dA = \iint_R 2\,dA = 2(\text{area of square}) = 2(1) = 2.$$

EXAMPLE: Using Green's Theorem to find the area. Find the area of the region bounded by the hypocycloid parametrized by $r(t) = (\cos^3 t)\mathbf{i} + (\sin^3 t)\mathbf{j}$ for t from 0 to 2π.

Answer: From the formula in the text,

$$A = \int_C x\,dy = \int_C x(t)\frac{dy}{dt}dt = \int_0^{2\pi} (\cos^3 t)(3\sin^2 t)(\cos t)\,dt = \int_0^{2\pi} 3(\sin^2 t)(\cos^4 t)\,dt. \text{ The}$$

evaluation of this rests on either a reduction formula for $\int \cos^6 x\,dx$, or you can chip away at it, with the equations $\sin^2 \theta = \dfrac{1 - \cos 2\theta}{2}$ and $\cos^2\theta = \dfrac{1 + \cos 2\theta}{2}$.

EXAMPLE: Prove Green's Theorem over the rectangle below:

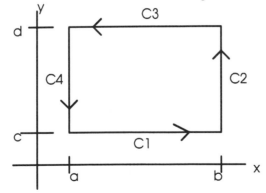

Green's Theorem states:

$$(*) \quad \iint_R \left(\frac{\partial N}{\partial x} - \frac{\partial M}{\partial y} \right) dA = \underbrace{\iint_R \left(\frac{\partial N}{\partial x} \right) dA}_{(2)} - \underbrace{\iint_R \left(\frac{\partial M}{\partial y} \right) dA}_{(3)}.$$

$$\underset{(1)}{}$$

Consider the second integral in (*): $\iint_R \left(\dfrac{\partial N}{\partial x} \right) dA = \int_c^d \int_a^b \dfrac{\partial N}{\partial x}\,dx\,dy = \int_c^d \left(\int_a^b \dfrac{\partial N}{\partial x}\,dx \right) dy$, and

$$\boxed{\int_c^d \int_a^b \frac{\partial N}{\partial x} dxdy = \int_c^d \left(N(b,y) - N(a,y) \right) dy}$$

Meanwhile, (from the left side of Green's Theorem), we look first at the integrals over the

vertical pieces of the curve: $\displaystyle\int_{C2} Mdx + Ndy + \int_{C4} Mdx + Ndy$.

On C2, let $\mathbf{r}(t) = b\,\mathbf{i} + t\,\mathbf{j}$, and on C4, let $\mathbf{r}(t) = a\,\mathbf{i} - t\,\mathbf{j}$. Then

$$\int_{C2} Mdx + Ndy + \int_{C4} Mdx + Ndy =$$

$$\int_{C2} M\frac{dx}{dt}dt + N\frac{dy}{dt}dt + \int_{C4} M\frac{dx}{dt}dt + N\frac{dy}{dt}dt$$

(where $\dfrac{dx}{dt} = 0, \dfrac{dy}{dt} = 1$ on C2 and $\dfrac{dx}{dt} = 0, \dfrac{dy}{dt} = -1$ on C4).

$$\int_c^d M(b,t)\cdot 0 + N(b,t)(dt) + \int_{-d}^{-c} M(a,-t)\cdot 0 + N(a,-t)(-1)dt =$$

Using the substitution $y = t$ on 1st integral, and $y = -t$ on 2nd

$$\int_c^d N(b,t)(dt) + \int_d^c N(a,-t)(-1)dt =$$

$$\int_c^d N(b,y)dy + \int_d^c N(a,y)dy \text{ and}$$

$$\boxed{\int_c^d N(b,y)dy + \int_d^c N(a,y)dy = \int_{C2} Mdx + Ndy + \int_{C4} Mdx + Ndy}$$

Now combining the boxed results,

$$\int_c^d \int_a^b \frac{\partial N}{\partial x} dxdy = \int_{C2} Mdx + Ndy + \int_{C4} Mdx + Ndy \,.$$

A similar effort shows a similar result is true for the third integral in (*) . Thus, we show

$$\int_C Mdx + Ndy = \iint_R \left(\frac{\partial N}{\partial x} - \frac{\partial M}{\partial y} \right) dA \,,$$

Green's Theorem, for a rectangular region.

Exercise! Can you use similar methods to prove Green's Theorem for the triangle below?

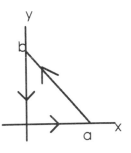

SECTION 15.5

"I started to talk to other mathematicians about proof, knowledge, and reality in mathematics and I found that my situation of confused uncertainty was typical. But I also found a remarkable thirst for conversation and discussion about our private experiences and inner beliefs"...Philip Davis and Reuben Hersh

SURFACE INTEGRALS

In this section, we examine the higher dimensional generalization of line integrals; the integral of a function or a vector field over a surface in 3-space. The graph of the function f, given as $w = f(x,y,z)$ would be in 4-space. We have

$$\iint_{Graph,\Sigma, \text{ of Surface}} g(x,y,z)\,dS = \iint_R g(x,y,f(x,y))\sqrt{[f_x(x,y)]^2 + [f_y(x,y)]^2 + 1}\,dA$$

EXAMPLE: Evaluate $\iint_\Sigma g(x,y,z)\,dS$ where g is given by $g(x,y,z) = z^2$, and Σ is the graph of the surface which is part of the cone $z = \sqrt{x^2 + y^2}$ between the planes $z = 1$ and $z = 2$

Answer: The function giving the surface is $f(x,y) = \sqrt{x^2 + y^2}$ and the notation $g(x,y,f(x,y))$ means: express g in terms of x and y. Thus $g(x,y,f(x,y)) = z^2 = x^2 + y^2$. Now surface area is given by:

$$\iint_{\Sigma} g(x,y,z)dS = \iint_{R} g(x,y,f(x,y))\sqrt{[f_x(x,y)]^2 + [f_y(x,y)]^2 + 1}\, dA =$$

$$\iint_{R} (x^2+y^2)\sqrt{\left(\frac{x}{\sqrt{x^2+y^2}}\right)^2 + \left(\frac{y}{\sqrt{x^2+y^2}}\right)^2 + 1}\; dA = \sqrt{2}\iint_{R}(x^2+y^2)\; dA\,.\ \text{We}$$

switch to polar coordinates:

$$\text{Integral} = \sqrt{2}\int_0^{2\pi}\int_1^2 r^2\; r\,dr\,d\theta = \sqrt{2}\int_0^{2\pi}\int_1^2 r^3\; dr\,d\theta,\ \text{which is easily evaluated.}$$

EXAMPLE: Evaluate $\displaystyle\iint_{\Sigma} g(x,y,z)dS$ where g(x,y,z) = x^2+y^2 and Σ is the graph of

the surface which is composed of. the part of the paraboloid $4 - x^2 - y^2$ above the xy plane
and the part of the xy plane lying inside the circle $x^2+y^2 = 4$.

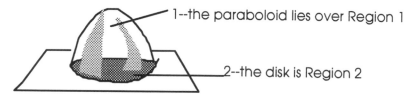

1--the paraboloid lies over Region 1

2--the disk is Region 2

Answer: On Region 1, which we will call R_1, f(x,y) $= 4 - x^2 - y^2$. On Region 2, or R_2,
f(x,y) = 0. And g(x,y,f(x,y)) = x^2+y^2 over both regions. So

$$\iint_{\Sigma} g(x,y,z)dS = \iint_{R_1} g(x,y,f(x,y))\sqrt{[f_x(x,y)]^2 + [f_y(x,y)]^2 + 1}\, dA +$$

$$\iint_{R_2} g(x,y,f(x,y))\sqrt{[f_x(x,y)]^2 + [f_y(x,y)]^2 + 1}\, dA$$

$$= \iint_{R_1} (x^2+y^2)\sqrt{[-4x]^2 + [-4y]^2 + 1}\, dA + \iint_{R_2} (x^2+y^2)\sqrt{0+0+1}\, dA$$

$$= \iint_{R_1} (x^2+y^2)\sqrt{16x^2 + 16y^2 + 1}\, dA + \iint_{R_2} (x^2+y^2)\, dA$$

$$= \int_0^{2\pi}\int_0^2 r^2\sqrt{16r^2 + 1}\; r\,dr\,d\theta + \int_0^{2\pi}\int_0^2 r^2 r\,dr\,d\theta$$

where the second integral presents no problems, and the first can be integrated with the substitution $u = 16r^2 + 1$, so $du = 32 \, r \, dr$. Then $\int r^3 \sqrt{16r^2 + 1} \, dr = \dfrac{1}{32} \int \dfrac{u-1}{16} \sqrt{u} \, du$.

EXAMPLE: Evaluate $\displaystyle\iint_{\Sigma} g(x, y, z) \, dS$ where $g(x,y,z) = x + 1$ and Σ is the tetrahedron below:

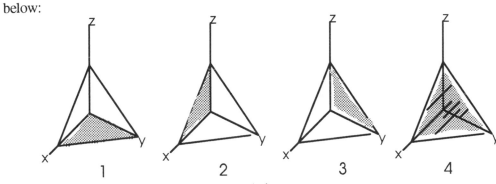

$$\begin{array}{cccc} 1 & 2 & 3 & 4 \end{array}$$

Hint: We sum 4 integrals: over Regions 1-4.

♦ On Region 1, $f(x,y) = 0$, with $g(x,y,f(x,y)) = x + 1$

♦ On Region 2, z varies on this face and z is not a function of x and y. So, we write $y = f(x,z) = 0$, where we use

$$\iint_R g(x, f(x,z), z)\sqrt{[f_x(x,y)]^2 + [f_z(x,y)]^2 + 1} \, dzdx = \int_0^1 \int_0^{1-x} (x+1)\sqrt{0 + 0 + 1} \, dzdx$$

on this region.

♦ On Region 3, we use a similar trick using $f(y,z) = 0$.

$$\iint_R g(x, f(x,z), z)\sqrt{[f_x(x,y)]^2 + [f_z(x,y)]^2 + 1} \, dzdx = \int_0^1 \int_0^{1-y} (1)\sqrt{0 + 0 + 1} \, dzdy \ .$$

♦ On Region 4, $f(x,y) = 1-x-y$, since z is determined by the plane $x + y + z = 1$. The integral contributed by this region is

$$\iint_R g(x, y, f(x,y))\sqrt{[f_x(x,y)]^2 + [f_y(x,y)]^2 + 1} \, dydx = \int_0^1 \int_0^{1-x} (x+1)\sqrt{1 + 1 + 1} \, dydx \ .$$

♦ The final result is the sum of these four integrals.

===

SECTION 15. 6

===

"God ever geometrizes"...Plato

INTEGRALS OVER ORIENTED SURFACES

The material in the next three sections, which close this guide, is fairly difficult. There are many criss-crossing analogies; many results may look vaguely familiar, enough to be somewhat confusing at first. Study the text carefully, until the material feels extremely clear to you.

✠ We have seen, from first year calculus, the integral along an *oriented* interval [a,b]:

$$\int_a^b f(x)\,dx .$$ The interval was oriented, because we went "from a to b".

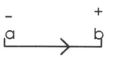

✠ We have seen the integral of a vector-valued function over an *oriented* curve C:

$$\int_C \mathbf{F}\cdot d\mathbf{r} .$$ We integrate the component of **F** in the direction of the curve, along the curve.

Recall $$\int_C \mathbf{F}\cdot d\mathbf{r} = \int_C \mathbf{F}\cdot\frac{d\mathbf{r}}{dt}\,dt .$$

✠ We have *not seen* (but are about to see) the integral of a function of **two** variables over the graph of an *oriented **surface***, S. If Σ denotes the graph of the surface, this integral is denoted

$$\iint_\Sigma \mathbf{F}\cdot\mathbf{n}\ dS \ .$$

In this integral, we integrate the component of F normal to the surface over the graph of the surface.

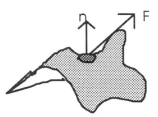

NOTE: Any integral of the form above is called a **"flux" integral**. In fluid dynamics and electro-magnetics, this integral measures the "in and out" effect of fluid through a surface.

We want to know how to evaluate this type of integral.

ORIENTING A SURFACE:

An orientable surface has an outward-pointing normal and an inward-pointing one. By the right hand rule, if the thumb points in the direction of the designated normal, then the curved fingers determines the orientation of little circles on the surface, as in the diagram below. These in turn, determine the orientation of the boundary curve C, of the surface.

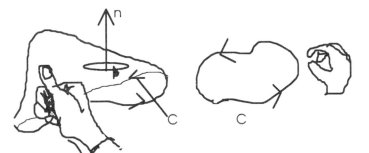

Curve C is oriented in the counterclockwise direction when we use the outward pointing normal

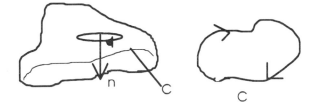

With the inward pointing normal we induce a clockwise orientation of C.

With this orientation scheme, the fluid flow in and out of a surface is measured as positive and negative.

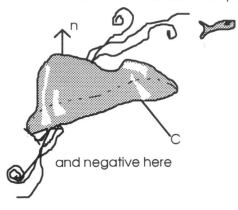

Fluid flow is measured as positive here

and negative here

FINDING THE NORMAL (old history!)

Recall that on a plane: $ax + by + cz = d$, the vector $a\mathbf{i} + b\mathbf{j} + c\mathbf{k}$ is *a* normal vector to the plane. (Why are there *many* normal vectors to a plane?) The equation of the tangent plane to the function $z = f(x, y)$ at $(x_0, y_0, f(x_0, y_0))$ is given by:

$$z = f(x_0, y_0) + f_x(x_0, y_0)(x - x_0) + f_y(x_0, y_0)(y - y_o),\text{or writing this implicitly,}$$

$$f(x_0, y_0) + f_x(x_0, y_0)(\mathbf{X} - x_0) + f_y(x_0, y_0)(\mathbf{Y} - y_o) - \mathbf{Z} = 0 .$$

Thus, a normal vector to the tangent plane is given by $f_x(x_0, y_0)\mathbf{i} + f_y(x_0, y_0)\mathbf{j} - \mathbf{k}$. If a vector is normal to the tangent planeof the surface, we say it is normal to the surface. When we want a unit normal vector, we divide the normal vector by its length, .

The upward pointing **unit normal** is $\mathbf{n} = \dfrac{-f_x\mathbf{i} - f_y\mathbf{j} + \mathbf{k}}{\sqrt{f_x^2 + f_x^2 + 1}}$ (the coefficient of **k** is positive)

The downward pointing **unit normal** is $\mathbf{n} = \dfrac{f_x\mathbf{i} + f_y\mathbf{j} - \mathbf{k}}{\sqrt{f_x^2 + f_x^2 + 1}}$ (change the sign).

WHAT IS THE MEANING OF A SURFACE INTEGRAL, $\displaystyle\iint_\Sigma \mathbf{F} \cdot \mathbf{n} \, dS$?

If $\mathbf{F} = M\mathbf{i} + N\mathbf{j} + P\mathbf{k}$, and Σ is the graph of the surface, and $\mathbf{n}$ is any normal, then

$$\iint_\Sigma \mathbf{F} \cdot \mathbf{n} \, dS = \iint_\Sigma \mathbf{F} \cdot \frac{\mathbf{n}}{\|\mathbf{n}\|} \; \|\mathbf{n}\| \, dA \text{ since } dS = \|\mathbf{n}\|dA .$$ **Thus we have the formulas:**

$$\iint_{\Sigma} \mathbf{F} \cdot \mathbf{n} \; dS = \pm \iint_{\Sigma} (-M f_x \; - N f_y \; + P) \, dA$$

where we use **+** if **n** is the **upward pointing normal** and - if **n** is the **downward pointing normal.**

EXAMPLE: Determine the orientation of the boundary of the surface comprised of 3 faces of the "interior" of the cube below , oriented by the normals in the picture.
Answer:

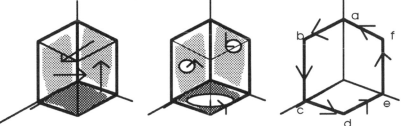

The boundary is pictured at far right. The orientation is described by a listing of the vertices traversed in order: a,b,c,d,e,f. Note that the spines along the axes are not part of the boundary of the surface. The entire surface of a cube would have no boundary.

EXAMPLE: Let $\mathbf{F} = y\mathbf{i} - x\mathbf{j} + z^2\mathbf{k}$. Find $\displaystyle\iint_{\Sigma} \mathbf{F} \cdot \mathbf{n} \; dS$ where Σ is the part of the cone

$z^2 = x^2 + y^2$ above the square pictured below left , with **n** directed upward.

Answer: The cone is given by $f(x,y) = \sqrt{x^2 + y^2}$. We find

$f_x = \dfrac{x}{\sqrt{x^2 + y^2}}, f_y = \dfrac{y}{\sqrt{x^2 + y^2}}$. And from the description of **F**, M = y, N = -x, P = z^2.

Now $\displaystyle\iint_{\Sigma} \mathbf{F} \cdot \mathbf{n} \; dS = \pm \iint_{\Sigma} (-M f_x \; - N f_y + P) \, dA$ where we use **+** for the upward

pointing normal.

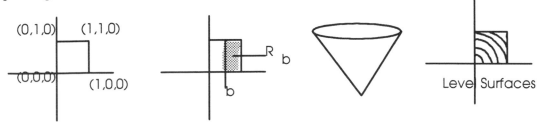

We use $\iint\limits_{\Sigma} \mathbf{F} \cdot \mathbf{n}\ dS = + \iint\limits_{\Sigma} (-Mf_x - Nf_y + P)\,dA$ (+, **because of upward normal**) =

$$\iint\limits_{R} (-y\ \frac{x}{\sqrt{x^2 + y^2}} + x\frac{y}{\sqrt{x^2 + y^2}} + x^2 + y^2)\,dA =$$

$$\iint\limits_{R_b} (x^2 + y^2)\,dydx = \lim_{b \to 0^+} \int_b^1 \int_0^1 (x^2 + y^2)\,dydx = 2/3\ (*)$$

(*) **NOTE:** The use of "b" with the limit simply makes us formally accurate when we integrate over the interior of a region. We won't use b and R_b in the future, however.

___

EXAMPLE: Let $\mathbf{F} = x\mathbf{i} + y\mathbf{j} + z\mathbf{k}$. Find $\iint\limits_{\Sigma} \mathbf{F} \cdot \mathbf{n}\ dS$ where Σ is part of the cylinder

$x^2 + z^2 = 1$ between the planes y = -2 and y = 1, and **n** is directed away from the y-axis.

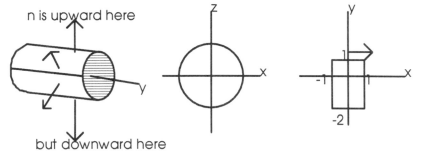

Answer: On the top half of the graph, we use an upward pointing normal, and on the lower half, the downward pointing normal. Then we form integrals over both halves and add.

On the top half, Σ_1, over region R_1, $f(x,y) = \sqrt{1 - x^2}$, and on the lower half, Σ_2, over region R_2, $(x,y) = -\sqrt{1 - x^2}$.

Meanwhile, on Region R_1, $f_x = \dfrac{-x}{\sqrt{1 - x^2}}, f_y = 0$. On Region R_2, $f_x = \dfrac{x}{\sqrt{1 - x^2}}, f_y = 0$.

Also M = x, N= y, M= z = f(x,y). We use

$$\iint\limits_{\Sigma} \mathbf{F} \cdot \mathbf{n}\ dS = \pm \iint\limits_{\Sigma} (-Mf_x - Nf_y + P)\,dA, \text{ and}$$

$$\iint\limits_{\Sigma} \mathbf{F} \cdot \mathbf{n}\ dS = \iint\limits_{\Sigma_1} \mathbf{F} \cdot \mathbf{n}\ dS + \iint\limits_{\Sigma_2} \mathbf{F} \cdot \mathbf{n}\ dS \text{ where}$$

(1) $\iint\limits_{\Sigma_1} \mathbf{F} \cdot \mathbf{n}\ dS =$

$\iint\limits_{R_1} \frac{-x(-x)}{\sqrt{1-x^2}} + \sqrt{1-x^2}\,dydx = \lim_{\varepsilon\to1} \int\limits_{-\varepsilon-2}^{0}\int\limits_{-2}^{1} \frac{1}{\sqrt{1-x^2}}dydx + \lim_{\varepsilon\to1}\int\limits_{0}^{\varepsilon}\int\limits_{-2}^{1}\frac{1}{\sqrt{1-x^2}}dydx$, where we

used ε's because the integral is improper. (The integral with respect to x gives $\sin^{-1}(x)$.)

(2) $\iint\limits_{\Sigma_2} \mathbf{F} \cdot \mathbf{n}\ dS =$

$\iint\limits_{R_2} \frac{x^2}{\sqrt{1-x^2}} - \left(-\sqrt{1-x^2}\right)dydx = \lim_{\varepsilon\to1}\int\limits_{-\varepsilon-2}^{0}\int\limits_{-2}^{1}\frac{1}{\sqrt{1-x^2}}dydx + \lim_{\varepsilon\to1}\int\limits_{0-2}^{\varepsilon}\int\limits_{-2}^{1}\frac{1}{\sqrt{1-x^2}}dydx$

The sum of (1) and (2) evaluated is 6π.

Questions Can you find the formula for the flux integral for any length pipe? What if the flow passed through the pipe? What answer would you expect if we used the downward pointing normal on the whole graph?

SECTION 15. 7

"A physical law must possess mathematical beauty."...Paul Dirac

STOKE'S THEOREM

Many results have been explored in these past sections:

Dimension	1	2	3
Area / Volume	$\int_a^b f(x)\,dx = $ area under a curve	$\iint\limits_R f(x,y)\,dydx$ volume under a surface	$\iiint\limits_D f(x,y,z)\,dzdydx$ "4 – d" volume under a solid
Special integrals	$\int_C F \cdot dr = $ flow along the curve	$\iint\limits_\Sigma F \cdot ndS = $ flux through the surface	**?**

We saw in the last section similarities between the integrals:

$$\int_C \mathbf{F} \cdot d\mathbf{r} \text{ and } \iint_\Sigma \mathbf{F} \cdot \mathbf{n}\,dS.$$

Stoke's Theorem tells us that if we choose just the right "**F**" on the right side, *we have an equality:*

$$\int_C \mathbf{F} \cdot d\mathbf{r} = \iint_\Sigma (\mathbf{curl\ F}) \cdot \mathbf{n}\,dS$$

How did we do this dimension leap? How did a double integral of one entity become a single integral of another? It's plausible--because Green's Theorem accomplished this. In fact, going back to first year calculus, the Fundamental Theorem of Calculus gave us

$$F(b) - F(a) = \int_a^b f(x)\,dx,$$

a result which achieves the deletion of an integral . Stoke's Theorem is an important result and a definite goal of this course. It is important on 3 levels: (1) it has geometrical and physical significance, (2) it has algebraic significance, and (3) on a practical level, it is extremely useful.

Conditions on the surface and boundary: Let Σ be an oriented surface with normal **n**, and finite surface area. Assume that Σ is bounded by a closed, piecewise smooth curve C, with the induced orientation of Σ. (Essentially: "nice" means nothing kinky)

Conditions on the function: Assume **F** has continuous partial derivatives at each non-boundary point of Σ. (Essentially: "nice"--we want things to be continuous so we can integrate them, and we plan to integrate partial derivatives.) Then :

$$\int_C \mathbf{F} \cdot d\mathbf{r} = \iint_\Sigma (\text{curl } \mathbf{F}) \cdot \mathbf{n}\, dS$$

In particular, if $\mathbf{F} = M\mathbf{i} + N\mathbf{j} + P\mathbf{k}$, then

$$\int_C M\,dx + N\,dy + P\,dz = \iint_\Sigma (\text{curl } \mathbf{F}) \cdot \mathbf{n}\, dS.$$

In other words we cross a dimension gap, and equate a line integral with a surface integral. In two dimensions, $\mathbf{F} = M\,\mathbf{i} + N\,\mathbf{j}$, and *this theorem reduces to Green's Theorem.*

Green' Theorem states: $\displaystyle \int_C M\,dx + N\,dy = \iint_R \left(\frac{\partial N}{\partial x} - \frac{\partial M}{\partial y} \right) dA$,

where the integrand in the right side is, in effect curl **F** if we use $\mathbf{n} = \mathbf{k}$ (This makes sense, because this theorem applies to the xy-plane.)

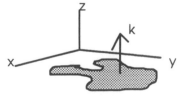

WHAT DOES STOKE'S THEOREM SAY, IN PHYSICAL TERMS?

To understand this, we will delve into ancient history to look even more closely to see how Stoke's Theorem is analogous to the Fundamental Theorem of Calculus.

The Fundamental Theorem told us that

> If f is continuous on [a,b], then there is an antiderivative of f, denoted F, and
>
> $$\int_a^b f(x)\,dx = F(b) - F(a).$$

The theorem says, essentially, that if area is rolled out as a function of x, then f(x) can be thought of as $\dfrac{dA}{dx}$, the **rate of change of area per unit of x**. And so : f is some function's derivative (namely, the area function's derivative).

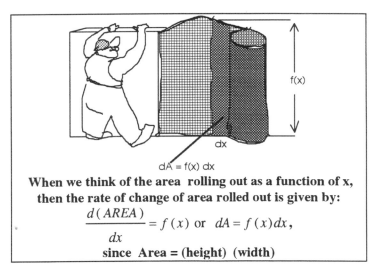

When we think of the area rolling out as a function of x, then the rate of change of area rolled out is given by:
$$\frac{d(AREA)}{dx} = f(x) \text{ or } dA = f(x)dx,$$
since Area = (height) (width)

So when we say:

$$\int_a^x \frac{dA}{dx}dx = A(x) - A(0)$$

we are saying that **an integral of a rate** demolishes an integral sign. And the cumulative change in area depends only on the area where we began and the area where we left off.

Now we are saying that $\displaystyle\int_C \mathbf{F}\cdot d\mathbf{r} = \iint_\Sigma (\text{curl } \mathbf{F})\cdot \mathbf{n}\, dS$. Since (curl **F**) · **n** represents the

rate of rotation of fluid per unit of surface area, then integrating over the surface *demolishes one of the integral signs* on the right side, and we end up with the *overall effect* of the fluid circulating around the boundary of the surface.

WHAT DOES THE THEOREM SAY, IN STRICTLY MATHEMATICAL TERMS?

The crucial idea is that under certain conditions, the mathematical quantity (curl **F**)· **n** is the derivative *(in some sense)* of "something". This may sound like hands flapping in the air,

but in fact, it is seen in higher level courses, that (curl **F**)· **n** = d (Mdx + Ndy + Pdz). It is important to note that conditions are imposed on both the function and the boundary.

WHY IS THE THEOREM IMPORTANT, AT A PRACTICAL LEVEL?

A line integral $\int_C \mathbf{F} \cdot d\mathbf{r}$ may look simple, but to evaluate it on a polygonal surface with 10 sides would mean evaluating 10 integrals. Stoke's Theorem equates integrals of this type with others that are possibly more easily solved. Furthermore, if a boundary is shared by two surfaces with the same orientation , we can integrate over the "easier" surface, since the integral depends only effects on the boundary. In the picture below, $\Sigma 1$ (which is a blob) and $\Sigma 2$ (which is a disk) share the same boundary with the same orientation. Therefore,

$$\iint_{\Sigma 1} = \iint_{\Sigma 2}$$

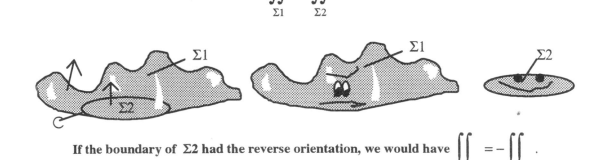

If the boundary of $\Sigma 2$ had the reverse orientation, we would have $\iint_{\Sigma 1} = -\iint_{\Sigma 2}$.

EXAMPLE: Use Stoke's Theorem to evaluate $\int_C \mathbf{F} \cdot d\mathbf{r}$ where $F = z^2\mathbf{i} - y^2\mathbf{j}$ and Σ is

the surface composed of 3 unit squares that we have seen before:

Answer: Rather than evaluate 6 line integrals, we use $\int_C \mathbf{F} \cdot d\mathbf{r} = \iint_\Sigma (\text{curl } \mathbf{F}) \cdot \mathbf{n} \, dS$ where our surface consists of the 3 faces. Let $\Sigma 1$ be the xy face, $\Sigma 2$ the yz face, and $\Sigma 3$ the xz face. Now

$$\text{curl } \mathbf{F} = \begin{vmatrix} \mathbf{i} & \mathbf{j} & \mathbf{k} \\ \dfrac{\partial}{\partial x} & \dfrac{\partial}{\partial y} & \dfrac{\partial}{\partial z} \\ z^2 & -y^2 & 0 \end{vmatrix} = 2z\mathbf{j}, \quad \text{Then on } \Sigma 1, \text{ where the normal is } \mathbf{k}, (\text{curl } \mathbf{F}) \cdot \mathbf{n} = 0, \text{ and}$$

similarly on $\Sigma 2$ where the normal is $\mathbf{i}$. On $\Sigma 3$ the normal is j, and $(\text{curl } \mathbf{F}) \cdot \mathbf{n} = 2z$. Thus

$$\int_C \mathbf{F} \cdot d\mathbf{r} = \iint_{\Sigma 2} 2z \, dz \, dx = \int_0^1 \int_0^1 2z \, dz \, dx = 1.$$

EXAMPLE: Use Stoke's Theorem to evaluate $\int_C \mathbf{F} \cdot d\mathbf{r}$ where

$$\mathbf{F} = \frac{1}{\sqrt{x^2 + y^2 + z^2 + 1}}(x\mathbf{i} + y\mathbf{j} + z\mathbf{k}) \text{ and C is the intersection of the paraboloid}$$

$2z = x^2 + y^2$ and the cylinder $x^2 + y^2 = 2x$.

Answer: By completing the square, we see that the cylinder is $(x-1)^2 + y^2 = 1$. The surfaces intersect on the plane $z = x$.

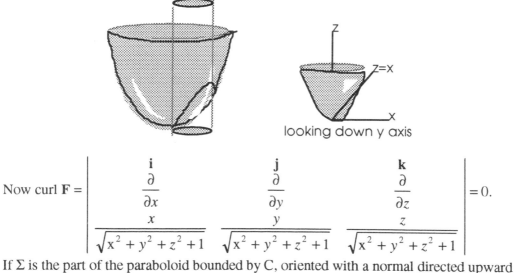

looking down y axis

$$\text{Now curl } \mathbf{F} = \begin{vmatrix} \mathbf{i} & \mathbf{j} & \mathbf{k} \\ \dfrac{\partial}{\partial x} & \dfrac{\partial}{\partial y} & \dfrac{\partial}{\partial z} \\ \dfrac{x}{\sqrt{x^2 + y^2 + z^2 + 1}} & \dfrac{y}{\sqrt{x^2 + y^2 + z^2 + 1}} & \dfrac{z}{\sqrt{x^2 + y^2 + z^2 + 1}} \end{vmatrix} = 0.$$

If Σ is the part of the paraboloid bounded by C, oriented with a normal directed upward (although it won't matter which normal we choose), we have $\int_C \mathbf{F} \cdot d\mathbf{r} = \iint_\Sigma (\text{curl } \mathbf{F}) \cdot \mathbf{n} \, dS =$

0. We did not even have to describe the graph of the region.

SECTION 15. 8

"The essence of mathematics is its freedom, said Cantor. Freedom to construct, freedom to make assumptions."...Philip Davis and Reuben Hersh in <u>The Mathematical Experience</u>.

THE DIVERGENCE THEOREM

We have seen the concepts of flow **around a curve**, and flow **through a surface**. What would be the generalization of flow, pertaining to a solid ? We speak of "divergence", the rate of mass flow per unit volume, at a point--the tendency of the fluid to "burst out" in all directions.

A **simple** solid region D is the solid region between the graphs of 2 functions on a simple region in the plane. The solid should have no sharp points (edges are 0.K.) or breaks .

DIVERGENCE

Given a vector field $\mathbf{F} = M\mathbf{i} + N\mathbf{j} + P\mathbf{k}$, the **divergence** of the field, denoted div $\mathbf{F}$, is the real number defined as

$$\boxed{\text{div } \mathbf{F} = M_x + N_y + P_z}$$

NOTE: Recall, when using the "del operator", $\nabla = \dfrac{\partial}{\partial x}(\ )\,\mathbf{i} + \dfrac{\partial}{\partial y}(\ )\,\mathbf{j} + \dfrac{\partial}{\partial z}(\ )\,\mathbf{k}$,

$$\boxed{\text{div } \mathbf{F} = \nabla \cdot \mathbf{F}}$$

We are about to imitate what we did with Stoke's Theorem and leap into another dimension.

The Divergence Theorem

Conditions on the region: Let D be a simple solid region whose boundary surface Σ is oriented by outward normal, **n.**

Conditions on the function: Let **F** be a vector field whose component functions have continuous partial derivatives on D. Then

$$\iint_{\Sigma} \mathbf{F} \cdot \mathbf{n}\, dS = \iiint_{D} \operatorname{div} \mathbf{F}(x,y,z)\, dV$$

NOTE: Once again we see that the integral of a rate demolishes the integral sign. What is particularly remarkable is the simplicity of the integral on the right!

EXAMPLE: Determine whether the region D is a simple solid region if D is the solid region bounded by the paraboloids $z = 1 - x^2 - y^2$ and $z = x^2 + y^2 - 1$.

Answer: The solid is the region between the graphs of 2 functions on a simple region in the $x = 0$, $y = 0$ and $z = 0$ planes. Hence, it is a simple solid region.

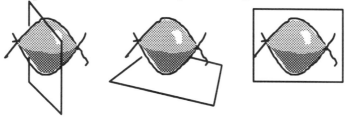

NOTE: Examples of simple regions are regions bounded by planes, inside spheres and ellipsoids, inside a cylinder and between planes, etc.

Examples of regions that are not simple are regions bounded by the surface $y = \sin x$ and the plane $y = 1$ as well as planes parallel to the x and z-axes, regions with pointy peaks, etc.

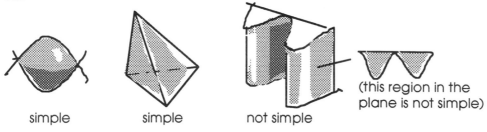

simple   simple   not simple   (this region in the plane is not simple)

EXAMPLE: Compute the integral $\iint\limits_{\Sigma} \mathbf{F} \cdot \mathbf{n} \ dS$ where $\mathbf{F} = yz\,\mathbf{i} + xy\,\mathbf{j} + xz\,\mathbf{k}$, and $\mathbf{n}$ is

the outward normal to the surface Σ, which is the boundary of the solid region in the first octant inside the cylinder $x^2 + y^2 = 1$ and between the planes z = 0 and z = 4.

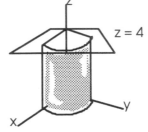

Answer: We know div $\mathbf{F} = M_x + N_y + P_z = 0 + x + x = 2x$. So:

$$\iint\limits_{\Sigma} \mathbf{F} \cdot \mathbf{n} \ dS = \iiint\limits_{D} \text{div } \mathbf{F}(x,y,z)dV = \iiint\limits_{D} 2x\ dzdydx = \int_0^{\pi/2}\int_0^1\int_0^4 2(r\cos\theta)\ r\ dzdrd\theta. \text{ The}$$

evaluation of this integral is straightforward.

EXAMPLE: Compute the integral $\iint\limits_{\Sigma} \mathbf{F} \cdot \mathbf{n} \ dS$ where $\mathbf{F} = x^2\,\mathbf{i} + y^2\,\mathbf{j} + z^2\,\mathbf{k}$, and $\mathbf{n}$ is

the outward normal to the surface Σ, which is the boundary of the solid parallelepiped with vertices (0,0,0), (1,0,0), (1,2,0), (0,2,0), (0,0,3),(1,0,3), (1,2,3), and (0,2,3).

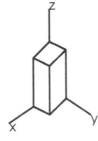

Answer: div $\mathbf{F} = M_x + N_y + P_z = 2x + 2y + 2z$. So:

$$\iint\limits_{\Sigma} \mathbf{F} \cdot \mathbf{n} \ dS = \iiint\limits_{D} \text{div } \mathbf{F}(x,y,z)dV = \int_0^1\int_0^2\int_0^3 (2x + 2y + 2z)\ dzdydx. \text{ Again the}$$

evaluation of this integral is straightforward.

REVIEW OF CONCEPTS

What is a vector field? Give examples from "real life" where vector fields occur. Can you sketch a vector field? How do you find the divergence and curl of a vector field? What theorems and results can you state about divergence and curl? (Do you know any physical applications of divergence and curl?)

What is a line integral? Where in "real life" might this occur?

What is a gradient of a function? What does it represent geometrically? What is a conservative vector field? What results do we get when the vector field **F** is the gradient of a function f?

What is a line integral? What is meant by the statement: "The line integral is independent of the path"? How would you prove that a line integral is independent of the path? How might you use Green's Theorem to evaluate a line integral?

What is the *Fundamental Theorem of Line Integrals*? In what way does it resemble the Fundamental Theorem of Calculus? Why is this a "fundamental" theorem? In what ways does it make our work easier? If **F** = grad f, can you state some results that follow? E.g., what can you say about the line integral being independent of the path, about the line integral about a closed curve, about curl **F**? Can you find curl **F**?

What is meant by a surface integral? A flux integral? What is an oriented surface and why is it important? Why do we have to worry about the orientation of a surface? What is induced orientation? Can you give an example of a surface that is not orientable? (See your text; the Mobius strip is such a surface.) What is a simple solid region?

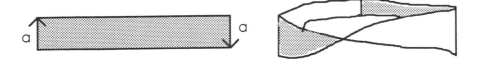

What is Green's Theorem? Stoke's Theorem? The Divergence Theorem? If asked to "verify" Stoke's Theorem for a given vector field and surface, would you be able to do this? (Show the hypotheses of Stoke's Theorem are met, and validate that the appropriate equation holds.) In what basic ways are all of these theorems related? In what ways are they different? Why are they important? What theorems relate a single integral to a double integral, and what does this theorem state? What theorem relates a double integral to a triple integral, and what does this theorem state? What is the geometric importance and the mathematical importance of these results? How can Green's Theorem be applied to, say, an annular (washer-like) region? In what way does Green's Theorem give us a short-cut in finding the area enclosed by a certain type of curve? Can you solve problems using these theorems?

Can You Identify These Integrals?

Say as much about these integrals as you can. Explain what they represent, how to
evaluate them, and what theory is associated with them.

a.) $\displaystyle\int_C f(x,y,z)\,ds$

b.) $\displaystyle\int_C \mathbf{F}\cdot d\mathbf{r}$

c.) $\displaystyle\int_C M\,dx + N\,dy + P\,dz$

d.) $\displaystyle\int_C \mathbf{grad}f \cdot d\mathbf{r}$

e.) $\displaystyle\int_C M\,dx + N\,dy = \iint_R \left(\frac{\partial N}{\partial x} - \frac{\partial M}{\partial y}\right)dA$ (whose theorem?)

f.) $\displaystyle\iint_\Sigma g(x,y,z)\,dS$

g.) $\displaystyle\iint_\Sigma \mathbf{F}\cdot \mathbf{n}\ dS$

h.) $\displaystyle\int_C \mathbf{F}\cdot d\mathbf{r} = \iint_\Sigma (\operatorname{curl}\mathbf{F})\cdot \mathbf{n}\ dS$ (whose theorem?)

i.) $\displaystyle\iint_\Sigma \mathbf{F}\cdot \mathbf{n}\ dS = \iiint_D \operatorname{div}\mathbf{F}(x,y,z)\,dV$ (what theorem?)

GOOD LUCK!

" Mathematics does have a subject matter, and its statements are meaningful. The meaning, however, is to be found in the shared understanding of human beings, not in an external nonhuman reality. "...Philip Davis and Reuben Hersh

960

ethos

ges

Sunday 9AM

446